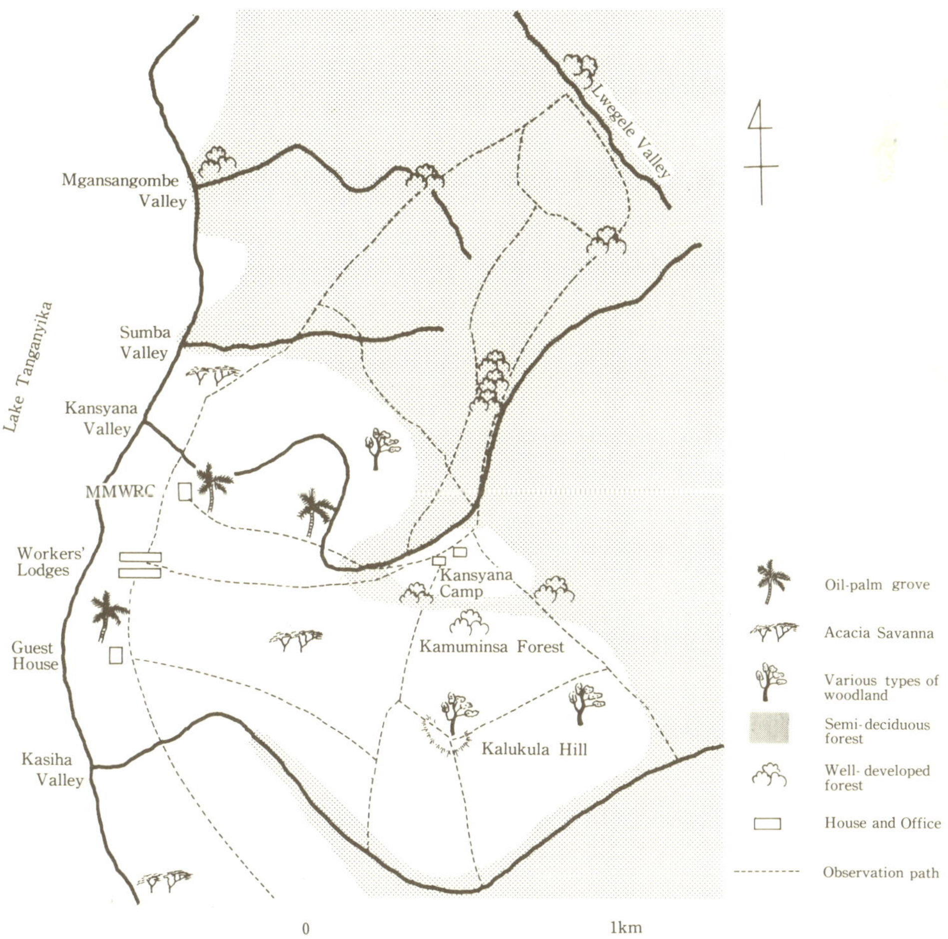
Mgansangombe
Valley
Lwegele Valley
Lake Tanganyika
Sumba
Valley
Kansyana
Valley
MMWRC
Workers'
Lodges
Guest
House
Kansyana
Camp
Kamuminsa Forest
Kalukula Hill
Kasiha
Valley
Oil-palm grove
Acacia Savanna
Various types of
woodland
Semi-deciduous
forest
Well- developed
forest
House and Office
Observation path
0
1km

THE CHIMPANZEES OF THE MAHALE MOUNTAINS

THE CHIMPANZEES OF THE MAHALE MOUNTAINS

Sexual and Life History Strategies

Edited by Toshisada Nishida

UNIVERSITY OF TOKYO PRESS

ISBN 4-13-068153-2
ISBN 0-86008-462-0
Printed in Japan

Dedicated to Professor Junichiro Itani
in commemoration of his retirement
from Kyoto University

Contents

Preface

This is the first English book on the chimpanzees of the Mahale Mountains, Tanzania. We have published many articles in English, but they appeared in a number of different academic journals, and many friends and colleagues have expressed a desire for easy access to our work in its entirety. This volume represents our effort to comply with the above requests. I am especially grateful to Professor Hosea Y. Kayumbo and Professor Richard Wrangham, who earnestly urged us to publish such a book.

We are very happy to commemorate the retirement of a great pioneer in primate studies, Professor Junichiro Itani, and the quarter-century anniversary of our Mahale study with this publication. Professor Itani contributed an article on his extensive surveys, and kindly advised us not to forget the Kyoto University African Primatological Expedition's tradition of extensive study from a bird's-eye view.

We are grateful to Mr. Mitsuo Inoue of the University of Tokyo Press for his editorial assistance. The publication of this book was supported by the Grant-in-Aid for Publication of Scientific Research Results from the Ministry of Education, Science and Culture of Japan.

October 12, 1989

Toshisada NISHIDA

List of Contributors

Toshikazu Hasegawa
Teikyo University, Faculty of Letters, Department of Psychology, 359 Ohtsuka, Hachiohji, Tokyo 192–03

Hitoshige Hayaki
Kyoto University, Faculty of Science, Department of Zoology, Kitashirakawa-Oiwakecho, Sakyo-ku, Kyoto 606

Mariko Hiraiwa-Hasegawa
The University of Tokyo, Faculty of Science, Department of Anthropology, Hongo, Bunkyo-ku, Tokyo 113

Michael A. Huffman
Kyoto University, Faculty of Science, Department of Zoology, Kitashirakawa-Oiwakecho, Sakyo-ku, Kyoto 606

Junichiro Itani
Kyoto University, Center for African Area Studies, 46 Yoshida-Shimoadachicho, Sakyo-ku, Kyoto 606

Kenji Kawanaka
Okayama University of Science, Faculty of Science, Laboratory of Anthropology, 1–1 Ridaicho, Okayama 700

Toshisada Nishida
Kyoto University, Faculty of Science, Department of Zoology, Kitashirakawa-Oiwakecho, Sakyo-ku, Kyoto 606

Yukio Takahata
Kyoto University, Faculty of Science, Department of Zoology, Kitashirakawa-Oiwakecho, Sakyo-ku, Kyoto 606

Hiroyuki Takasaki
Kyoto University, Faculty of Science, Department of Zoology, Kitashirakawa-Oiwakecho, Sakyo-ku, Kyoto 606

Shigeo Uehara
Sapporo University, Faculty of General Education, 3–7–3–1 Nishioka, Toyohira-ku, Sapporo 062

Acknowledgements

We owe our long-term study of Mahale chimpanzees to many people and institutions.

Financial support of our research has been provided mainly by Monbusho (Ministry of Education, Science and Culture) and JICA (Japan International Cooperation Agency). From the beginning of research in 1965, Monbusho has provided invaluable support through the Monbusho International Scientific Research Program. In the early periods T.N. was also supported for his research by the Takenaka Scholarship Foundation and the Wenner-Gren Foundation for Anthropological Research. In the spirit of international technical cooperation JICA sent wildlife experts trained in primatology to Mahale from 1975 to 1988. These experts collected the basic demographic data as Tanzanian wildlife research officers. JSPS (Japan Society for the Promotion of Science) sent H.T. to Mahale as a research scientist. We greatly appreciate the assistance extended by the officials of these institutions who facilitated our research.

We express our deep gratitude to the Tanzania National Scientific Research Council (now Tanzania Commission for Science and Technology), the University of Dar-es-Salaam, and the Serengeti Wildlife Research Institute for permission to conduct the research. We are especially grateful to Abduhl S. Msangi, Hosea Y. Kayumbo, and Karim N. Hirji who supervised our research and kindly provided invaluable comments; to Fivawo and A.E. Lyaruu who were most helpful in the administrative process involved in the granting of permission to conduct research; and to G. Malisa who helped us in various ways.

We are indebted to the Game (Wildlife) Division of the Ministry of Natural Resources and Tourism, and the Tanzania National Park Authorities for the cooperation they extended to the upkeep of the research station. We are particularly indebted to Raphael Jingu and the late Derek Bryceson for their efforts in the initial implementation of the JICA conservation project; to Fred M.R. Lwezaula and David Babu for the final establishment of the Mahale Mountains National Park; and to F.S. Njau, Gerald Bigurube, and Rehema Tibanyenda for their assistance in carrying out the research.

The Embassy of Japan has facilitated our research by extending assistance regarding communication, negotiation, and transportation. We thank, in particular, Nobuyuki Nakashima, Yasuhiro Inagawa, Yuriko Suzuki, Ten'yu Tsuda, Mitsuo Asaba, and Yasushi Kurokochi for the establishment and maintenance of the JICA project. Representatives of the Dar-es-Salaam office of JICA, especially Kin'ei Nishikawa and Kazuo Tanigawa, provided invaluable support for the chimpanzee conservation project. Kozo Tomita, who was the first representative of the Dar-es-Salaam office of JOCV (Japan Overseas Cooperation Volunteers), and his wife helped us during the early stages of the project when help was most needed.

We are grateful to the Regional and District Offices at Kigoma. Support was granted by the successive Regional Natural Resources Officers, Regional Game Officers, and in particular, Lyimo and A.I. Lema. Also, our thanks go to the commanders and crew of the S.S. Liemba for their efforts in assuring transportation on Lake Tanganyika.

The people of Mgambo, Nkonkwa, and Katumbi have always welcomed us warmly and have extended support during various stages of the research. The Arab merchants of Mgambo, in particular, Said Seif and Alofu, helped us to maintain our remote camp in the early period.

We should like to express our thanks to the acting directors, Erasmus Tarimo and Edeus Massawe, and the staff, H.I. Sariko, G.S. Mkombwe, S.N. Kimaro, and V. Mushi, of the Mahale Mountains Wildlife Research Centre for their long-standing cooperation and indispensable help.

Special thanks go to the former residents of Kasoje, who helped us establish our base camps, lent us boats, provided food, cut observation trails, provided plots for planting sugarcane and bananas, and helped us buy sugarcane and other commodities. The late Mwami Sobongo and his son Mzee Saidi accepted us as their family members. The late Mzee Jacobo and his relatives, Mzee Abdala Kakamba and his relatives, the late Mzee Seremani Majaga and his relatives, and all the other village people accepted us as "wenyeji" (residents).

We could not have carried out this long study without the dedicated help of our field assistants. Mtintimya (Ramadhani) Nyundo, nicknamed "Professor", is brilliant of finding animals. He was a skillful trapper and literally taught young "students", Jumanne Katensi, Moshi Bunengwa and Hamisi Bunengwa how to find chimps and identify them by individual names, the idiosyncracies of different chimps, the names of plants, and other important things.

We have been fortunate to have known many people with a variety of talents: the diligent and tenacious worker, Wamkele (Mohamedi) Seifu; the best fisherman and sharp-eyed observer, Moshi Hamisi; the tall, exceptional manager, Wangelo (Alimasi) Kasulamemba; the most skillful tracker, Kijanga (Rashidi) Hawazi; the Jack-of-all-trades, from tracking animals to brick-making, Ramadhani Kasakampe; the quick-tempered boss of the bosses, the late Issa Kapama Ally; the most famous drummer and dancer as well as a capable boat-driver, Athumani Katumba; the cheerful mediator, as eloquent as President Nyerere, Haruna Huseni; the faithful and most reliable farmer, Juma Kahaso; the diligent and hard-working, and most good-natured Sadi Katensi; the Hercules of Kasoje, Haruna Sobongo. We regret that space permits us only to mention the names of the shorter-term workers: Kabilambe Ismaili, Saidi Musa, the late Kabukula (Hawazi) Kasulamemba, Rashidi Kitopeni, Matumula Nyundo, Yassini Kiyoya, Omali Kabaju, and Hamisi Sadi.

We express special thanks to Bill McGrew, Jane Goodall, Richard Wrangham, and Geza Teleki for their comments on our study. These have stimulated new interests, expanded our scope, and improved our methods.

We thank the many colleagues, who studied the chimpanzees of Mahale, and who benefited us by influencing our thought on chimpanzees, and by helping camp management. These include Dick and Jenipher Byrne, Anthony Collins, Kevin Hunt, Hidemi Ishida, Kohsei Izawa, Takayoshi Kano, Masato Kawabata, Satoshi Kobayashi, Bill McGrew, Ken'ichi Masui, Akio Mori, Kohshi Norikoshi, Rogath Olomi, Takashi Torigoe, Takahiro Tsukahara, and Caroline Tutin.

Field workers who stayed at Kasoje to study the local people, fish, or insects, rather than chimpanzees, also contributed to the management and development of our camp. They included Reizo Harako, Michio Hori, Makoto and Hideko Kakeya, Hiroya Kawanabe, Tetsuo Kuwamura, Tetsuya Narita, Kenji Takamura, Jun Takeda, and Jiro Tanaka. In particular, the Kakeyas contributed much to the initial development of the Kansyana Camp.

We also obtained support from residents of Dar-es-Salaam or Kigoma including scientists such as Shunya Hino and Toshimichi Nemoto, businessmen such as Toshio Inomata and Koichi Kobayashi (JETRO), and journalists such as Masataka Ito. We also are greatly indebted to our families for everything from housekeeping to data collection, and would like to thank Hatsuko Kawanaka, Haruko Nishida, Hitomi Takahata, Kazumi Takasaki and Moyo Uehara.

Plant specimens were identified by N. Gachathi, M.G. Gilbert, J.B. Gillett, M.A. Hanid, J. Kahurananga, E.P.K. Kayu, S.P. Kibuwa, and G.W. Sangai of the East African (now Kenya) Herbarium; by L.B. Mwasumbi and R.C. Wingfield of the University of Dar-es-Salaam; and by D. Bridson, P.S. Green, R.M. Polhill, B. Verdcourt, and J.J. Wood of the Royal Botanic Gardens at Kew.

Animal specimens were identified by B. Bolton (ants, bees, and wasps), S. Bacchus and W.A. Sands (termites), J.M. Ritehil (orthopteras), A.F.

Stimson (reptiles and amphibians), Michael Grant (ungulates), P.D. Jenkins (shrew), and Jean Ingles (rodents) of the British Museum (Natural History), and by Duff-Mackay (mammals) of the Corryndon Museum (now the National Museum of Kenya).

The research was facilitated by the support by Jiro Ikeda, Masao Kawai, Shiro Kondo, Yuichi Ono, and many other colleagues. T.N. expresses his deepest gratitude to Naotsune Watanabe, Hitoshi Watanabe, Kazuro Hanihara, Keiichi Omoto, and Banri Endo of the Department of Anthropology, the University of Tokyo, for supporting him while he was in Africa. T.N. also is extremely indebted to Yoshiko Endo for her devoted secretarial work from 1977 to 1988. K.K. is grateful to Yoshimasa Kamaki, Jiro Ikeda, Hiroaki Kobayashi, Shuichi Kameda, Kinya Yasui, and Yuzuru Hamada of Okayama University of Science for their kind support. S.U. is grateful to Takahiro Miyara, Saburo Hirao, and Noriko Takamatsu of the Faculty of General Education, Sapporo University for their kind support. T.H. thanks Yoshitaka Umeoka, Hiroaki Niki, Tadasu Oyama, and Hiroto Katori for their support and encouragement.

We must here give special mention of T. Kano's long-lasting friendship and cooperation in our research. He has helped us send students to Mahale while he has been busy as the organizer of the Pygmy Chimpanzee Project.

Of course, this book would never have come into the world without the pioneering theoretical and practical work of Kinji Imanishi and Jun-ichiro Itani. Our quarter-century of research faced many financial, political, and man-power difficulties; it was almost discontinued many times. On such occasions, Professor Itani assisted us effectively. We would like to dedicate this book to Itani-san for his pioneering studies and continuous encouragement.

THE CHIMPANZEES OF THE MAHALE MOUNTAINS

1

A Quarter Century of Research in the Mahale Mountains: An Overview

Toshisada NISHIDA

Department of Zoology, Faculty of Science, Kyoto University

The purpose of this chapter is twofold: to discuss certain aspects of our Mahale research which may be unfamiliar to non-Japanese primatologists, and to provide an introduction to the whole book, especially the research background of the subsequent chapters. First, the history and rationale of our research are discussed. This is followed by descriptions of the study sites and methods. Third, an overview of the past studies at Mahale is presented. This is necessary since there have been many publications on the chimpanzees of Mahale. An exhaustive list of publications in English and in Japanese is given for readers who would like to obtain further information regarding each topic.

I. HISTORY OF RESEARCH ORGANIZATION

1. "Prehistory" of the Research at Mahale

V.L. Cameron, who arrived in "Ras Kungwe" on the 23rd of March, 1874, was the first European to see and describe this wild and picturesque mountain range (Cameron, 1877). In 1876, H.M. Stanley explored Lake Tanganyika along the eastern shoreline, and vividly described the landscape and resident people (Stanley, 1878). His description suggests that he camped at the mouth of the Lubulungu River in June, 1876 (Fig. 1.1, 1.2).

There are some records from the first half of this century by "Weissen Vaters" ("White Fathers" sent by the order of Roman Catholic missionaries) and district officials about the social structure, history, and legends of the Bende-Tongwe people (Avon, 1915 –16; von Majerus, 1911, 1915–16; Kirschstein, 1937; Hatchell, 1941; Grant, 1949).

Between the 1920s and 1930s, Grant, Loveridge, and Pakenham collected some birds in the Kigoma District, but practically nothing was published. In 1940 and 1941, Moreau (1943), using a grant arranged by

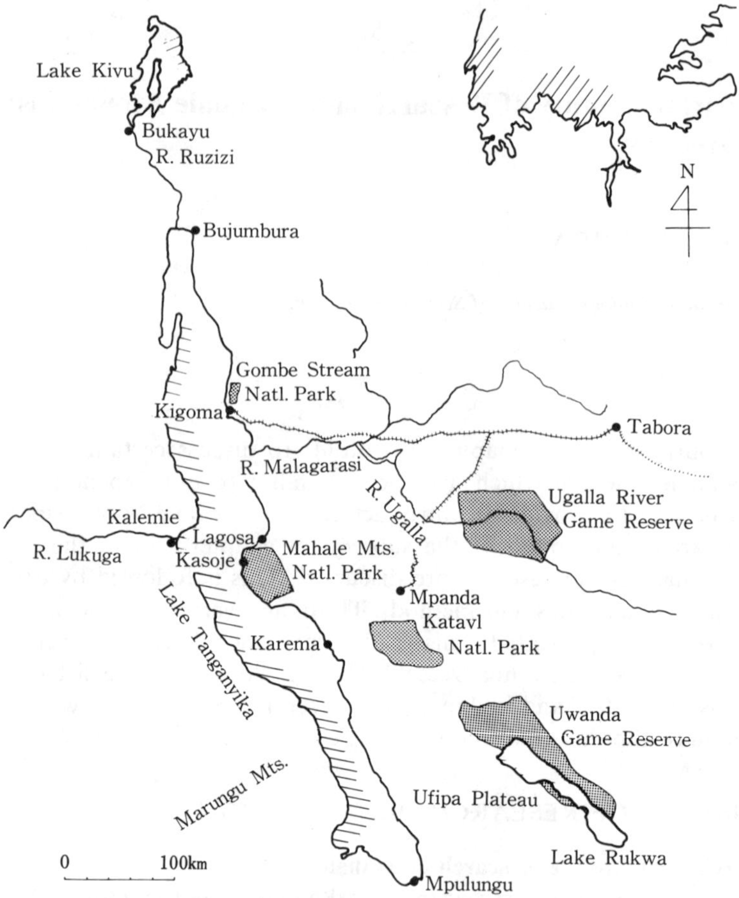

Fig. 1.1 Map of western Tanzania, showing game reserves and national parks.

the British Museum, sent an African collector, Salimu Asumani, to the eastern side of Lake Tanganyika. Salimu visited several localities, especially Mahale and the Ufipa scarp forests, and obtained 600 specimens of birds, mostly montane species, from which 15 new forms were described (see below).

The highest peak of the massif, Mt. Nkungwe, was climbed first by C.C. O'Hagan (1940) in 1939 and second by J. Procter (1958) in 1956 (Fig. 1.3). Both collected plants and left valuable descriptions of vegetation types around the Mahale Mountains.

Dollman (1935) exhibited the damaged skull of a chimpanzee, which had been bought by B.W. Savory of the Political Service, at a meeting of the Linnean Society of London. He became the first person to report

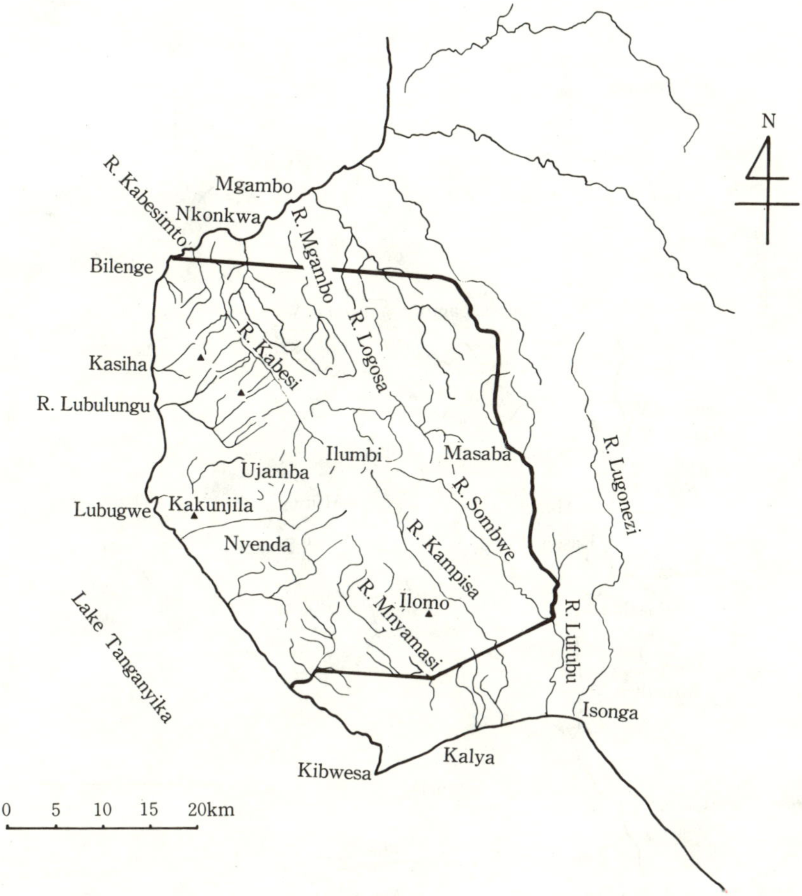

Fig. 1.2 Map showing the Mahale Mountains National Park and major rivers.

to the scientific community that chimpanzees inhabit the Mahale Mountains. Subsequently, Hatchell (1941), Moreau (1942), McConell (1945), and Grant (1946) reported on the distribution of chimpanzees around Mahale. It was pointed out by Kano (1977) that, judging by his extensive study (Kano, 1972), Moreau's (1942) description proved to be surprisingly correct, even though Moreau himself had never visited Mahale.

In spite of this early work, the area was probably unexplored by professional scientists until the late 1950s. In 1958 and 1959, the Oxford University Exploration Club sent expeditions ("University of Oxford

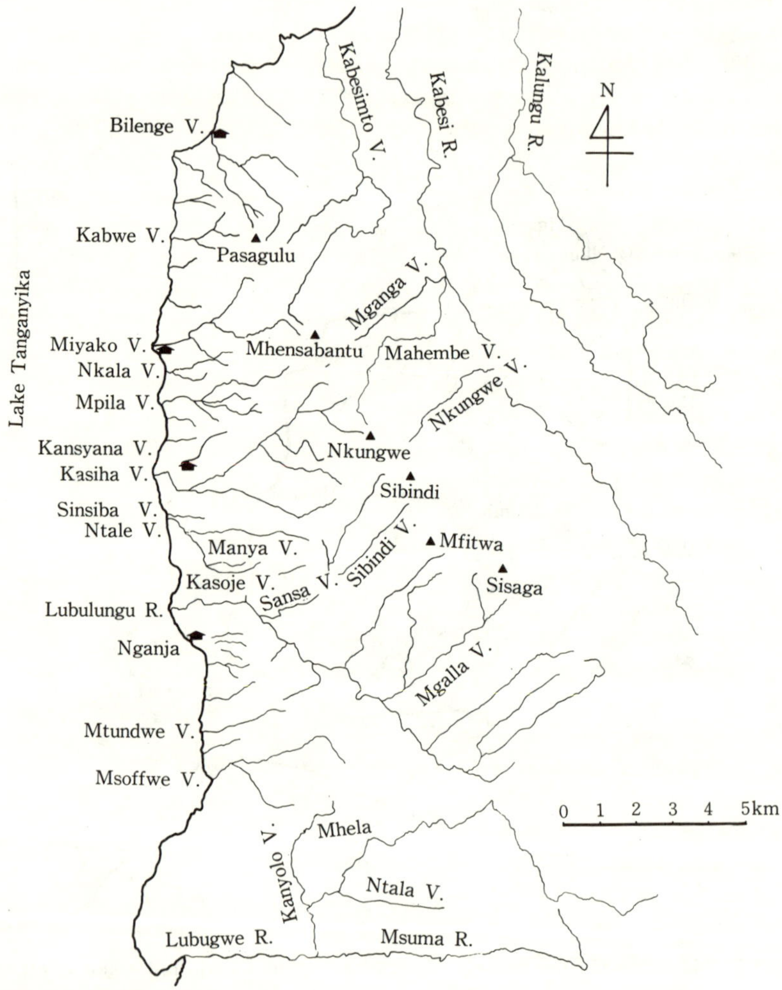

Fig. 1.3 Map of the study area.

Tanganyika Expedition") to the Mahale Mountains. These were led by P.L. Simkin in 1958 and by T.E. Stevens in 1959, and were made up of young scientists who were studying zoology, parasitology, ornithology, botany, geology, psychology, and anthropology. They stayed in the Mahale area for three months (July—September) each year, with their base camp at Kibwesa in 1958 and at Sinsiba (Kasoje) in 1959.

Their publications include lovely introductory notes (Simkin, 1958, 1959a), brief summary reports (Davis, 1959; Juniper, 1959; Simkin and Juniper, 1961), a more conclusive summary report (Stevens, 1962), an experimental study of cognition among African children (Davis, 1961), notes on little-known or new species of dragonflies (Pinhey, 1959), molluscs (Verdcourt, 1961; Pain and Verdcourt, 1962), and birds (Ulfstrand, 1960a, b, c; Ulfstrand and Lamprey, 1960), a paper on migration patterns of kites and eagles (Ulfstrand, 1960d), short reports on geology (Wadsworth, 1963) and on vegetation, fauna, and geography (Stevens, 1962). Three new genera and fourteen new species of spiders were described (Cooke, 1964). Unpublished information includes a floral list, and an important description of the customs of the Batongwe people (Simkin, 1959b).

The expedition also collected 305 bird specimens, several hundred fish from the streams draining into Lake Tanganyika, 4,000 specimens of spiders, and more than 4,800 botanical specimens. Twenty living epiphytic orchids were flown back to England. More than 7,000 feet of 16 mm cine film was also exposed during the expedition, focusing on the fieldwork and on village life (Davis, 1959; Juniper, 1959; Simkin and Juniper, 1961).

The expedition revealed affinities with western Africa in the flora and fauna of the Mahale Mountains. Strangely enough, there was little mention of the chimpanzees of Mahale; Simkin and Juniper (1961) wrote of their mountain camp that "chimpanzees, red colobus monkeys and bushbabies were frequently heard in the gallery forests, while leopards and hyenas often contributed to the cacophony of night noises." Simkin (1959a) referred to the chimpanzees' habit of "raiding Africans' beehives near Mgambo," of which he was told by the then District Commissioner at Kigoma. (This story was still common in Bugansa and the Nyenda Plateau even in the 1970s).

2. The Organization of the Japanese Study of Apes

Japanese study of African great apes in the natural environment began in 1958, when Kinji Imanishi and Junichiro Itani ("the Gorilla Expedition" sponsored by the Japan Monkey Center) traveled around equatorial Africa searching for a good field-site for a long-term study of wild gorillas (Imanishi, 1958). The study met with initial success (Kawai and Mizuhara, 1959; Imanishi, 1960; Kawai, 1961), which, unfortunately, could not be maintained because of political conditions in Zaire. As a result, Imanishi changed the focus of study from gorillas to chimpanzees (Itani, 1961).

In 1961, with financial support from the Ministry of Education, Science and Culture (Monbusho), Imanishi organized the Kyoto University Africa

Primatological Expedition (KUAPE). The expedition originally studied wild chimpanzees at Kabogo Point along the shore of Lake Tanganyika. Chimpanzees proved to live in small, flexible groups (Azuma and Toyoshima, 1961–62; Itani, 1964; Nishimura and Azuma, 1977), but they were not often seen. Therefore, Itani moved the base camp to the Kasakati Basin in 1963.

At Kasakati, useful studies of food ecology (Izawa and Itani, 1966; Kawabe, 1966; Suzuki, 1966, 1969) and of grouping and ranging patterns (Izawa, 1970, 1977; Kawabe, 1977; Suzuki, 1977) of non-provisioned chimpanzees were completed. The chimpanzees, however, proved difficult to habituate to observers.

Itani, who took over the expedition in 1965, decided to establish two more camps, at Filabanga and at Mahale. Three methods of habituating chimpanzees were attempted: "habituation" by T. Kano at Filabanga, "impressing" by K. Izawa at Kasakati, and "provisioning" by T. Nishida at Mahale (Itani, 1967).

Filabanga was selected as one of the stations because in 1965 Itani and Suzuki (1967) observed as many as 47 chimpanzees traveling in a single file, which enabled them to hypothesize that "a large-sized group" may be a social unit of chimpanzees. Kano attempted to habituate wild chimpanzees by contacting and remaining with the group, but without giving them any incentives. Subsequently, the home range of Kano's study group proved too large (being estimated at 200–400 km^2) to habituate the animals (Kano, 1971a). As a result, he began to study patterns of chimpanzee distribution (Kano, 1971b, 1972, 1977).

Izawa obtained a male infant chimpanzee, Brucy, in Nairobi and attempted a unique method of demonstrating to wild individuals friendly interactions between human observer and chimpanzee. Izawa thought that if Brucy was allowed to enter the group of wild chimpanzees, he also would be accepted by them. Unfortunately, Brucy continued to be fearful of his wild counterparts and what might have been one of the first efforts to return a home-raised chimpanzee to the wild did not bear fruit (Izawa, 1975). However, using individuals that he recognized as a base, Izawa did demonstrate the existence of a large-size social unit in unprovisioned chimpanzees (Izawa, 1970).

At the foot of the Mahale Mountains, Nishida, with the assistance of the village people of Kasoje, established a large sugarcane plantation in elephant grass bush to attract chimpanzees. Beginning in March of 1966, chimpanzees began to visit the sugarcane plantation regularly. The K Group chimpanzees were habituated to human observers by the middle of 1966 (Nishida, 1967, 1968, 1973a). It being economically infeasible to maintain many field stations, three camps (Kabogo, Kasakati, and Filabanga) were closed permanently in 1967. However, the research and life amidst the African wilderness left an unforgettable memory in the first generation of chimpologists (Itani, 1970, 1977b). KUAPE continued to send researchers intermittently to Mahale until 1974.

In 1975, the Japan International Cooperation Agency (JICA), a major Japanese aid agency, began to send primatologists to Mahale as Game (Wildlife) Research Officers. JICA's goal was to further international cooperation in nature conservation and was a response to the Tanzanian government's keen interest in a proposal by Itani and Nishida that the Mahale Mountains be maintained as a reserve of flora and fauna, especially chimpanzees.

Thus, the "Mahale Mountains Chimpanzee Research Project" began the collection of basic data on chimpanzees, the ultimate aim being the conservation of this rapidly disappearing species of ape. Longitudinal studies included climatological monitoring (Appendix), fecal analysis, and analysis of mother—offspring relationships, as well as demographic observations (Nishida, 1981, 1982). At the same time, the Tanzanian government established the Kasoje Chimpanzee Research Station (KCRS), which has cooperated closely with the Japanese experts. This project enabled Japanese researchers to monitor two chimpanzee unit groups continuously until early 1988. In 1979 KCRS was reorganized as one of the centers of the Serengeti Wildlife Research Institute (SWRI), and was newly christened the Mahale Mountains Wildlife Research Centre (MMWRC).

Almost all of the former residents of the Mahale Peninsula were transferred to two Ujamaa villages, Nkonkwa and Kalya, by the Government in August 1974 (Nishida, 1978). Agriculture in the Kasoje area was banned by the MMWRC in 1977. In 1985, the Mahale Mountains National Park was finally established as the first "foot-walking" National Park of Tanzania (Nishida, 1985).

Since 1979, the University of Tokyo Ape Expedition to Africa, led by Nishida and funded by the Japanese Ministry of Education, Science and Culture, has been sending students to Mahale every two years. Researchers from the Universities of Stirling, Edinburgh, St. Andrews, Michigan, and Dar-es-Salaam have also contributed to chimpanzee studies at Mahale.

3. Other Japanese Studies

All the Japanese studies, other than those of chimpanzees, have been done by KUAPE or people associated with KUAPE.

The Batongwe tribe has been studied in terms of subsistence strategies (Kakeya, 1976), ethnotaxonomy (Itani, 1977a; Kakeya, 1978), settlement patterns (Hata, 1968; Itani, 1984a), house-building (Suzuki, 1972), and honey-collecting activities (Takeda, 1976). Their social life was described by Suzuki (1972), Itani, Nishida, and Kakeya (1973), Nishida (1973a), and Itani (1977b). Kawabata (Kawabata and Doi, 1972; Kawabata, 1975) studied the fish fauna near Kabogo Point. More recently, the Kyoto University freshwater fish research team lead by H. Kawanabe has studied ecology and reproductive strategies of cichlid fish at the Myako Camp (Hori, 1985a, b, c, 1987; Kawanabe, 1983; Kuwamura, 1985a, b, 1986a, b, 1987a, b; Narita, 1983; Narita, Mulimbwa, and Mizuno, 1985; Takamura, 1983a, b, 1984, 1987).

II. RATIONALE AND IMPLICATIONS OF THE APE STUDY

1. The Origin of the Human Family

Imanishi had long been interested in the origins of animal societies (Imanishi, 1951). His successful studies of Japanese macaques stimulated his interest in the origin of human society, and of the human family in particular (Imanishi, 1961). From an ethnographic survey, he extracted four criteria for the definition of a human family. These were: the presence of a within-family incest taboo, family exogamy, the presence of community (some social bond among neighboring families), and the division of labor by sex.

He reasoned that clues to solving questions regarding the origin of the human family would be provided by two types of research. The first focused on the social structure of the great apes ("primatological approach"), the second on the social structure of current hunter-gatherers ("anthropological approach"). Thus, two base camps were established in Tanzania: one was at Kabogo Point as already mentioned, the other was near Lake Eyasi for the study of the Hadza (Tomita, 1966). Tanaka's ecological study on the Central Kalahari San (Tanaka, 1969, 1980) also took place under Imanishi's influence.

2. Phylogeny vs. Social Structure

The unexpected result of the study was the discovery that the social structure of chimpanzees is radically different from that of gorillas. Imanishi had hypothesized that family-like groups of great apes, maintaining peaceful relationships with each other, would be the ancestral types of the human family. "Familoid" was the term coined by Imanishi for the hypothetical pre-human-family structure exemplified by the gorilla harem group (Donisthorpe, 1958; Kawai and Mizuhara, 1959). I still vividly remember Itani's reaction to this postulate. In July 1966, I arrived at Kigoma Port on the steamship S.S. Liemba, where Itani, who had just arrived from Japan, came to meet me. I shouted from onboard, "No familoid in a large-sized group (unit-group)!" Itani replied, "It cannot possibly be the case!"

The incongruence between phylogeny and social structure posed a grave methodological difficulty. If our two closest relatives display different social structures, how can we judge which is closer to the social structure of our common ancestor?

When, in 1966, the social unit of chimpanzees was shown to differ from that of gorillas, Imanishi had already retired from Kyoto University. He seemed greatly disappointed at the result. Recently, he made a final comment on the problem, "We cannot reconstruct the protohuman society from the study of wild apes" (Tachibana, 1987).

For us, this problem must be dealt with if we wish to continue our long-term studies. Social structure has proved not to be a conservative trait, but is comparatively easy to change. Thus, our field studies must be motivated by a different rationale from that originally proposed by Imanishi. Field studies may be justified in several ways.

First, field work may provide working hypotheses or models which otherwise would not have been produced. For example, Itani (1966) proposed a "preband" hypothesis in which the human family crystalized out of a community, similar to the "large-sized-group" of the chimpanzee.

Second, field studies may facilitate comparison of aspects of human and great ape social structure. For example, the discovery of female transfer (Schaller, 1963; Nishida and Kawanaka, 1972) drew attention to the similarity of the "patrilocal" (or male philopatry) system among African great apes and humans.

Third, we should aim, through our field work, at ethological and ecological comparisons of great apes and humans rather than at the direct reconstruction of protohominid society. Differences should be considered to be just as important as similarities. For example, Nishida (1970) pointed out that food-sharing among chimpanzees is different from that of hunter-gatherers in several respects. Thus, food-sharing among chimpanzees does not voluntarily occur without begging, there is no predetermined food-sharing site such as a camp, and no food "exchange" occurs between individuals, including between the sexes.

Although the discovery of female transfer made us more optimistic, further progress was due to the common sense recognition that what we should seek from the field study is an understanding of the relationships between ecological, ethological, morphological, and phylogenetic variables. The recognition that each species has its own unique evolutionary history allowed the inclination toward "the reconstruction paradigm" to be corrected.

Finally, with the increasing use of a behavioral ecological paradigm, our studies have moved closer to natural history. We study the chimpanzee as "another unique species" (Foley, 1988) rather than as a special cousin of the human species. This line of research produces less biased results and is more successful in contributing to a real understanding of the evolutionary positions of chimpanzee and man (Nishida, 1987a).

Thus, although the research work at Mahale was originally heavily biased toward "sociological" study under Imanishi's influence, now many researchers are examining a wide range of chimpanzee behavior; from posture and locomotion to food ecology to vocal communication.

3. Implications of Fieldwork on Chimpanzees

There are some implications of chimpanzee research that are unique in natural history. This is due to the fact that the chimpanzee is one of the closest relatives of our species. The chimpanzee can be used as a model of mankind or of .the common ancestor of great apes and humans. In the former case, chimps have been used, with great practical success, as human models in the fields of medicine, psychology, and demography. The latter model is useful when studying themes of less immediate practical value, such as the reconstruction of human ancestors. The most important point is that research on chimpanzees helps provide the biological background for human behavior.

1) Biological Background of Human Social Behavior

Chimpanzees and humans are very closely related. It is very likely that their common ancestor possessed some behavioral patterns which are common to both living humans and chimpanzees. Because any such shared trait has been possessed by humans from prehistoric times, it is likely to be well-programmed and less flexible. For example, the following social behavioral patterns are known to be present in both the chimpanzee and our own species: strong tendency to avoid incest, long-lasting mother-offspring relationship, male philopatry, strong antagonism among groups, cooperation among males, development of reciprocal altruism, triadic awareness, alliance fickleness strategy, revenge system, sex difference in political behavior, and so on (Goodall, 1971, 1986; de Waal, 1982, 1984, 1989; Nishida, 1983a; Nishida *et al.*, 1985).

Such information would help us understand which behavioral tendencies are relatively easy or difficult to change, knowledge that may be important if we want to reform institutions or customs to avoid disputes. This line of reasoning provides a more balanced view of humanity to the still widely supported view that humans are born as a blank slate.

Field study also will provide hypotheses to examine the kinds of natural environments in which such shared behavior evolved.

2) Models and Public Education

Medical Models: Chimpanzees in captivity are used for the production of vaccines against the human diseases (such as hepatitis) that are not infective to other animals (e.g., Honjo, 1985).

Psychobiological Models: Chimpanzees are also used for the study of human learning ability, psychobiological disorders, and social development.

For example, field study has shown that adult female apes only rarely experience sexual cycles accompanied by menses; instead they are usually pregnant or lactating. Socio-psychological disorders seen among some modern women may partly be explained by their having many menstrual cycles (Short, 1976).

Demographic Models: Chimpanzees in nature give birth only about every 5 years. Continuous breast-feeding which inhibits ovulation is likely to make postpartum amenorrhea continue for 3 – 4 years. Gatherer-hunters also show a long period of postpartum amenorrhea: the Kalahari San, for example, give birth every four years (Howell, 1979). Chimpanzees and gatherer-hunters also are similar in that mortality is very high in the first year of life. Demographic study of chimpanzees may help us understand the selection pressures acting on prehistoric human populations, thus aiding in the formulation of modern demographic policies.

Public Education: Chimpanzees, through their presence in zoos, play a major role in public education, especially the teaching of evolution.

3) Formulation of Apt Conservation Policies

Chimpanzees have rapidly been dwindling in number, mainly because of environmental destruction by human beings. The total wild population in known localities is about 15,000 and is estimated maximally at about 100,000 (Teleki, 1987). The latter figure may be exaggerated as it is based on the assumption that many chimpanzees still live in a vast, unexplored area, e.g., the Zaire River Basin. Even if the latter figure is correct, it is not large enough for us to become complacent, since it represents the total population throughout Africa. Field studies make it possible for us to establish scientific conservation measures by clarifying population densities, ranges, the rate of population increase or decrease, etc.

4) Improvement of Upkeep of Captive Colonies

Demand for chimpanzees continues to exceed supply, and resources continue to decrease in the natural environment. Since the removal of apes from the wild should be opposed, it has become important to establish reproductive colonies that can provide the necessary number of animals for research in captivity. It has been understood only recently that rearing in a group similar to the social unit found in the natural environment, which is the most humane treatment of chimpanzees, is also the most productive rearing method (de Waal, 1982; Yoshihara, 1987). Field studies provide the data regarding the social environment (such as the age and sex composition) as well as the physical conditions (such as food, temperature, and humidity) which are most important in keeping chimpanzees healthy and productive.

5) Screening of Medicinal Plants

Although wild plants are a natural treasury of materials for medicine, only 2% of the 300,000 flowering plants on the earth have been chemically analyzed (Myers, 1979). It will be impossible to analyze all the species of plants, and, even if it were possible, one would require a vast amount of money and time. However, medicinal plants used by chimpanzees may provide an effective means of screening potentially useful plants (see below).

6) Source of Intellectual Pleasure

The search for human evolutionary origins is based on curiosity which is common to mankind. Many people are interested in chimpanzees as objects of comparison with themselves. The contributions of field studies to popular books, photos, and cinema provide the public with inexhaustible pleasure and intellectual stimulation.

III. THE MAHALE MOUNTAINS

1. Physiognomy

The Mahale Mountains (latitude 6°S, longitude 30°E) are located in the largest promonotory on the eastern shore of Lake Tanganyika. Running in a straight line from north-west by west to south-east by east, Mahale is characterized by numerous prominent peaks which tower above the clouds (Fig. 1.2). Mahale is often referred to as the Nkungwe Mountains, from the name of the paramount peak (2,460 m above sea level) of the massif. Damp air blowing from the lake causes considerable cloud and mist development and supports both the extensive montane forests above 1,800 m and the concentration of gallery forests (Kasoje Forest) at the north-western foot of the mountains. Numerous valleys dissect the mountains, some supporting permanent streams that flow into the lake. The eastern side of the massif is covered mostly with much drier vegetation such as tall *Brachystegia* woodlands (Fig. 1.3).

Mahale can be regarded as a "forest island" in the midst of the miombo woodland which extends over a vast area of western Tanzania and eastern Zaire. The area is exceedingly isolated, the mountains forming an uplifted island between the lowlands of east and west Africa. The nearest mountain masses, the Uganda-Kive-Burundi highlands to the north, and the Ufipa-Nyasa highland to the south, are several hundred kilometers distant (Fig. 1.1). Moreau (1943) studied the montane and primary evergreen forest avifauna of the area. Since Mahale is geographically equidistant from the northern and southern highlands, it would be expected that "northern" and "southern" species would be equally represented in the region. However, the bird fauna of Mahale contains ten times as many northern as southern species. Thus, Mahale has 21 "northern species", but only two "southern species". Only five of the 21 "northern" species occurring in Mahale, but both of the southern species, are represented by endemic races (Moreau, 1943).

Moreau (1943) obtained the following interesting conclusion from his own zoogeographical study and from geological evidence collected by others (see Fig. 1.1).

"The Malagarasi Depression may be of very old tectonic origin, but it is only recently rejuvenated, and in the Depression the present Malagarasi valley is being cut as a narrow erosion trench. The Karema Gap, on the other hand, represents an old rift valley, actually older than the Lake Tanganyika rift valley, by which it is crossed obliquely, its western end being represented by the Lukuga Gap at Albertville (Kalemie). Thus we may envisage a stage in the past history of this part of Africa when the Kivu-Urundi highland extended unbroken southwards past the site of Kigoma and across a yet uncut Malagarasi depression to the Kungwe-Mahare area, but was already separated from the Ufipa-Nyasa-Rhodesia-Marungu highland by Karema-Lukuga rift."

The geological history of Lake Tanganyika and its surroundings and the biogeographical implications of that history were more recently summarized by Fryer and Iles (1972).

2. Flora

Mahale is characterized by its extremely diversified vegetation. The vertical zonation and horizontal heterogeneity of plant types in the area are caused by local micro-differences in humidity, temperature, rainfall, and soil conditions. The type of forest which is most important for the survival of the chimpanzee is a "semi-deciduous" or "semi-evergreen" gallery forest, which is present from 780–1,300 m. From olden times, gallery forest has been the main site of slash-and-burn agriculture by the Bantu-speaking Batongwe tribe. This has resulted in the extermination of the virgin forest except for a small patch called "isigo lyetabami" where dead headmen of the villages are buried. This small patch of high forest suggests that the climax stage consists of *Albizia glaberrima, Xylopia parviflora, Pseudospondias microcarpa, Parkia filicoidea, Julbernardia seretti*, and *Pycnanthus angolensis*. Ecological succession is believed to proceed from abandoned field to elephant grass bush and *Harungana-Ficus* forest, and then through *Pycnanthus* forest to the climax *Albizia-Xylopia-Julbernardia* forest.

Higher altitude forest, which appears above 1,500 m, contains various plant communities. Many trees are covered with Spanish moss (*Usnea*). The forest develops especially along streams and at higher altitudes also on steep slopes near the main peaks. Major constituents are *Parinari excelsa, Anthonotha noldeae, Croton megalocarpus, Ficalhoa laurifolia*, etc. Tree ferns (*Cyathea*) are seen on the riversides. The highest parts of the mountains are sometimes covered with almost pure stands of hollow-stemmed "Alpine bamboo" (*Arundinaria alpina*).

Brachystegia woodland occupies the major part of the Mahale area with the exception of the western forested slopes. This genus exhibits vertical zonation: *B. longifolia* flourishes in the lower basin of large rivers, which is replaced by *B. bussei* at the height of 800 –1,000 m. Between 1,200 and 1,700 m, *B. manga* and *B. spiciformis* increase, while on the higher slopes, *B. boehmii* becomes dominant.

In the lower rugged terrain, *Diplorhynchus-Combretum* open woodland is dominant. At the higher altitudes of the western slope (1,200–1,400 m), *Uapaca kirkiana* forms conspicuous pure stands and provides comfortable evergreen shade.

The high-altitude montane savanna is characterized by scattered trees of *Parinari curatellifolia, Erythrina abyssinica, Ricinodendron heudelotti*, and *Terminalia mollis*. Beautiful flowers of *Protea* catch one's eyes. The highest parts of Mahale are mostly covered by montane grassland and miscellaneous flowering plants, an everlasting chrysanthemums *Helichrysum* spp. with white, yellow, and red varieties being the major constituent of the flower garden. Finally, the eastern hinterland of the Mahale Massif is extensively covered with pure stands of the solid-stemmed *Oxytenanthera* bamboo.

More detailed description of the vegetation for the Mahale area is given by Nishida (1972, 1977a), Nishida and Uehara (1981), Nishida *et al.* (1981), Anonymous (1985), and Collins and McGrew (1988). Nishida and Uehara (1981) compiled a vernacular (Sitongwe) and Latin dictionary of plants, which include more than 550 species.

3. Fauna

The animals of western Tanzania have different origins; savanna adapted species of east Africa share the area with savanna and woodland adapted species of south Africa, and forest adapted species of western Africa or the Congo Basin (Carcasson, 1966; Itani, 1977c; Kielland, 1978). Mahale is a biogeographically interesting region; its animals and plants show affinities to those of western rather than eastern Africa, while remnants of older faunas and floras are also preserved in the area (Davis, 1959; Juniper, 1959; Stevens, 1962).

Mahale is the richest area of Tanzania in west African fauna. Of particular interest are the ground squirrel (*Protoxerus*) (Davis, 1959), brush-tailed porcupine, Angolan black-and-white colobus, Livingstone's turaco (*Tauraco livingstonii*), and crowned hawk-eagle (*Stephanoaetus coronatus*). The Angolan black-and-white colobus of Mt. Nkungwe is, very likely, a heretofore unknown distinct subspecies (Nishida *et al.*, 1981). The subspecies of crested guinea-fowl of Mahale (*Guttera edouradi sethsmithi*) has not been recorded in western Tanzania, except at Mahale. This subspecies is distributed over northern Zaire, Uganda, and Kenya, west of the Rift Valley (Ulfstrand, 1960c). New species and subspecies of butterflies endemic to Mahale have been recently described (Kielland, 1978). Many species of forest-living butterflies of the genus *Charaxes* are most impressive. Several species of butterflies which were previously found only along the east coast of Tanzania and Kenya, and, nowhere in between, were discovered in Mahale may indicate that, at one time, the biota was more homogeneous throughout Tanzania (Kielland, 1968).

The presence of some mammals in the area was only recently recorded by the Japanese team. These include blue duikers, Sharpe's grysbok, bushy-tailed mongoose, banded mongoose, red-legged sun squirrel, and other rodents (*Praomys* and *Mus* (*Leggada*)). Some of these sightings represent considerable extensions of known ranges (Itani, 1977a; Hasegawa and Hiraiwa-Hasegawa, 1980; Nishida, 1985, unpublished). A list of mammals recorded at Mahale appears in Table 1.1.

Preliminary ecological studies have been conducted of red colobus (Nishida, 1972), yellow baboons, red-tailed monkeys, blue monkeys, and vervets (Takahata, 1980; Uehara, Hiraiwa-Hasegawa, and Nishida, unpublished data). Bats, shrews, and many rodents still remain to be identified. In addition to butterflies, ants, and termites, fruitflies have been collected extensively (Takada and Uehara, 1987; Okada, Asada, and Kawanaka, 1988). The fauna of the Tongweland including Mahale, from mammals to insects, with vernacular (Sitongwe) names, has been extensively described by Itani (1977a). A list of mammal, bird, and fish fauna was published in Anonymous (1980).

Table 1.1 List of mammals in the Mahale Mountains National Park[a].

Family and Latin name	ORDER	English name	References[b]
	INSECTIVORA		
Macroscelidae			
Rhynchocyon cirnei		Chequered elephant-shrew	2
Crocidura monax		White-toothed shrew	6
	CHIROPTERA		
	bats not yet identified		
	PRIMATES		
Lorisidae			
Galago crassicaudatus		Thick-tailed galago	2
Galago senegalensis		Lesser galago	3
Cercopithecidae			
Cercopithecus aethiops		Vervet monkey	1
Cercopithecus ascanius		Red-tailed monkey	
Cercopithecus mitis		Blue monkey	1
Papio cynocephalus		Yellow baboon	1
Colobus angolensis		Angolan black-and-white colobus	2
Colobus badius		Red colobus	1
Pongidae			
Pan troglodytes		Chimpanzee	1
	PHOLIDOTA		
Manidae			
Manis temmincki		Cape pangolin	2
	RODENTIA		
Muridae			
Mus tenellus			6
Praomys jacksoni		African soft-furred rat	6
Cricetidae			
Cricetomys emini		Giant rat	2
Sciuridae			
Protoxerus stangeri		Giant forest squirrel	2
Heliosciurus rufobrachium		Red-legged sun squirrel	5
Thryonomyidae			
Thryonomys swinderianus		Cane rat	2, 3
Spalacidae			
* *Heliophobius* sp.		Mole rat	2
Hystricidae			
Atherurus sp.		Brush-tailed porcupine	2
Hystrix galeata		Crested porcupine	2, 3
	CARNIVORA		
Canidae			
Canis adustus		Side-striped jackal	2
Lycaon pictus		Wild dog	5
Mustelidae			
Aonyx capensis		Cape clawless otter	2
Lutra maculicollis		Spotted-necked otter	2
Mellivora capensis		Ratel	2

Family and Latin name ORDER	English name	References[b]
Viverridae		
Genetta genetta	Common genet	3
Genetta tigrina	Large-spotted genet	2
Viverra civetta	African civet	2
Ichineumia albicauda	White-tailed mongoose	2
Bdeogole crassicauda	Bushy-tailed mongoose	3
Mungos mungo	Banded mongoose	3
Hyaenidae		
* *Crocuta crocuta*	Spotted hyaena	2
Felidae		
Felis sylvestrics	African wild cat	3
Felis serval	Serval cat	2
Panthera pardus	Leopard	2
Panthera leo	Lion	5
TUBULIDENTATA		
Orycteropodidae		
Orycteropus afer	Aardvark	2
PROBOSCIDEA		
Elephantidae		
* *Loxodonta africana*	African elephant	2
HYRACOIDEA		
Procaviidae		
Heterohyrax brucei	Yellow-spotted dassie	3
* *Dendrohyrax arboreus*	Tree dassie	3
PERISSODACTYLA		
Equidae		
* *Equus bruchelli*	Grant's zebra	3
ARTIODACTYLA		
Suidae		
Phacochoerus aethiopicus	Bushpig	2
Potamochoerus porcus	Warthog	2
Hippopotamidae		
Hippopotamus amphibius	Hippopotamus	2
Bovidae		
Tragelaphus scriptus	Bushbuck	2
Sylvicarpa grimmia	Bush duiker	2
* *Hippotragus equinus*	Roan antelope	2
Cephalophus monticola	Blue duiker	4
Raphicerus sharpei	Sharpe's grysbok	4
Oreotragus oreotragus	Klipspringer	2
Syncerus caffer	African buffalo	2

[a] Itani (1977a) provides a more extensive list of mammals living in the overall Tongweland including Mahale. However, he rarely mentions whether particular animal species occurred in Mahale.

[b] 1. Kano, 1971b; 2. Nishida, Uehara and Nyundo, 1979; 3. Anonymous, 1980; 4. Hasegawa and Hiraiwa-Hasegawa, 1980; 5. Nishida, 1985; 6. Unpublished data

* Not found in the Kasoje area.

IV. CONSERVATION

The value of the Mahale Mountains and the proposals and process relating to the reserve development have been described above, as well as in Hasegawa and Nishida (1984), Anonymous (1980), Nishida (1985) and Wrangham (1980). The scientific importance of the hinterland, the Ugalla area in particular, has also been mentioned (Itani, 1977d; Moore, 1986; Nishida, 1989a).

Thanks to the conservation measures taken by the Tanzanian government since 1975, some animals which were never or rarely seen before have appeared on the scene. These include African wild dogs, warthogs, leopards, buffaloes, and lions (Nishida, 1985).

V. RESEARCH ON CHIMPANZEES

1. Arrangement of Research Camps

The intensive study of chimpanzees has been conducted in the Kasoje area of the Mahale Mountains National Park. During 8 months of the initial study, Nishida lived at Kasiha, renting a room from a village chief (Nishida, 1973a). There are now three research camps in Kasoje: at Bilenge, Myako, and Kansyana (Fig. 1.3). Camps of Bilenge and Myako are on the shore of Lake Tanganyika. The former has served as the base camp for the study of B Group, and the latter for K Group. Kansyana Camp (6° 07′S, 29°44′E, about 820 m above sea level) is about 1 km inland from the shore of Kasiha (Fig. 1.4). Kansyana is within the overlapping area of the K and M Groups. It has been used for the study of both these unit-groups, depending on their seasonal range shifting (Nishida, 1968; Nishida and Kawanaka, 1972). In addition, Nganja camp was used in 1976 –1979 for the study of M Group and N Group.

2. Method

K and M Groups were provisioned and habituated at Kansyana in 1966 and 1968 respectively. Since then, and until recently, they had been supplied with sugarcane and banana. Provisioning allowed human observers to approach chimpanzees in close proximity in their natural habitat. At first, food was provided only at Kansyana ("fixed provisioning"), but after 1971 a small amount of sugarcane was also carried into the bush, for distribution when encountering chimpanzees ("mobile provisioning"). This allowed the keeping of an attendance record in the dense bush (Nishida, 1979b).

Since 1975, artificial feeding has been reduced because heavy feeding apparently influences the ranging behavior of animals. This occurred

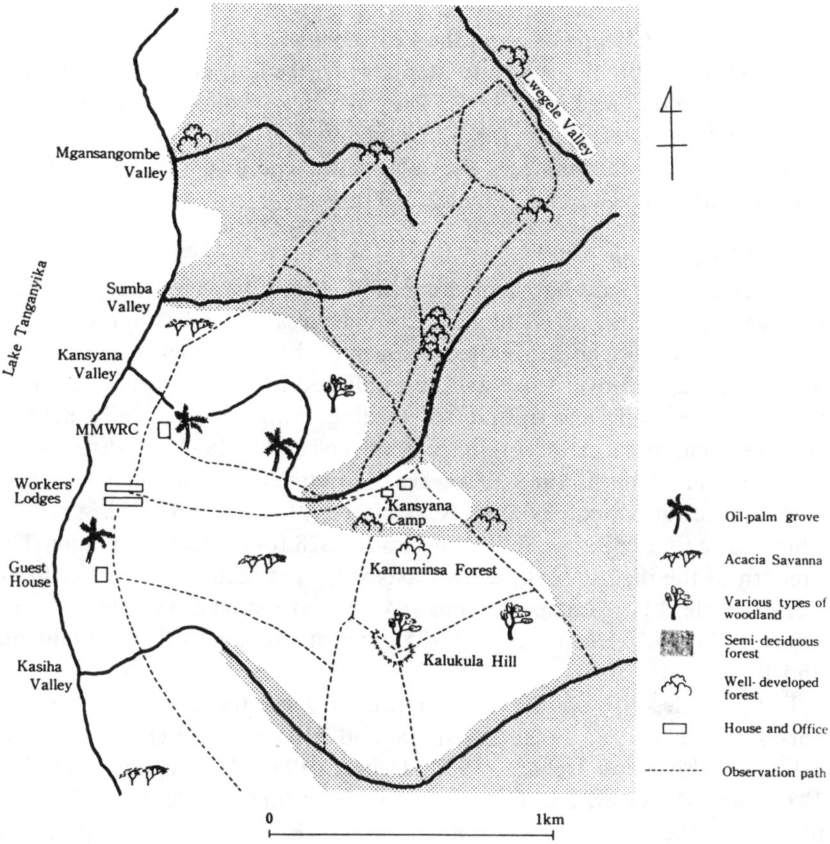

Fig. 1.4 The sketch map showing the headquarters of management and research and their surroundings.

also at Gombe (Wrangham, 1974; Goodall, 1983). However, provisioning at Mahale only matched the large extent seen at Gombe during the earliest period. Provisioning has been drastically reduced since 1981 and completely abandoned since 1987, for fear that poachers might take advantage of this method and because of the possibility of the transmission of human disease to chimpanzees.

In the earlier periods, with only one or two observers present, *ad libitum* sampling was the most profitable method of observation in terms of the quantity of data collected. With the increase of observers (sometimes up to five scientists and as many assistants in the bush) focal sampling has become feasible and is used as the major sampling method in our studies.

VI. ECOLOGY

1. Influences of Chimpanzees on the Environment

Chimpanzees break twigs and stems when feeding on plant parts and when making day and night beds. They swallow seeds of fruits, large and small, and transport them for a considerable distance. Since evacuated seeds germinate, chimpanzees act as a major seed dispersal agent (Takasaki, 1983a; Takasaki and Uehara, 1984).

2. Feeding Behavior

No extensive study based on quantitative data has been published on the feeding ecology of Mahale's chimpanzees. Nor has an extensive vegetation map of the study area been published. Ecological data which have accumulated over time will be published in the next publication of this series. So far, descriptions of feeding, ranging, and daily activity rhythms remain at a preliminary level (Nishida, 1974, 1976a, 1977a; Nishida and Uehara, 1983). However, a detailed list of foods eaten by chimpanzees at Mahale is already available (Nishida and Uehara, 1983). This shows that chimps eat 198 species (of 328 food-types) of plants. The breadth of the diet at Mahale contrasts with that seen at other localities. For example, 141 plant species and 184 types are eaten at Gombe (Wrangham, 1975), and 156 species and 205 types at Bossou, Guinea (Sugiyama and Koman, 1987).

Detailed descriptions have been provided on the natural history of ant-eating (Nishida, 1973b; Nishida and Hiraiwa, 1982) and termite-eating (Nishida and Uehara, 1980; Uehara, 1982; Collins and McGrew, 1985, 1987; McGrew and Collins, 1985). In particular, Uehara (1982) has illustrated the various tactics adopted by chimpanzees in response to differences in life cycles and species of termites.

3. Body Weight

Body weights of chimpanzees fluctuate seasonally, probably in response to changes in food supply. Surprisingly, a comparison of body weight showed that female chimpanzees of Gombe are significantly lighter than those of Mahale or of eastern Zaire (Uehara and Nishida, 1987).

4. Meat-eating and Relations with Larger Animals

In addition to at least 25 species of insect food, chimpanzees of Mahale have been reported to feed on at least four species of birds and 12 species of mammals (Nishida and Uehara, 1983). Recent observations added four more species to the list of vertebrate food: civet (Huffman, Nishida, and Uehara, unpublished), blue monkey (Uehara, unpublished), chequered elephant shrew (Takasaki, unpublished), and guinea fowl (Kawanaka, unpublished). This list suggests a dietary breadth comparable to that of protohominids.

Formerly, it was suggested that chimpanzees of Gombe and Mahale differed in methods of capture and in the particular species which were frequently eaten (Nishida, Uehara, and Nyundo, 1979). Recent study, however, demonstrated that these apparent differences were influenced by the size and composition of study groups and by sex differences in hunting behavior. Thus, larger parties that include more adult males are likely to hunt red colobus monkeys, while adult females on their own tend to "pick up" juvenile duikers (Nishida, 1980c, 1981; Takahata, Hasegawa, and Nishida, 1984). Red colobus monkeys not only are the major prey of chimpanzees (Wrangham, 1975; Busse, 1977; Kawanaka, 1982a; Norikoshi, 1983; Takahata, Hasegawa, and Nishida, 1984; Boesch and Boesch, 1989), but it is also the rule rather than the exception that two to four individuals are captured at once when "colobus-hunting" occurs (Goodall, 1986; Nishida *et al.*, 1984). This situation is similar in both Gombe and Mahale.

The discovery that chimpanzees sometimes scavenge mammals (Hasegawa *et al.*, 1983) may well have a major impact on the interpretation of early hominid sites, such as the "Hippo site" at Lake Turkana, Kenya. Earlier assertions that nonhuman primates do not scavenge (e.g., Teleki, 1973) have been disproved.

More surprising, perhaps, is the discovery that male chimpanzees killed a leopard cub, which was dragged out from a rocky cave (Hiraiwa-Hasegawa *et al.*, 1986; Byrne and Byrne, 1988). This observation showed that the elimination of a "future" predator or competitor is not a uniquely human trait.

5. Sex Differences in Subsistence Activities

From fecal analysis, Uehara (1984) demonstrated that M Group adult females eat more wood-boring ants than do adult males. Moreover, when adult males are in larger groups, they eat insects less frequently (Uehara, 1986). This sex-related dietary difference reflects not only a difference in the dietary requirements of each sex, but also sex differences in the extent of social activities. Thus, when there are many adult males in a unit-group, they must spend more time than adult females in social grooming to form and maintain alliances (Uehara, 1986, but see Chap. 10). Similar data were presented for Gombe's chimpanzees by McGrew (1979).

Hiraiwa-Hasegawa (1989) found that sex differences in feeding appears at an early age. Thus, female infants and juveniles spend more time feeding on ants than do males. Young females also build daytime beds more frequently than do their male counterparts.

VII. ETHNO-MEDICINE

At both Gombe and Mahale, chimpanzees were seen to swallow leaves of *Aspilia* (Asteraceae) slowly, and without chewing, in the early morning. Moreover, whole leaves were typically evacuated intact (Wrangham and Nishida, 1983).

Since the consumption rate was so slow, the nutritional value of the leaves was questioned. In fact, people of tropical Africa used to ingest parts of *Aspilia* as medicine (Watt and Breyer-Brendwijk, 1962), and chemical analysis revealed that *Aspilia* leaves contain a strong antibiotic "Thiaruburine-A" (Rodriguez *et al.*, 1985). *Aspilia* is consumed by apparently healthy individuals, both male and female, and old and young. Moreover, its ingestion increases, at least at Mahale, in the rainy season when intestinal nematodes increase (M. Kawabata, unpublished). It is likely, therefore, that chimpanzees use the plant as a vermicide.

Subsequently, Takasaki and Hunt (1987) and Nishida (Newton and Nishida, in press) found leaves of *Lippia pulicata* Baker, *Commelina diffusa* Burm., and *Ficus exasperata* Vahl being ingested in a similar way (that is, swallowed slowly without being chewed in the early morning). Although these leaves have not yet been chemically analyzed, it is interesting that some other *Lippia* species are also used as a medicine by African people (Watt and Breyer-Brendwijk, 1962) and that at least one species of *Ficus* is used by Amerindians as an antihelminthic agent (Caldecott, 1987).

At times other than early morning, and in the evening in particular, chimpanzees chew the leaves of at least *Aspilia* and *Ficus* rapidly and in great quantity as they feed on the leaves of other food plants. Therefore, the "medically effective elements" of the leaves are likely to be concentrated only in the early morning. The curious manipulation of the leaf with the tongue suggests that the active elements are absorbed through the "buccal route" as is nitroglycerine (Newton and Nishida, in press).

Recently, Huffman and Seifu (1989) observed an adult female, who was apparently suffering from an intestinal disorder, ingest a fair amount of juice from the pith of young shoots of *Vernonia amygdalina* Del. (Compositae). Other healthy chimpanzees present including her unweaned son and friends never did so. Preliminary analysis of the plant reports that the methanol extract shows strong anti-bacterial activity (Ohigashi *et al.*, 1989).

These types of study would be useful in many parts of central Africa. The new field of "chimpanzee ethnomedicine" provides an effective means of "screening" medically useful plants.

VIII. SOCIAL STRUCTURE

1. Population Structure

Chimpanzees live in a unit-group (or "community"; Goodall, 1983), which is a multi-male, multi-female, semi-closed group consisting of 20 – 105 animals (Nishida, 1968; Hiraiwa-Hasegawa, Hasegawa, and Nishida, 1984).

The Mahale Mountains support the highest population density of the eastern race of chimpanzees in Tanzania. Kano (1972) estimated the total population to be 2,000, of which almost 30% live around Mahale. A detailed description of the population is given in Chap. 3.

2. Age-sex Class and Female Reproductive Cycles

Wild chimpanzees mature slowly. The age-sex classes of chimpanzees have been defined by Nishida (1977b) and Hiraiwa-Hasegawa, Hasegawa, and Nishida (1984). Great individual differences exist in the growth rate and in relationships between mother and infant (Hiraiwa-Hasegawa, 1986).

Female sexual cycles and reproductive profiles have been given by Hasegawa and Hiraiwa-Hasegawa (1983). Anovulatory sexual swelling has been observed in lactating females (Takasaki *et al.*, 1986). Estrous synchrony was found among adult females (Nishida, 1977a, 1979a), although quantitative analysis was lacking.

3. Grouping and Ranging Patterns

Group size has been determined accurately for only two unit-groups. Twenty-seven individuals were present in K Group in 1967 (Nishida, 1968) and 105 in M Group (Hiraiwa-Hasegawa, Hasegawa, and Nishida, 1984) in 1980. Within a unit-group (and probably within a local population) the number of adult females generally exceeds that of adult males, while immature animals roughly equal adults in number (Hiraiwa-Hasegawa, Hasegawa, and Nishida, 1984). A unit-group usually splits into many temporary subgroups or parties. Mothers and infants are virtually the only stable units that continue over a month; however, a unit-group shows a certain degree of integrity. For example, most chimpanzees in both K and M Groups usually move seasonally and in concert from north to south or from south to north. Although individuals often appear to move rather independently to observers following individual chimpanzees, they nevertheless often synchronize their activities and have the same general destination. From a bird's eye view, movement of a unit-group may be ameba-like, each part seems to move randomly over time, but the whole body gradually moves to one destination. For M Group, subgroups consisting of more than 50 individuals are frequently observed at least when *Saba florida*, a large fruit, is in season. In some seasons, almost all members of a unit-group are concentrated within 1 km^2 while searching locally abundant food resources such as *Garcinia* fruits (Nishida, 1974).

Adult males and cycling females usually range over longer daily distances than do lactating mothers. However, the range covered in the course of a year is similar in all members of a unit-group, although adult males more frequently do boundary patrol. In other words, in the short-term, adult males cover larger areas than do females with dependent offspring, but on a long-term basis, the ranges of both are almost the same. Therefore, members of a unit-group can be said to share a common home range (see Chap. 4).

4. Inter-unit-group Relationships

Unit-groups avoid and are antagonistic to one another (Nishida and Kawanaka, 1972). A unit-group usually monopolizes its own home range, though overlap is extensive between some unit-groups such as K Group and M Group. Party size and the number of adult males present determine the result of clashes between unit-groups. Adult males of different unit-groups are particularly antagonistic and sometimes fight and display against each other (Nishida, 1979b). However, a 7-year-old male (Kawanaka, 1982b) and a 13-year-old male (Hasegawa, unpublished) of K Group were each seen interacting peacefully with a few adult males of M Group in the overlapping area between the two unit-groups. Moreover, estrous females move from one unit-group to another as mentioned below.

5. Female Transfer

Most females emigrate from their natal unit-group to one of the neighboring groups when they reach adolescence (Nishida and Kawanaka, 1972; Kawanaka and Nishida, 1975; Nishida *et al.*, 1985, see Chap. 3). The process of transfer is gradual (Uehara, 1981).

Young, immigrant females are usually welcomed by adult males but typically elicit aggressive responses from adult and adolescent females. New female immigrants, therefore, tend to follow adult males from whom they seek protection against hostile resident females. Furthermore, "secondary immigrants" who have spent many years together in another unit-group tend to associate together and to form coalitions against resident females. When a new female immigrant is the subject of aggression by resident females, adult males frequently rush to the scene, where they defend the immigrant females and then punish the resident females (Nishida and Hiraiwa-Hasegawa, 1985; Nishida, 1989b). When females enter their new unit-group, they are usually in estrus and mated with almost all resident males within a few days (Nishida, unpublished). When a new female immigrant, accompanied by her infant son, was not fully in estrus, the mother and infant were attacked by both males and females (Nishida and Hiraiwa-Hasegawa, 1985). Some females change their residential group more than twice, and a few females return to their natal group (Nishida *et al.*, 1985).

Males usually remain in their natal group from birth until death. However, there have been three cases of parous females permanently transferring between groups, accompanied by their juvenile sons (Kawanaka, 1981; Nishida *et al.*, 1985; Takahata and Takahata, 1989).

6. Infanticide and Cannibalism

Two cases of between-group (Nishida, Uehara, and Nyundo, 1979) and five cases of within-group (Norikoshi, 1982; Kawanaka, 1981; Nishida and Kawanaka, 1985; Takahata, 1985; Masui, 1986) infanticide have been recorded. Four cases of within-group infanticide were followed by cannibalism. Furthermore, at least four other infants are suspected to have been victims of infanticide (Hiraiwa-Hasegawa, Hasegawa, and Nishida, 1984; Nishida, 1989b).

Within-group infanticide is an evolutionary puzzle because the males who are the real or suspected killers may have fathered the victimized infants. Since adult males of different unit-groups are enemies, it is possible that killer adult males mistakenly identified their victims as offspring of males of neighboring unit-groups. This hypothesis is supported by the fact that the core areas of all the mothers of infanticide victims lay in the overlapping area of two unit-groups (Nishida and Kawanaka, 1985; Takahata, 1985).

IX. SOCIAL RELATIONSHIPS AND COMMUNICATION

1. Ethogram

No attempt at an exhaustive listing of ethogram in the chimpanzees of Mahale has been conducted. Preliminary listing is found in Nishida (1970). More recently-found patterns of behavior have been described by Nishida (1987b) and in other articles (see below).

A detailed analysis of communication patterns and a comparative analysis of vocal and gestural communication among great apes were conducted by Mori (1982) and Mori (1983), respectively.

2. Male-male Relationships

Adult males frequently tend to gather together and to reciprocate grooming. They show stronger social bonds than those among adult females or between adult males and females (Nishida, 1968, 1970; Kawanaka, 1984). Male bonding probably functions to facilitate the defense of their communal territory (Wrangham, 1975, 1979a; Nishida, 1979b).

Adult males tend to form a linear dominance hierarchy (Nishida, 1970). However, if the unit-group includes many adult males, it is often difficult to determine the dominance relationships among all individuals (Bygott, 1974). Thus, Hasegawa and Hiraiwa-Hasegawa (1983) could divide M Group's 10 males into only 4 classes: alpha, high-ranking, middle-ranking, and low-ranking (see also Chaps. 7 and 8).

Males appear reluctant to submit to higher ranking individuals. Adult males rarely pant-grunt to other males, except to the alpha male, and adolescent males rarely do so to other adolescent males (Hayaki, Huffman, and Nishida, 1989; Kawanaka, 1989, see Chap. 9). Apparently, they do not want to display overtly their subdominant status.

Coalitions are formed among adult males, making dyadic relationships unclear. "Basic rank" (Kawai, 1958) or "formal dominance" (de Waal, 1982) must be determined between two males when there are no other chimpanzees close at hand. However, it is sometimes difficult in the natural habitat to observe interactions between some pairs of males in the absence of other males. This may not be accidental; some males may deliberately avoid meeting particular rivals except in the company of their allies (Nishida and Hiraiwa-Hasegawa, 1987).

Alpha status confers a great reproductive advantage, in terms of both copulation and nutrition, on the male. (Nishida, 1983a, in press; Hasegawa and Hiraiwa-Hasegawa, 1983). To maintain his position, an alpha male needs the assistance of his coalition partners, while a rival male also needs his own allies to outrank the alpha male. This situation gives allies of both rivals "political" power. Such alliances give individuals the opportunity to outmaneuver higher-ranking males (Nishida, 1983a).

Food-sharing, food-snatching, or mating do not always occur along the lines predicted by the dominance hierarchy (Nishida, 1970, 1977a; Takahata, Hasegawa, and Nishida, 1984). This represents a most interesting aspect of chimpanzee sociality, and one which resembles that of human beings (Itani, 1988).

There have been two cases of defeated alpha males who remained estranged from the "male cluster" of a unit-group. These males spent most of their time alone at the periphery of the group's territory (Nishida, 1983a; Takahata, 1986).

3. Adult Male-adult Female Relationships

Most lactating females usually roam within their own narrow core-areas and rarely interact with adult males (Nishida, 1979b). When wide-ranging adult males meet with non-estrous females, they sniff their genitals, apparently checking their sexual state (Nishida, 1984). Adult males groom adult females more often when they are in estrus than when they are not (Nishida, 1986b). Females in estrus follow males, usually moving in a large party over a wide range.

4. Adult Female-adult Female Relationships

It is difficult for observers to ascertain dominance relationships among females, since they rarely meet. However, most habituated adult females show a linear dominance hierarchy (Nishida, 1979b, 1989b). Dominance rank among females is highly correlated with age and tenure in a group. Females sometimes unite and form temporary coalitions against a new female immigrant. Adult females also may unite against an older adolescent male who has directed charging displays at them, or against a young adult male who bullies one of them. New female immigrants who have given birth tend to associate closely with each other (Nishida, unpublished). Rarely (and not always effectively in terms of physical support), long-term coalitions are formed between immigrant females who emigrated from the same unit-group (Nishida, 1989b).

Nulliparous adult females sometimes select, follow, and groom one of the resident mothers. This "babysitter relationship" is not based on attraction to the mother, but to the infant (Nishida, 1983b).

5. Adult-immature Relationships

Although adult males rarely interact with infants, they do sometimes groom, transport, play with, and protect them (Nishida, 1983b). As juvenile and adolescent males become older, adult males display increasing hostility toward them. This is especially true in sexual contexts (Hayaki, 1985a). Adolescent males must always be careful of adult males, so as not to elicit aggression from them (Nishida, 1980a; Hayaki, 1985a). They perform the pant-grunt greeting to adult males in an exaggerated manner when they notice the slightest sign of arousal in them (Hayaki, Huffman, and Nishida, 1989, see Chap. 9).

Fig. 1.5 The *Brachystegia bussei* woodland on a hill and semi-deciduous forest in a valley (the Myako valley). (Photo by T. Nishida)

Fig. 1.6 Wedding ceremony of Batongne people in the Kosiha village, 1969. (Photo by T. Nishida)

Lactating females are usually indifferent to infants other than their own, but nulliparous females often seek contact with infants and engage in alloparenting behavior (Nishida, 1983b; Uehara and Nyundo, 1983).

Male orphans tend to follow particular adult males while female orphans follow particular adult females when an elder sibling is unavailable (Hayaki, 1988). Preferred adults are often past-prime. In one case a 5-year-old female orphan followed her adolescent brother (Hayaki, 1988).

6. Mother-offspring Relationships and Development

Only the outlines of the mother-offspring relationship has been published so far (mostly in Japanese; Nishida, 1977b, 1979a). Chapters of this book (see Chaps. 13, 14, and 15), therefore, are among the first analyses of mother-infant relationships based on the systematic collection of data.

Although the sample size used was small, it appears that females tend to mature socially earlier than males. This is inferred from the earlier ages at which a daughter begins grooming her mother, engaging in mutual grooming with her mother, and begins grooming others (Nishida, 1988).

A few females have been seen to transfer to another unit-group, leaving their weanling sons in the original group. Two females transferred under these circumstances temporarily (Kawanaka, 1981; Nishida *et al.*, 1985) and one female did so permanently (Hiraiwa-Hasegawa and Hasegawa, 1988).

Association partners become differentiated for males and females as they grow older (Hayaki, 1988). Males leave their mothers and associate with adult males, while females frequently continue to associate with their mothers into early adolescence.

7. Mating Strategies

Sexual relationships among chimpanzees are variable; two types are opportunistic and restrictive mating. In opportunistic mating, an estrous female may mate with many males in succession. Younger and less dominant males tend to engage in opportunistic mating more frequently. Higher-ranking and older males are more likely to show restrictive mating patterns (Hasegawa and Hiraiwa-Hasegawa, 1983). High-ranking males, particularly an alpha male, sometimes display "possessiveness" (Tutin, 1979) toward particular females in estrus, preventing other males from mating with "their" females, especially when ovulation is likely to occur (Hasegawa and Hiraiwa-Hasegawa, 1983). On the other hand, subdominant males sometimes show "safari behavior" (Tutin, 1979), taking their target females towards the periphery of the group's range in order to monopolize her (Nishida, 1983a).

Alpha status confers enormous reproductive advantage to a male, particularly when the number of adult males in the group is small (Nishida, 1981, 1983a; Hasegawa and Hiraiwa-Hasegawa, 1983). The number of adult males in a party influences male mating strategies (McGinnis, 1979; Hasegawa and Hiraiwa-Hasegawa, 1983).

Female choice plays some role in the coordination of mating. First, a female transfers spontaneously to another unit-group, not because she is driven away. Second, she may not acquiesce to a male who surreptitiously solicits her for a "safari", or she may break a consortship by attracting other males through loud cries (McGinnis, 1979; Nishida, 1981). However, if an estrous female is possessed by a powerful alpha male, she has almost no alternative but to mate only with him. A consort male may spend a considerable portion of his time closely watching his mate, at the expense of feeding and resting (Nishida, in press).

Females change their mating strategies according to the stage of their life cycle. For example, new female immigrants and young adult females tend to move promiscuously and mate with less dominant males (Hasegawa and Hiraiwa-Hasegawa, 1983; Takasaki, 1985; Hasegawa, 1989, see Chap. 5).

X. INCEST AVOIDANCE

No review of incest avoidance among Mahale's chimpanzees has been published. Mother-son copulation was often observed during the final process of weaning (N = 3 pairs; Nishida, 1981, unpublished).

There is only one case of a sexually mature son mating with his mother (Nishida, unpublished). In this instance, an 11-year-old male, TB, forcibly mated with his mother who was fishing arboreal ants in a tree. Intromission occurred, but the mother screamed in protest and drove him away. The son fell to the ground without ejaculating, continued screaming loudly, and never tried to mate with her again. Three past-prime mothers mated with many adult males, but never with their sons. In view of the lack of observation of this behavior at Mahale, Goodall's (1986) description of many examples of incest between mothers and their sexually mature sons is surprising.

It is less likely that father-daughter or brother-sister incest occurs, because females usually emigrate during adolescence. Furthermore, even if this type of mating does take place, it is not likely to result in a conception, since females have a long-lasting period of adolescent sterility (see Chap. 3). The simultaneous presence of an estrous adult female and her adolescent brother in M Group was observed only once, and they were not seen to mate (Kawanaka, unpublished).

XI. COGNITION

Research on cognition in nonhuman primates is difficult under the uncontrolled conditions of the natural environment. No systematic study of chimpanzee cognition has been conducted thus far. However, a few suggestive episodes reflecting social intelligence have been described. These include manipulation and "allegiance fickleness" (Nishida, 1983a), deception (Nishida, 1981, 1988), and cognitive mapping (Nishida, 1981). Hayaki (1985b) described an example of "imaginary social play," in which an adolescent male played with a branch as if he were playing with a friend.

Although admittedly speculative, the collection of such anecdotes would make it possible to formulate more rigorous methods of observation, description, and analysis of such episodes (Whiten and Byrne, 1986, see also Chap. 16).

XII. CULTURE

Field studies have shown that behavioral patterns of chimpanzees differ greatly across populations. Goodall (1973a) made the first comparisons between Gombe and Mahale. Subsequently, massive evidence of cross-cultural differences has accumulated.

1. Subsistence Activities

Chimpanzees display conspicuous differences in their dietary repertoire (McGrew, 1983; Nishida *et al.*, 1983; Goodall, 1986). Insect-eating is especially variable; Gombe chimpanzees feed on driver ants but neither *Camponotus* nor *Crematogaster* ants, while Mahale individuals feed on the latter two, but never on driver ants. Techniques in feeding on leaves and fruits are also likely to differ between populations (Nishida *et al.*, 1983).

With regard to tool use, chimpanzees at Gombe (McGrew, 1974) and at Bossou (Sugiyama, Koman, and Sow, 1988) have an elaborate method of using a long wand to fish for driver ants, while Mahale's chimpanzees do not have such a technique, although they occasionally use a long wand for expelling animals (Nishida and Hiraiwa, 1982) or extracting a bird nestling (Takasaki, unpublished) from the hole of a tree. Sponges for drinking water are manufactured from leaves at Gombe, but this behavior has been seen only once at Mahale (Takasaki and Tsukahara, unpublished) and therefore is not an established local culture there. It is also possible that tool-using techniques and hunting behavior differ strikingly between eastern and western chimpanzees (Boesch and Boesch, 1989, in press).

It was once asserted that responses to water differ between Gombe and Mahale (Nishida, 1980b). However, the Gombe data collected by McGrew (1977) have proved to be less clear-cut and may be an artifact of his rather short-term research period (Goodall, pers. comm.). Thus, the differences between Gombe and Mahale with regard to water have become increasingly blurred.

2. Gestures

McGrew and Tutin (1978) have pointed out that the "grooming-hand-clasp" posture of Mahale's chimpanzees is not found at Gombe. The function of this symmetrical face-to-face posture used in mutual grooming was interpreted as guaranteeing the mutuality of grooming by insuring that the same body parts are groomed and preventing cheating (Nishida, 1988).

Nishida (1980a) has described a courtship gesture named the "leaf-clipping display", which has not been seen in other parts of Africa, except at Bossou, Guinea, where the gesture was not ritualized as a courtship display (Nishida, 1987b). Recently, Nishida (1987b) described another courtship display unique to Mahale, while Hiraiwa-Hasegawa (1989) has pointed out that the greeting gesture of young Mahale chimpanzees is delicately different from that of Gombe.

3. Transmission of New Behavior

It appears that new and useful behavioral patterns are rarely invented by chimpanzees. The observation of new habits by primatologists in the field is even less likely. Most observations of the propagation of new behavioral patterns has occurred among free-ranging nonhuman primates when there was some form of human intervention (McGrew and Tutin, 1978).

Mahale's chimpanzees began to feed on newly available cultigens such as mangoes, lemons, and guavas (Takasaki, 1983b; Takahata *et al.*, 1986) after the resident humans were transferred to "Ujamaa villages" (see above). Adult females rather than juveniles sometimes initiated the new behavior, in contrast to the general belief that adult nonhuman primates are inherently conservative.

XIII. SUMMARY

1. The history of the Japanese study of wild chimpanzees is summarized. It was begun by Imanishi and Itani, who were interested in reconstructing the society of early hominids and especially in the origin of the human family. Along this line, earlier studies were concentrated on the social

structure, which led to major discoveries on the social unit, female transfer, and antagonistic relationships between unit-groups. With the introduction of behavioral ecology in the late 1970s, the framework and scope of research was extended and the research methods were elaborated. Our research station was also internationalized and opened to non-Japanese researchers. Our work contributed to the establishment of the Mahale Mountains National Park in 1985. Provisioning was abandoned so that the research would not be at odds with nature conservation.

2. The natural environments of the Mahale Mountains are briefly explained. Mahale's unique biogeographical status lies in the affinity of its flora and fauna with west Africa.

3. The achievements of the Mahale Mountains Chimpanzee Research Project are briefly outlined, with an exhaustive list of publications in English.

Acknowledgements

I thank W.C. McGrew for his constructive comments on the preliminary manuscript and for helping to build up an "exhaustive" list of publications relating to Mahale. I am grateful to T. Hasegawa, M.A. Huffman, M. Kawabata, K. Kawanaka, H. Takasaki, T. Tsukahara, and S. Uehara for the permission to use unpublished data. I am also grateful to S. Uehara and H. Takasaki for their invaluable comments and Y. Ikeda for drawing maps.

2

Safari Surveys of the Vegetation and the Chimpanzee Groups in the Northern Half of the Mahale Mountains

Junichiro ITANI

Center for African Area Studies, Kyoto University

I. INTRODUCTION

On September 14, 1965, at Ititiye Camp in the Kasakati Basin, five members of our study team—Akira Suzuki, Kosei Izawa, Takayoshi Kano, Toshisada Nishida, and I—met to discuss our strategies for future chimpanzee research (Fig. 2.1). Ititiye Camp, located inland 30 km east of Lake Tanganyika, was our second camp; the first had been set up at Kabogo. Although Suzuki, Izawa, Munemi Kawabe, and I had spent one year and nine months in the Kasakati Basin, we had not succeeded in habituating any chimpanzees to human presence and observation.

Because Suzuki and I were scheduled to leave Ititiye in late September for Japan via the Budongo Forest in Uganda, the main objective of the meeting was to determine the remaining three researchers' tactics. Before this meeting, we had spent two weeks on two preliminary survey trips. On one trip Suzuki and I traversed the Masito Plateau, east of Kasakati, to reach Tumbatumba, 45 km east of Ititiye, and took a westerly route along the escarpment north of the Lugufu Basin to Ititiye Camp. The other trip was made by Izawa, who traveled south along the lakeshore to make the first preliminary observation of the chimpanzees of Kasoje, on the western foot of the Mahale Mountains.

These survey trips revealed two promising sites, and we made plans accordingly. Izawa would remain in the Kasakati Basin and complete his ongoing study. Kano would set up a front camp in Filabanga and attempt to approach the group of 43 chimpanzees seen by Suzuki and me (Itani and Suzuki, 1967). Nishida would set up a camp at Kasoje and concentrate on provisioning the chimpanzees, which were reported to appear in the local people's cultivated fields.

Izawa continued his research in the Kasakati Basin and closed Ititiye Camp in April 1967. Although he did not succeed in habituating chim-

Fig. 2.1 Meeting at Itiye Camp on September 14, 1965. From left, T. Nishida, T. Kano, A. Suzuki, K. Izawa, Juma Ramadhani, and Khalphani Juma. (Photo by J. Itani)

panzees, he published papers on groups visiting the basin (Izawa, 1970, 1977).

Kano lived at Filabanga for 11 months and tracked the chimpanzees there. In this habitat, which was even drier than the Kasakati Basin, the chimpanzees had much larger ranging areas than first expected, and the conditions for observation were not favorable (Kano, 1971a). In September 1966 I helped Kano close Filabanga Camp, and the two of us extended our survey to the Ugalla River in the east (Itani, 1977d, 1979). After this, Kano engaged in an extensive survey of the distribution, population, and environment of chimpanzees on the eastern shore of Lake Tanganyika (Kano, 1971b, 1972).

Nishida finally succeeded in provisioning a group of chimpanzees at Kasoje. His individual identifications revealed the composition and outline of unit-groups. He described in particular detail the social relationships and behaviors of a group he named K Group (Nishida, 1968, 1970). Later, together with K. Kawanaka, he attempted to habituate M Group, which ranged south of K Group, and revealed the relationship between the two groups (Nishida and Kawanaka, 1972; Kawanaka and Nishida, 1975).

Izawa's survey trip to Mahale took him 17 days: he left Ititiye on August 26 and returned on September 11. Although we saw the Mahale Mountains from afar when we began our chimpanzee research in Kabogo in 1961, his trip was only our second approach to the area after Azuma's preliminary trial in 1962. During eight days in Mahale, Izawa saw numerous chimpanzee beds in trees along the lakeshore between Bilenge and Kasiha and saw the animals themselves four times on the foothills of Mahale. He reported it likely that there were at least three groups of chimpanzees. In comparison with the chimpanzees of Kasakati, those of Mahale, probably because they ranged near areas of human habitation, were relatively unafraid of humans, and the local people were likewise indifferent to the chimpanzees. The local people's tradition prohibited killing chimpanzees, and Izawa had the impression that provisioning them with sugarcane would be feasible. On the same trip he attempted to climb Mt. Nkungwe upstream along the Kasiha River, but retreated because of the lack of food. On his way back from Kasoje, he spent four days investigating the area from Nkungwe Bay to Mt. Kakungu, where he heard chimpanzee calls twice. He concluded that Mahale was the most promising of the areas.

It was on October 12, 1965, that Nishida visited Kasoje. The progress of his work and his success in provisioning the chimpanzees in May 1966 are detailed in one of his books (Nishida, 1973a).

This is the way our chimpanzee research at Mahale, and subsequently in the Mahale Mountains National Park, began.

One year after the meeting at Kasakati on September 25, 1966, I visited Mahale. Accompanied by two Africans, I walked west from Mpanda to reach the Mahale Mountains. I may be the only researcher who approached Mahale not from the lake but by an inland route. Later I attempted several extensive safaris around Mahale.

At Mahale, after their initial success in provisioning the chimpanzees, our colleagues continued intensive studies of K and M Groups. During my own short visits to Mahale, I often planned several-day safaris. This chapter deals with the data on chimpanzees collected during such safaris. However, data recorded at places too remote from the K and M Groups (e.g., the Nyenda Plateau, 25–30 km south of Kasoje), among those expected to have little social interaction with the K and M Group chimpanzees, are not included.

Kano (1971b) reported in his study on the distribution of primates in the area south of the Malagarasi River, within Kigoma District, and in the northwestern part of the Mpanda District. Nishida also made a number of safaris, and other researchers who have studied at Kasoje collected data from the local people. Some of these data are briefly summarized in Chapter 3. In this chapter, I would like to limit the data to those from my own seven safaris.

Year after year the research has progressed. It is a great pleasure that a book such as this one, which was undreamed of at the onset of our chimpanzee study, can be published. However, there are still many unanswered questions about relationships among unit-groups. Furthermore, our tradition of wandering over a large area and posing questions seems to have gone out of fashion among the younger generation of researchers. Although the data presented here are incomplete, I believe that they will be of some use.

II. STUDY AREA AND METHODS

The seven safaris primarily covered the vicinities adjacent to the home ranges of K and M Groups, but also included a little larger area (Fig. 2.2). The Mahale Mountains extend from north-northwest to south-southeast, bounded by Lake Tanganyika to the west and the hilly country covered with *miombo* woodland and *Oxytenanthera abyssinica* thicket to the east. It is likely that the distribution of chimpanzees at Mahale is discontinuous from Mt. Karobwa to the northeast and from the Wansisi Mountains to the east. Although Kano (1971b: Fig. 10) connected the distribution between Mahale and Karobwa, there is little positive information from the area in between. The chimpanzees in the Mahale Mountains, therefore, are considered to be isolated in the area extending about 60 km north-northwest to south-southeast and 25 km from east to west. The area covered here is its northern half.

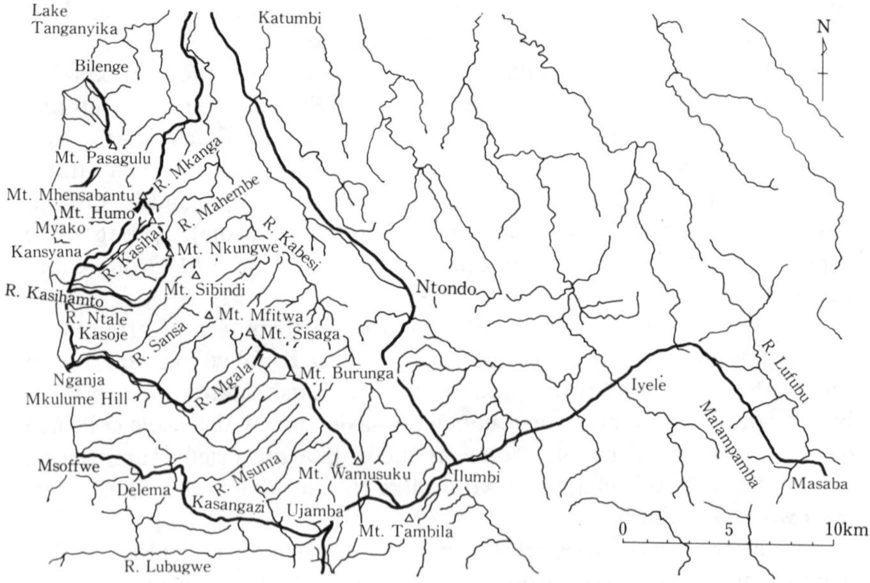

Fig. 2.2 The study area, and the routes of the seven safaries.

Although direct encounters with chimpanzees were rare, their calls were heard on many occasions. In addition, as indirect evidence of the chimpanzees' presence, data on beds, food remnants, and feces and information from the local inhabitants were recorded. The seven safaris reported here were made between 1966 and 1979. The eviction of local inhabitants in favor of a subsequently planned national park in the area, under former President Nyerere's *Kijiji Cha Ujamaa* policy, had not been completed at that time, and information on the chimpanzees was available from the inhabitants who remained.

I was accompanied on the safaris by 2–4 Tongwe workers, who also provided me with invaluable information on the geography, place names, plants, and so on. We traveled with a minimum of equipment and camped for a week in no-man's land. We walked with a 1/50,000 scale map which, without contours, showed only the stream lines, mountain ridges, and peaks.

The 1955 edition of the 1/50,000 map published by the Tanganyika Department of Lands & Survey, which I used for the safaris, and the 1983 edition of the 1/50,000 map with contours published by the Tanzania Ministry of Lands Survey and Mapping Division differ considerably on the heights of mountain peaks. The present chapter follows the 1983 edition for the heights of mountains, and as a result it may differ from the other chapters in this volume on such figures. For example, the peak of Mt. Nkungwe is indicated as 2,460 m a.s.l. in Chapter 1, which follows the 1955 edition, while in this chapter it is 2,520 m. The measurement taken by our altimeter in 1979 indicated that the peak of Mt. Nkungwe is 2,490 m, which lies between the two figures.

III. RESULTS

1. Across the Mahale Mountains (September 22–27, 1966)

On September 22, 1966, heading from Mpanda to the west-southwest, we spent the night at J. Kielland's house in Sibwesa (30°45′E, 06°31′S) 35 km from Mpanda, and on the 23rd he gave us a ride to Mt. Kafulungu. From there Isa Rajabu, Mohamedi Ismaili, and I went on foot to the hamlet called Mwina (30°15′E, 06°16S).

On the 24th we began our approach to Mahale on foot. We walked 25 km west to reach Masaba (30°04′E, 06°12′S). On the 25th we walked north along the ridge of Malampamba, lying east parallel to the main ridge of Mahale, passed the hamlet of Iyele (30°01′E, 06°12′S) between the two ridges, and climbed up the steep slope until 23:00. We could not reach Ilumbi, so we bivouacked.

The vegetation and fauna en route are summarized as follows. The area between Sibwesa and Mwina was interwoven with the sources of the east-flowing Katuma, the north-flowing Lwegele, and the south-flowing Msenguse, Lwega, and Nyamuzi Rivers. There were few villages. Pure stands of *Julbernardia globiflora* and *Isoberlinia angolensis* were dominant in the vegetation. The bamboo thicket of *Oxytenanthera abyssinica*, which is thought to be a secondary vegetation due to human disturbance, was also found here and there. Although the vegetation between Mwina and Masaba differed little from the above, towards the west meager riverine forests appeared, in contrast with the bare riverbanks before Mwina. Along the main stream of the Lugonezi River there was a fully developed evergreen riverine forest.

Across the Isabika River, a tributary of the Lufubu River, to its west bank, *Julbernardia* disappeared. Besides the *Isoberlinia* woodland and bamboo thicket, the savanna vegetation of *Diplorhynchus condylocarpon* and *Combretum collinum* appeared in patches. On the hill at Malampamba, west of Masaba, *Oxytenanthera* bamboo was most dominant. The source area of the north-flowing Lagosa River was also completely covered with bamboo up to the middle of the steep slope leading to Ilumbi, where it completely disappeared and was replaced with pure stands of *Brachystegia spiciformis*.

The large mammals we observed in the area were two herds of buffaloes, on the riverbank of the Katuma and on a foothill of Mt. Kafulungu, and three giraffes in a woodland on the west bank of the Syankala River. The local inhabitants said that giraffes and sable antelopes do not cross the Lugonezi River to the west. Feces and other signs of elephants were prominent to the west of Masaba. (Kielland said that when he bivouacked near a hamlet called Katente upstream on the Musenguse around 1956, he could not sleep all night owing to the noise made by herds of elephants continuously passing from south to north.) On the slope of Mt. Kafulungu we saw zebra feces. No chimpanzee beds were found.

We visited a hamlet called Mhambwe, upstream on the Lwega, to hire a porter for the journey up to Masaba. Kielland said he saw signs of chimpanzees on Mt. Ninangombe and Mt. Mwesi (alt. 1,937 m) north of Masaba. From the summit of Mt. Kafulungu we commanded a view of Mt. Mwesi. The top of Mwesi was covered with montane forest, in contrast with Kafulungu, which was covered with grass. Kielland said, "As the government of Tanzania set up a camp at Mwesi for refugees from Burundi, probably the chimpanzees moved somewhere else."

Kielland also said that chimpanzees inhabit the Wansisi Mountains, 40 km southeast of Mwesi. Later Kano surveyed the Wansisi area and concluded that the 500 km^2 range is a chimpanzee habitat completely isolated from the others and that it is the southern limit of chimpanzee distribution in Tanzania (Kano, 1972).

There was a small hamlet named Mwina on the Nyamuzi River, a tributary of the Lwega River. Kapipi Kabango, about 60 years old, said that there had been no chimpanzees in the area for a number of years. On the west bank of the Lugonezi River, covered with evergreen riverine forest, there was a hamlet called Kabungu. Mukulo Senlekuo, an old man, said that many years ago chimpanzees had come there. Probably they were temporary visitors from Mahale; and if that is the case, his remark was the earliest information on the chimpanzees of Mahale. Ramadhani Sinundo of Masaba said, "Chimpanzees come to a hill called Kakula west of Masaba." However, Kasumuni Liyoba, who had lived for 14 years at the hamlet of Iyele upstream on the Lagosa 10 km northeast of Masaba, said he had never heard chimpanzee calls there. The area was covered with bamboo thicket.

Chimpanzees may dislike *Oxytenanthera* bamboo thicket. Mahale chimpanzees were also said to travel down to the Lufubu and near the Lugonezi. Although the area is probably only temporarily visited by Mahale chimpanzees, it may be regarded as the eastern limit of their distribution. It is interesting that the chimpanzee's habitat borders that of the giraffe, with the Lugonezi River serving as a boundary.

On September 26, it did not take even half an hour to walk to Ilumbi from the campsite. The house of the most renowned *mwami* (chieftain) of the area, Lukandamila Liyoba, was in a *Brachystegia spiciformis* woodland. Ilumbi, about 1,700 m a.s.l., commanded a view of the woodland to the east, where I had walked. To the west, across the shallow valley of the Kabesi, Mt. Nkungwe, Mt. Sisaga, and other peaks of the Mahale main ridge were seen. In the afternoon, climbing up the main ridge slope, we reached the northern saddle point of Mt. Tambila. I left Isa and Mohamedi there, and without my pack reached the summits of Mt. Wamusuku (alt. 2,340 m) and Mt. Burunga (2,360 m), but I had to give up the summit of Mt. Sisaga as darkness fell just before we reached the top. Returning to the saddle point, we climbed down to a hamlet called Ujamba, on a crest branching west from the main ridge, and spent the night there. Ujamba was located at the highest elevation of all the hamlets in Mahale. It was estimated at 2,000 m a.s.l.

On the 27th, after traversing a slope to the north from Ujamba, we descended the narrow ridge to the hamlet of Kasangazi, which was also a *mwami*'s residence. The valley of the Msuma immediately below Kasangazi is a tributary of the Lubugwe River. Across the Msuma, on its west bank, we passed a hamlet called Kapala. After heading west across a plateau, we climbed down a steep slope to Msoffwe on the shore of Lake Tanganyika. We went on north by canoe and landed at Nganja at 18:30. At 20:25 we arrived at Nishida's camp at Kansyana.

Although the area upstream of the Kabesi River was thought to have been covered with riverine forest, the area had been mostly turned into cultivated fields. The steep slope up to the main ridge was covered with tall grasses, with scattered *Erythrina abyssinica, Protea* sp., *Syzygium guineense*, etc. As we climbed higher, the grasses became shorter and were eventually replaced with montane grassland. From the saddle point on the main ridge to Mt. Wamusuku and Mt. Burunga, the vegetation was a mosaic of montane grassland, montane forest, and *Arundinaria alpina* bamboo thicket.

Behind the hamlet of Ujamba was a deep montane forest. To the north of the hamlet, there was a large maize field in the forest. Fallows were covered with herbaceous bush of nettles and *Periploca linearifolia.* Although the plateau between Kasangazi and Msoffwe was covered mostly with secondary vegetation like *Hyparrhenia* sp., occasionally patches of *Uapaca kirkiana* and *Parinari curatellifolia* were seen. The slope down to Msoffwe was a woodland of *Brachystegia* spp.

Lukandamila Liyoba said, "Chimpanzees come along the Kabesi River. Some 10–13 chimpanzees form a group, and if two such groups meet, they make much noise." Probably he referred to meetings of two parties of a single unit-group. On the main ridge we did not hear any calls or see any beds. At Ujamba, I was informed that chimpanzees ranging from Kabambi Mtipula came this far only occasionally to eat maize stems when the plants become tall. The informant guessed that they came from over the main ridge in the direction of the Kabesi River.

Mwami Lubaka Kaliyafilime at Kasangazi, about 60 years old, said: "Chimpanzees have lived there since early times, and people never kill them. They do not come from the lakeshore but live here, and occasionally visit Ujamba or the main ridge. They are frequently seen ranging from the Msuma River to the Lubugwe River. They do not eat the ears of maize, but they eat the stems of maize and sugarcane." At Kapala, too, though no call was heard on the day we were there, we heard that chimpanzees often came.

Ramadhani Ismaili, who walked in the rear when climbing down the steep slope to Msoffwe, heard chimpanzee calls in a valley a little upslope of Msoffwe. We found seven beds nearby. Heading to Kasoje by canoe, we counted 235 beds in *miombo* trees along the lake. Undoubtedly there was a group of chimpanzees around Msoffwe.

A single bushbuck seen near Kapala was the only large mammal witnessed between Ilumbi and Msoffwe. However, we saw many traps of various kinds on the way, for leopards, honey-badgers, bushbucks, rock hyraxes, and so on. The importance of hunting to the mountain-dwelling Tongwe people has been described by Kakeya (1976), and Itani (1977a) has described their trapping techniques.

2. Around Mt. Nkungwe (September 29–30, 1966)

Since my arrival at Kansyana was one week behind schedule, I had to shorten my stay at Kasoje. However, I did not want to give up the planned climbing of Mt. Nkungwe with Nishida. The climb was started on the 29th. Since Nishida (1973a) has written the details of this trip, here the description is limited to the outline of the route and information on the distribution of chimpanzees. Our ascent of Mt. Nkungwe was the fourth recorded: the first was by O'Hagan via the Kasiha Valley in 1939, the second by Procter via the Kabesi River in 1946, and the third by the Oxford Expedition in 1958–59. Many unsuccessful attempts by others, including S. Azuma and Izawa, had been recorded. Although its altitude is only 2,520 m, Mt. Nkungwe's slopes and thick bush cover did not allow an easy approach to its summit.

Nishida planned a route from Kansyana to Mt. Nkungwe via Mt. Mhensabantu and Mt. Humo. Part of the trail up along the ridge going northeast from Kansyana to Mt. Mhensabantu had been cleared in June. Thanks to this prior preparation, the climb, which had taken O'Hagan four days, took us only two days.

Nishida and I departed Kansyana at 8: 20 on the 29th, accompanied by three men, Omari Kabule, Mohamedi Saidi, and Juma Mkwituwa. We reached the ridge of Mhensabantu at 9: 45 and its summit at 14:20. At 16:20 we arrived at the saddle point between Mt. Mhensabantu and Mt. Humo and looked in vain for water. We reached the summit of Humo at 17: 00 and took a trail off the main ridge. Just before sunset, we found water dripping between rocks in the montane forest clinging to the northeastern slope of Nkungwe, and there we bivouacked.

On the morning of the 30th we started climbing at 8:50 and reached the saddle point between Humo and Nkungwe at 9:20. After two hours of climbing in thick bush we arrived at the summit of Nkungwe (2,520 m). The main ridge extends to the southeast—Mt. Sibindi, Mt. Mfitwa, and Mt. Sisaga. Between Nkungwe and Sibindi lies a deep gap. We chose the southwest ridge on the south of the Kasihamto Valley leading to Mt. Kahoko for a descent. After much struggle in the bush we arrived back at Kansyana Camp at 23:30.

We did not have enough time to study the vegetation, which is described in more detail in the account of my second climb of Mt. Nkungwe (Section 7, below). Here I mention only that on the way down there was a pure stand of *Ficalhoa laurifolia* at 1900 m a.s.l., at the source of the Kasihamto.

During these two days, we heard chimpanzee calls six times. On the morning of the 29th, at 6:00, we heard a call in the upstream direction of the Kansyana Valley. Shortly after starting from Kansyana, at 8:20, we heard another call from almost the same place. At 9: 40, a little before reaching the southwestern ridge of Mt. Mhensabantu, we heard a third call from northwest below.

Nishida said that the provisioned K Group had not been around Kansyana for a few days and that unhabituated chimpanzees were around the feeding site. These animals were later named M Group, and the home ranges of K and M Groups were found to be overlapping in the Kansyana area (Nishida and Kawanaka, 1972). The first three calls we heard that day were probably from the same group; however, it is uncertain whether they were from K or M Group.

A little before reaching the *Philippia-Hypericum* zone on the Mhensabantu ridge, we heard a fourth call, from the Kasiha Valley to the southeast. The right bank of the Kasiha is almost entirely cliff rather than slope. The call was coming from the bottom of the valley, and it was obviously from neither K nor M Group. On reaching the lower limit of the montane forest we heard a fifth call, again from the Kasiha Valley. The fourth and fifth calls were probably from the same group.

On the 30th, after reaching the summit of Nkungwe, we climbed down the southwestern ridge leading to Mt. Kahoko. We heard the sixth chimpanzee call from the Ntale River due west of Kahoko. If we consider that the first, second, and third calls were of K Group, the sixth was of M Group, and the fourth and fifth were of another group, the distribution of the calls can be reasonably explained. Nishida (1973a) stated that this call came from the direction of the Sansa River. If this is the case, we have to consider the presence of a fourth group. Later the group that emitted the fourth and fifth calls was named H Group. On the evening of the 29th, we found chimpanzee feces on a rock on the ridge just before reaching the summit of Humo. This fact indicates that the group which was calling in the Kasiha Valley—which we called H Group—occasionally crossed over the main ridge of Mahale.

On the saddle point between Mhensabantu and Humo, we heard strange calls, which the Tongwe guides told us were those of *nkamba* (*Colobus angolensis*). On our way back, at a rocky place, we heard the same calls, and a little later found hairs and a mandible of *Colobus angolensis*. We collected the mandible. Nishida (1973a) inferred that it was killed by a leopard. However, on the basis of what we learned later, it is also possible that it was killed by chimpanzees.

The climbing of Mt. Nkungwe was not the end of my safari in 1966. I went to Mgambo by boat and went upstream along the Lwegele River, which flows into Nkungwe Bay. After tracking chimpanzees in the Mibanga area, I crossed the Lugufu Basin and returned to Ititiye Camp in Kasakati. The circular route starting from Ititiye on September 1 and returning there on October 8 totaled 450 km.

3. Again to Ilumbi (June 6–9, 1971)

Five years later I again visited the Mahale Mountains. The aim of the visit this time was to conduct an ecological anthropological study among the mountain-dwelling Tongwe people. On June 6, 1971, Makoto Kakeya, Hideko Kakeya, Haruna Huseni, Alimasi Kasulamemba, Saidi Musa, and I went to Msoffwe from Kasoje. From Msoffwe, we passed Ujamba, crossed the Mahale main ridge, and visited Ilumbi. We went north along the Ilumbi ridge to reach Katumbi Village on Nkungwe Bay. Although it was a short safari, taking only four days, it supplemented the information on the vegetation and distribution of chimpanzees in the Mahale Mountains.

The hamlet of Msoffwe was deserted. As on the previous safari, there were many chimpanzee beds in the *miombo* woodland on the lakeshore around Msoffwe. Undoubtedly Msoffwe was one of the core areas of a chimpanzee group, whose range will be discussed below.

After climbing a steep slope, we reached a plateau. We spent the first night at Delema, the first hamlet we came across. The vegetation on the plateau indicated that villages had been built and deserted there many times over the years. The plateau is about 1,500 m a.s.l. Other than the forests scattered near waters, the area was covered with 2 m-high *Hyparrhenia* sp. People at Delema said that these patched forests were frequently visited by chimpanzees. The plateau is the source of the Lubugwe and Lubulungu Rivers. On the 7th we passed Makusu Hamlet, where people were said to hear calls of chimpanzees coming up along the Lubulungu River.

We again passed Kapala and Kasangazi, where I had visited five years ago. We obtained no information other than that there were more chimpanzees on the Nyenda Plateau to the south. After climbing the steep ridge of Kasangazi, at 17:00 we reached Ujamba, where I again met with friendly people.

On the 8th we crossed over the main ridge and climbed down to Ilumbi. The notable tree species were *Parinari excelsa, Anthonotha noldeae, Ricinodendron hendelotii*, and *Croton megalocarpus*. The montane grassland was beautiful, with many kinds of flowers in bloom just after the rainy season.

We spent the night at Ilumbi, where we got no new information on chimpanzees. On the 9th we walked north on the ridge of Ilumbi. The woodland dominated by *Brchystegia spiciformis* consisted of *Parinari curatellifolia, Uapaca kirkiana,* and *Garcinia huillensis*, most of which bear fruit favored by chimpanzees. Although we stopped at Ntondo, the hamlet was empty. Without further information on chimpanzees, we descended to Katumbi on the lakeshore. On the way, we heard no chimpanzee call and saw no beds, food remnants, or feces.

4. To Mt. Sisaga (June 18–21, 1971)

On June 11, 1971 Nishida arrived at Kasoje with his family. On the 18th, Nishida, Makoto Kakeya, Mohamedi Seifu, Mauldi Seifu, Sadi Katensi, and I left Kasoje by boat. At 9:00 we started our safari up the Lubulungu River. Before long we heard chimpanzee calls from the north, on the foot of Mt. Kahoko. The calls were from M Group. Passing the junction with the Sansa River, flowing from the east, we went up along the main stream of the Lubulungu. At a pool called Songosiba, there were new footprints of a chimpanzee in the sand. Upstream from the pool was a waterfall, and further upstream was another pool, called Gwamwami, followed again by another waterfall upstream. We climbed on the right bank. The vegetation along the main stream was thicket without developed forest. As it was difficult to climb along the valley, we detoured it on the east bank and again approached the main stream. At 18:00 we found a site appropriate for camping. We heard no chimpanzee call.

On the 19th we started at 9:00. After half an hour's walk we reached the fork of the Lubulungu and the Mgala Rivers. We walked on the south bank of the Mgala River. *Sapium ellipticum* and *Anthocleista schweinfurthii* were prominent in the area. *Erythrina abyssinica* and *Ricinodendron heudelotii* were occasionally seen. Gradually the vegetation was replaced by bush mixed with *Hyparrhenia* sp., through which walking was difficult. This was said to be the site of Mwami Nyundo's fields, deserted about fifty years earlier. In the thicket *Steganotaenia araliaceae, Canthium crassum*, and *Harungana madagascariensis* were prominent. The altitude was thought to be about 1,500 m. We neither found any evidence of chimpanzee habitatation nor heard any call. Cutting through the bush, we reached the montane grassland at 15:00. *Aloe* sp. and *Helichrythum* were seen. Off the ridge the thicket was dense. *Arundinaria* bamboo thicket began to cover part of the ridge. At 17:00 we descended to the source of the Mgala River and camped there.

On September 20 we returned to the ridge that we had passed the day before. Climbing through the montane forest, we reached the main ridge. There two beehives were placed in a tree; a man of Ntondo to the east of Mahale was said to have set them. We passed through a saddle point covered with bamboo, and at 11:00 we reached a small peak. Leaving all our gear there, we headed northwest for Mt. Sisaga. The steep slope of 1 km (estimated by sight) was a tough climb after the descent to a deep saddle point. At the saddle point we saw a trap for catching leopards and a pitfall for trapping bushbucks. On our way up we found a chimpanzee bed in a bamboo thicket at about 2,350 m a.s.l. The bed had been made less than one month before. This bed evidenced a chimpanzee's presence at one of the highest altitudes ever recorded.

Through the bamboo thicket, clinging to a rock covered with grasses, we reached the summit (alt. 2,474 m). The summit was covered with thickets and grasses. It commanded magnificent views of Mt. Mfitwa and Mt. Sibindi just in front and Mt. Nkungwe beyond them. A fireplace in which the mountain-dwelling Tongwe people enshrined the mountain spirits was buried in the grasses at the summit. At 13:40, a sudden outburst of chimpanzee calls came from a forest about 2,300 m a.s.l. in the source area of the Kabesi River to the north. The chimpanzees must have been alerted by seeing us on Sisaga. The outbursts were heard three times in a row. Although we could not see them with binoculars, it is most likely that the chimpanzees crossed the main ridge along the montane forest.

We departed the summit at 14:00 and returned to the point where we had left our gear. Ascending and then descending the main ridge, we passed to the east of Mt. Wamusuku. Although the distance to the rock north of Mt. Tambila was only 12 km in a straight line, cutting through the bamboo thicket and bracken bush was difficult. After sunset the southeasterly wind was cold. At 23 : 10 we finally arrived at Ujamba, where we spent the night. Unfortunately all the men were away, and we were unable to get any new information on the chimpanzees. On the 22nd we climbed down the Mahale Mountains and conducted our first study of the Nyenda Plateau to the southwest. However, the data from the Nyenda Plateau are not included here because the area is too distant from that of the Kasoje study groups.

5. Mt. Pasagulu (January 28–30, 1974)

Mt. Pasagulu (alt. 1,580 m) is the northernmost peak of the Mahale Mountains. It had been known from the early stages of the chimpanzee study at Mahale that a unit-group of chimpanzees live around this peak, partly overlapping its range in the Takata Valley with K Group to the south (Nishida and Kawanaka, 1972; Kawanaka and Nishida, 1975). On January 28, 1974, Shigeo Uehara, Haruna Huseni, Mohamedi Seifu, Sadi Katensi, and I left Kasoje for Mt. Pasagulu. The long, hard rain became mixed with hail, and we waited at Myako Camp until it cleared. We landed at Bilenge at 15:00 and immediately started climbing. We crossed over the Bilenge River to the south bank. The area is covered with *miombo* open forest comprising many *Brachystegia bussei* trees, among which we counted several chimpanzee beds. Besides *B. bussei*, there were *B. spiciformis, Uapaca kirkiana, Syzygium guineense, Strychnos innocua*, and patches of *Oxytenanthera abyssinica*. We chose a campsite at 1,250 m a.s.l. for the night. We heard no chimpanzee call.

On the 29th, it rained until 15: 00, when Uehara, Mohamedi, and I started climbing. *Ricinodendron hendelotti, Gardenia joris-tonantis, Hymenocardia acida*, and *Oxytenanthera* bamboo were sparsely scattered. Although we reached the peak just in front of Pasagulu, visibility was low owing to the thick cloud cover. We postponed the attack on Pasagulu to the next day, and returned to the campsite. We neither saw any sign of the chimpanzees' presence nor heard any call. Mohamedi and Haruna informed us that an old man named Jungan Kalike said there was a Tongwe village near the summit of Mt. Pasagulu.

On June 30 the sky was clear. Climbing the peak we had reached the day before, we realized that the summit of Mt. Pasagulu was just behind and to the left of the peak. We followed the left slope of the front peak directly to the summit, which we reached at 11: 05. There were neither trees nor bamboos, only grasses. A troop of 17 olive baboons (*Papio cynocephalus*) crossed the ridge to the south. On the summit we heard a *Colobus angolensis* call from the forest at the source of the Kabwe River. The altitude was estimated at 1,350 m, which is probably the lowest recorded for the habitation of *C. angolensis*. We descended along the ridge stretching to the southwest, collecting plant specimens as we went. On the way we found no evidence of chimpanzees' presence.

6. Upstream on the Kasiha River (September 3–5, 1976)

On September 3, 1976, Kohshi Norikoshi, Mohamedi Seifu, Ramadhani Nyundo, and I attempted to climb along the Kasiha River. As already mentioned, the same route was attempted by Izawa in 1965, without success. Also, as mentioned in Section 2, I heard chimpanzee calls of neither K nor M Group from the southwestern ridge of Mt. Mhensabantu. The aim of this safari was to check on their presence. The Kasiha River has its source at the saddle point between Mt. Humo and Mt. Mhensabantu, and flows southwest. This main stream merges on the left bank with a tributary from Mt. Nkungwe and with another flowing down from the southwestern ridge of Nkungwe. Along these two tributaries I had seen unclimbable waterfalls on the return route from Nkungwe in 1966. Therefore I planned to proceed along the main stream.

At 9:40 we set out from Kansyana Camp. At 9:55 we entered the Kasiha Valley and went upstream in the riverbed. At the mouth of the valley we found a few feces of chimpanzees, of either K or M Group. After an hour's walk, the valley became a typical V-shaped one, with a series of pools and waterfalls. We climbed alternately on the right and left sides of the stream. There were many otter feces, and we saw a boomslang. The usual species of wagtail usually seen at Kasoje is *Motacilla aguimp*, but here we saw *M. alba*. There was no evidence of chimpanzees' presence. At 11:58, we reached the junction with the first tributary, which flows into the main stream as a waterfall. In front of us was a chain of three waterfalls. After bypassing them, we rested and came across feces of leopard. The distance we had walked so far would have been only 2 km in a straight line. Up to this point, the riverbed was clear, without developed riverine forest. We thought it strange that we saw no evidence of chimpanzees' presence.

At 13:10 we set out again, bypassing three more waterfalls to the right and left. While climbing up a steep slope beside one of the falls, we found an earthen pottery shard. Ramadhani said the impressed patterns on its surface differed from those used today. We were surprised to see that people had inhabited even such steep slopes in the past; such settlement, however, would not contradict Tongwe oral traditions and Stanley's (1878) descriptions. At 13:53 we passed the junction of the second tributary, 2.5 km upstream from the mouth of the valley. The valley became narrower, and the riverside trees taller. The bottom of the valley looked dark.

At 14:08, we found a chimpanzee bed for the first time on the trip. The bed, 2–3 months old, was on top of a tree 12–13 m high. This site was 3 km away from the feces we had seen on the same day downstream, and 300 m higher in altitude. Although we could not identify the species of the riverside trees, undoubtedly the site was already within the lower limit of the montane forest zone. Surface water disappeared in the stream as we climbed higher. Along the valley we found feces about two weeks old, containing seeds of *Harungana madagascariensis*. There was a mixed group of *Colobus badius* and *Cercopithecus ascanius*. At 15:35 we reached the junction of another tributary flowing down from the saddle point between Mt. Nkungwe and Mt. Humo. We continued going up the main stream. It began to rain. At 17:20, on a slope a little elevated from the bottom of the valley, we found a place where we could bivouac for the night.

Here we found two old chimpanzee feces containing seeds of *Ficus capensis*. The northern riverside was bare rock; trees, including a tall *Cynometra* sp., were found only on the southern bank. The moon shone through a narrow strip of sky. We heard no chimpanzee call.

On September 4 we set out at 9: 00. As we climbed steep slopes, we dislodged stones, which fell into the valley below us. We came across a troop of *Colobus badius*. There were stands of *Cordia millenii* and *Polyscias fulva*. We clung to an almost vertical cliff below the peak of Mt. Humo. We proceeded using vines and dwarf shrubs as our footholds. The rock was wet with dripping water, and we all got soaked. We saw olive pigeons (*Columba arquatrix*), which are rare in the lowland. At 11:45 we found a chimpanzee's bed, and another at 12:30. Remnants of fruit of *Aframomum* sp. eaten by chimpanzees about 10 days earlier were found. A troop of *Cercopithecus mitis* crossed the valley on the other side. At 13:45 another remnant of *Aframomum* sp., consumed the daybefore, was found. At 13:55 we counted eight old beds in trees growing on this cliff. Cutting through the bush and seeking footholds, we continued climbing. We saw another remnant of *Aframomum* and *Ficus capensis* fruits from the day before. Finally, at 14:30, we reached the main ridge and grassland on the summit of Mt. Humo.

Walking northward along the ridge of Mt. Humo, we reached a waterhole at the source of the Mkanga River, a tributary of the Kabesi. It was said that Nishida and Mohamedi found it in April 1976. We bivouacked on the ridge and saw a red-winged starling (*Orychognathus morio*), a bird not found in the lowland. At 17:10 we heard a chorus of chimpanzees' calls from the lower area of the Mahembe River, another tributary of the Kabesi, south of the Mganga. Near the bivouac point we found five new chimpanzee beds and three feces 2–3 days old, containing seeds of *Cordia milleni*. There were many stands of *Cordia* laden with fruit. We saw a flock of 5–6 violet-backed starlings (*Cinnyricinclus leucogaster*). At 18:30 we heard chimpanzee calls from the same place as before. The group seemed to be large. At 19:20 we heard another chorus.

On September 5 at 7:30, chimpanzees chorused at the same place as the night before. At 7:45 we heard another chorus. At 8:08, 8:28, and 8:50 we heard the third, fourth, and fifth choruses echo in the valleys. Along the main ridge, we reached the summit of Mhensabantu at 10:00. We descended along the declining narrow ridge stretching due west from the summit to Myako Camp, where we arrived at 14:50. En route we neither saw any sign nor heard any calls from chimpanzees.

7. Again to the summit of Mt. Nkungwe (August 21–23, 1979)

In August 1979 I attempted my second climb of Mt. Nkungwe. This time I was accompanied by Kisaburo Tokuda, Toshikazu Hasegawa, Mariko Hiraiwa-Hasegawa, J. Magombi (then Park Warden of Ruaha National Park), Mohamedi Seifu, and eight others. Before the climb, a trail had been cleared to the southwestern ridge of Mt. Mhensabantu, and the trip was not difficult. This time, since we had an altimeter, we could record the altitudinal transition of vegetation with considerable precision.

On August 21, we started climbing from Kansyana Camp. At 11:15 we reached the ridge (alt. 1,170 m). The vegetation showed that we were still within the Kasoje Forest. At 1,250 m the stand of *Cynometra* sp. became sparse, but the semideciduous forest comprising *Vitex* sp., *Maesa lanceolata, Pericopsis angolensis, Parkia filicoidea*, etc., continued for a while. At 1,360 m we entered the moist forest zone, where Spanish moss (*Usnea gonioides*) was dangling from tree branches. At 1,510 m on the ridge, where the forest disappeared for a while, bushes of *Dodonaea viscosa, Phillipia* sp., and *Hypericum* sp. appeared intermittently. In the montane forest at 1,580 m we saw *Podocarpus* sp., *Croton megalocarpus*, etc. We recorded *Protea* sp. at 1,615 m, and tree ferns at 1,705 m. At 1,735 m we came across a troop of *Colobus badius*. At 1,810 m *Parinari excelsa* appeared. At 18: 20 we reached the summit of Mhensabantu (alt. 2,080 m).

At 19: 05 we heard chimpanzee calls from the steep western slope of Mt. Humo toward the Mahembe River. At 19: 25 we reached the campsite east of Humo. There were several chimpanzee beds in trees just below the campsite. In the night we heard chimpanzee calls twice from a forest upstream on the Mahembe River. Here again we found old earthen pottery. Why had people lived on such a steep mountain slope at an elevation higher than 2,000 m a.s.l.? Ramadhani Nyundo and others called these people of the past, whose life style undoubtedly affected the area's vegetation, "Wanyankungwe" (the Nkungwe people).

At 9: 00 on the 22nd we headed for Nkungwe, leaving our gear at the campsite. We proceeded along the main ridge, thickly covered with brackens (*Pteridium aquilinum* and *Pteris catoptera*), *Dodonaea viscosa*, and tangling *Periploca linearifolia*. From the saddle point to the south of Humo we climbed the slope of Nkungwe. At 2,160 m we entered the montane forest, in which there were many trees with breast height diameters of 60 –80 cm. The timberline was at 2,310 m. At 13: 30 we reached the summit of Nkungwe (2,520 m), whose altitude was indicated as 2,490 m by our altimeter. On the peak to the south across a deep gap, there was a mixed group of about 30 *Colobus angolensis* and 12–15 *C. badius*. This observation is detailed in Nishida *et al.* (1981). *C. badius* ranges from the lowland Kasoje Forest to the main ridge of Mahale, just like the chimpanzee, while *C. angolensis* limits its distribution to the montane forest. Notably their relationship is, as seen above, partly sympatric and not competitive.

At 15:15 we set out on our return route. At 16:17 an outburst of chimpanzee calls came from the forest along the Mahembe River, a tributary of the Kabesi. At 16: 42, on the summit of Humo, we heard another outburst. At 17: 45 we arrived at the campsite. After sunset we heard no more chimpanzee calls.

On the morning of the 23rd, the chimpanzees we had heard call the evening before were quiet. We started at 8:55 descending along the main ridge stretching north from the summit of Mhensabantu. This ridge was much clearer and dryer than the southwestern ridge of Mhensabantu. Around 2,000 m, *Phillipia* sp. and *Erythrina abyssinica* were seen. At 1,760 m *Uapaca kirkiana* and *Brachystegia boehmii* appeared. At 1,640 m *Usnea gonioides* dangling from tree branches disappeared, and *Brachystegia spiciformis* became dominant. At 15:10 we turned onto the ridge branching to the east. At 1,510 m there were fresh signs of buffalo. At 1,405 m there was a stand of *Brachystegia microphilla.* At 13: 07, we descended to the Kabesi River. Walking along animal trails on the main foothill on the west side of the valley covered with a beautiful open forest, we reached Kalolwa Village on the lakeshore. En route no chimpanzee bed was seen. From Kalolwa we used a boat and reached Myako Camp at 18: 00.

IV. DISCUSSION

Although I simplified and trimmed my notes on the safaris, the above description remains long. All the information gathered on chimpanzees is included. One of the areas not covered by the seven safaris is the 7 km along the main ridge between Mt. Nkungwe and Mt. Sisaga, which includes the peaks of Mt. Sibindi and Mt. Mfitwa. Mt. Sibindi was reached by Nishida and Kawanaka in 1969. Other areas not covered include that along the Kabesi River and the Mukulume Hills south of M Group's range. Information on the Mukulume area is given in greater detail by researchers who stayed at Kasoje for a longer period of time (see Chap. 3).

Only brief descriptions of the vegetation in this area have been published so far. Nishida (1968) classified the vegetation around the Kasoje area into three types of forest and four types of open land, and Nishida and Uehara (1981) classified it into a total of 14 types: 3 types of forest, 10 types of grassland or woodland, and 1 type of montane grassland. Secondary vegetation due to human disturbance develops in various forms, in particular at deforested places, as Nishida and Uehara (1981) noted and as I confirmed after repeated visits to the area not only through direct observations but also through indirect information provided by the Tongwe workers who accompanied me on the safaris. Here I classify the vegetation roughly into three types: primary vegetation, secondary

Table 2.1 The vegetation types in the Mahale Mountains.

Present classification	Classification in Nishida and Uehara (1981)
I. Primary vegetation	
1. Lowland forest	A–3
2. Montane forest	A–2
3. *Miombo* open forest	B–1, 3, 4, 7
4. Riverine forest	A–3
5. Montane grassland	C–1
II. Secondary vegetation	
1. *Diplorhynchus-Combretum* savanna	B–2
2. *Hyparrhenia* grassland	B–10
3. Bracken bush	B–10
III. Likely secondary vegetation	
1. *Oxytenanthera* bamboo thicket	B–6
2. *Arundinaria* bamboo thicket	A–1
3. *Phillipia-Hypericum* bush	B–8, 9

vegetation, and likely secondary vegetation. These were further subdivided into 5, 3, and 3 types respectively, giving a total of 11 types (Table 2.1). In addition to these, areas near human habitation develop a few secondary types of vegetation—e.g., elephant grass (*Pennisetum*) grassland, *Imperata* grassland, woodland dominated by *Acasia*—but these are ignored here. In the right column of Table 2.1, the corresponding classification by Nishida and Uehara (1981) is also given.

The lowland forest (I–1) has been described in detail elsewhere (Nishida, 1968; Nishida and Uehara, 1981), and the formation of the forest itself has been shown to be greatly affected by seed dispersal by the chimpanzees (Takasaki, 1983a; Takasaki and Uehara, 1984). Here I mention only that the upper limit of this vegetation type is about 1,300 m. The montane forest (I–2) can probably be further divided into upper and lower zones. The boundary is around 2,000 m. The distinction is that in the upper zone the dominant tree species is *Parinari excelsa*. Between the lowland forest (I–1) and the montane forest (I–2), *Phillipia-Hypericum* bush (III–3) occasionally develops: it corresponds to the *Protea-Ricinodendron* montane savanna (B–8) and *Dombeya-Erythrina* montane savanna (B–9) described by Nishida and Uehara (1981). Although type III–3 is well developed in the Ruwenzori Mountains, showing a clear zonation, at Mahale it appears only as a narrow belt at 1,300 – 1,600 m. In a drier area, such as on the northern ridge of Mhensabantu, this vegetation zone is seen between the montane forest (I–2) and the *miombo* forest (I–3). Its upper limit is as high as 2,000 m. This type is included in category III because it is the zone most easily disturbed by humans (for example, it is swept by forest fire every 2–3 years) and rarely preserves its original vegetation.

Although what is meant by *miombo* is not so clear, here not only the typical *Brachystegia, Julbernardia*, and deciduous *Isoberlinia* species but also those having the same life forms—*Pterocarpus angolensis, Pericopsis angolensis, Combretum molle, Uapaca kirkiana, Parinari excelsa, Monotes elegans* (e.g., B–3, 4, and 7 in Nishida and Uehara, 1981)—are included. However, it is likely that the pure stands of *Uapaca kirkiana* near the hamlet of Makusu on the plateau up Msoffwe have been artificially preserved.

A typical riverine forest (I–4) is observed only along the Lugonezi River and upstream into the steep slopes of Mahale. Its component species are often *Ficus* spp., and this vegetation type resembles the lowland forest (I–1). Upstream on rivers, the vegetation sometimes becomes a pure stand of *Cynometra* sp.

The montane grassland (I–5) seen at Mahale, though not as typical as that seen at the Ruwenzori Mountains or the Virunga Mountainshas, many beautiful flowering herbs, e.g., *Aloe* spp., *Helichrysum* spp., etc. The grass height is low. Being a summit phenomenon, this vegetation type, though it does not cover a large area, is seen on all the peaks over 2,000 m in the main ridge. As most summits of Mahale are believed to be the residences of the clan guardians or ancestor spirits of the Tongwe people, these places are protected from human disturbance except for occasional fires that reach them from the lowland.

The *Diplorhynchus-Combretum* savanna (II–1) is thought to be secondary vegetation of the primary *miombo* open forest transformed by human influence. Although the Mahale area does not provide a typical example of this vegetation, the low savanna woodland of *Brachystegia longifolia* may be included in this type. Neither *Diplorhynchus condylocarpon* nor *Combretum collinum* exceed 3 m in height.

Hyparrhenia grassland (II–2) and bracken bush (II–3) both cover the ground abandoned after cultivation. Once it is covered with these vegetation types, succession stops for a long time. Although the *Hyparrhenia* grassland covering the eastern ridge of Sisaga has already been invaded by shrubs, *Hyparrhenia* is still dominant after the desertion of fields more than half a century ago. *Hyparrhenia* occurs from the lowland to the lower boundary of the montane forest zone, while bracken becomes dominant in the montane forest zone. On the main ridge in particular, thickly growing brackens (*Pteridium aquilinum* and *Pteris catoptera*) are covered with vines of *Periploca linearifolia*, which makes walking extremely difficult. This vegetation type covers a large area on the western slopes of Nkungwe and Mhensabantu. Probably the whole area was used as maize fields for the Tongwe people.

It is likely that the *Oxytenanthera* bamboo thicket (III–1) and the *Arundinaria* bamboo thicket (III–2) are secondary vegetation types corresponding to the *miombo* woodland and montane forest. In addition, on the main ridge of Sisaga, intrusion of *Arundinaria* into the montane forest and its coexistence with the montane forest tree species are often observed. The physiognomy of the main ridge is a beautiful harmony of patchy mosaics of montane forest, bamboo thicket, and montane grassland. However, it should be noted that this is the result of human activities in the past (Fig. 2.3).

The vegetation types favored by chimpanzees are the primary vegetation types I–1, 2, 3, and 4. K and M Groups live in the lowland forest, while the groups in the Kasakati Basin live in the riverine forest and *miombo* woodland. Also the present study has shown that there are groups which mainly live in the montane forest. The vegetation types II and III are not the main habitats of chimpanzees. In other words, even the Tongwe people, who subsist in harmony with nature, have eventually degraded the chimpanzees' habitat.

Figure 2.4 plots all the points of information on chimpanzees obtained in the seven safaris. The information is still insufficient for drawing a map of the distribution of chimpanzee unit-groups in Mahale. However,

Fig. 2.3 The northwestern slope of Mt. Tambila on the main ridge of the Mahale Mountains. A beautiful mosaic of montane grassland and montane forest, whose lower fringes are invaded by *Arundinaria* bamboo bush, a secondary vegetation on deserted cultivated fields. (Photo by J. Itani)

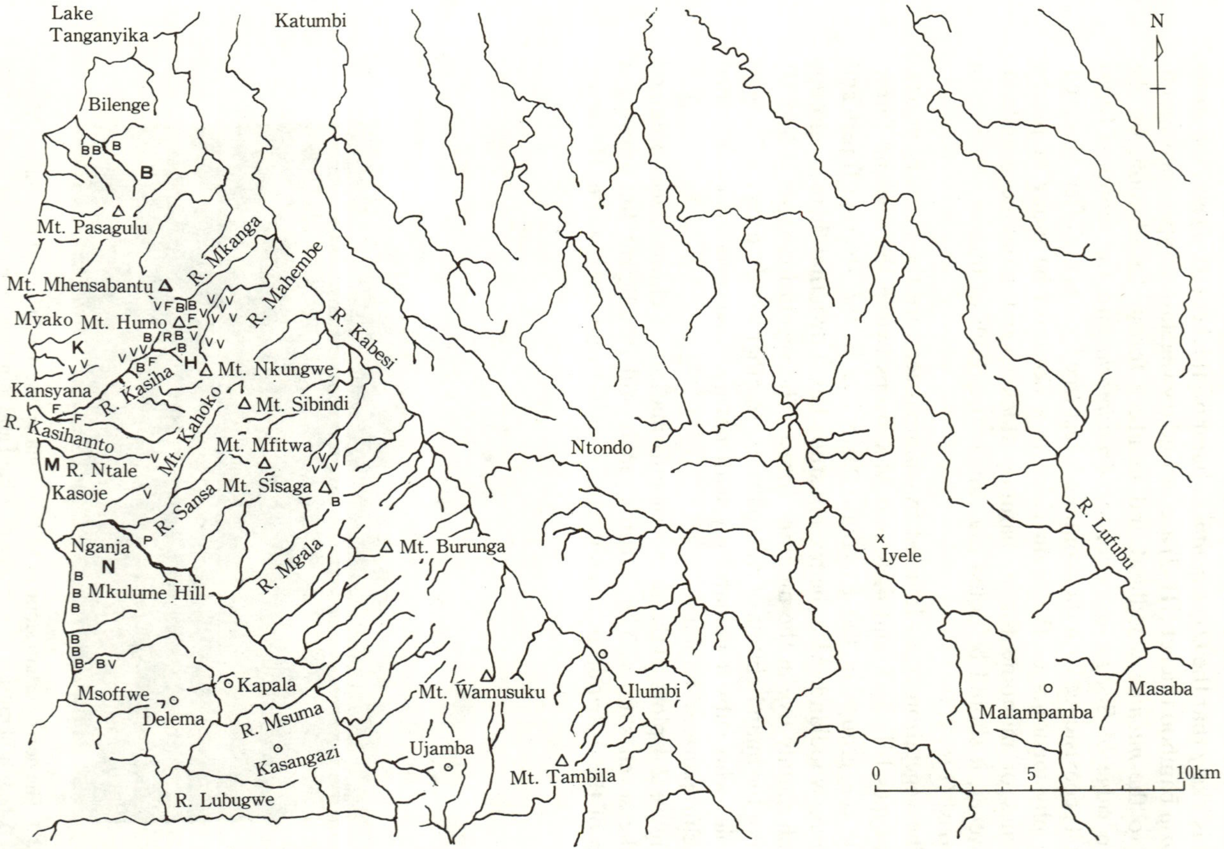

Fig. 2.4 The sites of chimpanzee signs and information.
○: Positive information on chimpanzees from villagers; ×: negative information on chimpanzees from villagers; B: beds; F: feces; P: foot prints; R: food remnants; V: vocalization. Large capitals B, K, M, N, and H are the names of unit-groups.

some comments may be made with regard to the study groups.

First, the home range of H Group does not overlap with those of K and M Groups. As described above, the signs of H Group were prominent only higher than the Kasiha River's fork with its second tributary flowing down from Mt. Nkungwe. The fork was elevated 1,200 m a.s.l. and was about 3 km away from the signs left by K or M Group. Higher than that, their food remnants, feces, and beds were almost continuously distributed, as if surrounding the peak of Humo, and crossed over the main ridge. H Group's home range extends to the riverine forests of the Mganga and the Mahembe Rivers, tributaries of the Kabesi. Of the six calls heard on the climb of Mt. Nkungwe in 1966, two, the fourth and fifth calls, were from H Group. The pant-hoots we heard on the Mahembe as we reached the main ridge after climbing along the Kasiha River in 1976 and the calls heard at the same place in 1979 were also theirs.

On the summit of Mt. Sisaga, too, chimpanzee calls were heard from the Kabesi River side. Probably the calls were not from H Group. In 1969, when Nishida and Kawanaka climbed Mt. Sibindi, they heard chimpanzee calls near the summit and concluded that a group of chimpanzees crossed over the main ridge (Kawanaka, personal communication). It is uncertain whether the calls they heard were from H Group, from the same group whose calls Nishida and I heard to the east of Sisaga, or from another group. The observed locations of H Group, which were clear on all my three climbs on the main ridge between Nkungwe and Mhensabantu, makes it likely that they have a home range that straddles the main ridge rather than running oblong from north to south. Such a home range must be dependent on the montane forest, in contrast with a lowland forest-dependent planar home range such as that of K or M Group.

In this line of inference, it is plausible to assume 4–5 unit-groups along the main ridge from Nkungwe to Ilumbi, i.e., the source area of the Kabesi River. The southernmost group has a home range including Ilumbi and Ujamba. Although they do not use the lowland forest, they cross the main stream of the Kabesi to the *miombo* open forest for fruits of *Parinari curatellifolia, Garcinia huillensis*, and *Uapaca kirkiana*, and they sometimes visit the riverine forest in the east. We have not obtained enough data on such mountain-dwelling chimpanzees. An important question for research in the future is how the vegetational differences in habitats affect the transfers of females between unit-groups.

North of the Takata River, two groups were suspected to be present at first (Kawanaka and Nishida, 1975), B Group having its home range around Bilenge near the lake and P Group having its home range around Mt. Pasagulu. At present, however, it has been confirmed that there is only one group ranging both areas (McGrew and Collins, 1985; Collins and McGrew, 1988; see also Chap. 3). Although at one time we discussed the possibility that B and H Groups could be the same group, it is now certain that they are different groups. B Group depends on the *miombo* open forest (I–3) and riverine forest (I–4) rather than on the lowland forest and montane forest.

N Group, ranging in the Mukulume area south of the home range of M Group, is also dependent on the *miombo*-riverine forest mosaic vegetation. The chimpanzees who left many beds in *miombo* trees along the lake between Msoffwe and Nganja must have been in this N Group. It is also plausible to assume that another *miombo*-riverine forest-dependent unit-group exists to the south, around Lubugwe.

From a panoramic point of view this chapter drew out three types of chimpanzee habitat with regard to major vegetation types in the northern half of Mahale, in particular in the vicinities of the home ranges of K and M Groups. The lowland forest used by K and M Groups is unique to the Kasoje area, and Mahale is one of the rare localities where montane forest is developed in western Tanzania. In other words, the most common chimpanzee habitat in western Tanzania is the *miombo*-riverine forest mosaic. Kabogo, Kasakati, Makolongo, and Filabanga are all this type of chimpanzee habitat. By contrast, the western bank of the Ugalla River almost completely lacks riverine forest and may be called simply a "*miombo* type." In such a habitat, I proposed a seasonal migration of chimpanzees over a large area (Itani, 1979).

It is clear that the primary vegetation types are very important for chimpanzees, and it is worth emphasizing the importance of ecological studies of the montane forest-dwelling chimpanzees.

V. SUMMARY

1. This paper describes the vegetation and the distribution of chimpanzee groups in the northern half of the Mahale Mountains on the basis of data and information collected on seven safaris in the area.

2. The vegetation types are categorized into three major types: primary vegetation, secondary vegetation, and likely secondary vegetation. These are further subdivided into 5, 3, and 3 types respectively, giving a total of 11 types.

3. It is likely that there are 4 –5 unit-groups (including H Group) along the main ridge of the Mahale Mountains, in addition to the known unit-groups, B, K, M, N, and L Groups, along the lakeshore.

4. There are three types of chimpanzee habitats in the area: (1) lowland forest type, (2) montane type, and (3) *miombo* open forest-riverine forest

type. For all habitats the primary vegetation types are the most important.

Acknowledgements

I am grateful to all who contributed to this paper in various forms.

3

Demography and Reproductive Profiles

Toshisada NISHIDA, Hiroyuki TAKASAKI, and Yukio TAKAHATA

Department of Zoology, Faculty of Science, Kyoto University

I. INTRODUCTION

Demographic data on great apes provide an invaluable model for reconstructing demographic trends in early hominids (Teleki, Hunt, and Pfifferling, 1976). Although collection of this type of data has increased recently, demographic studies of great apes are still in the early stages of development (Galdikas, 1981; Harcourt, Fossey, and Sabater-Pi, 1981; Goodall, 1983; Hiraiwa-Hasegawa, Hasegawa, and Nishida, 1984; Kano, 1987; Sugiyama, 1989). In particular, detailed long-term demographic data on wild chimpanzees are scant, besides Mahale, the only other source for such data is Gombe (Teleki, Hunt, and Pfifferling, 1976; Goodall, 1983, 1986). Data from Mahale are particularly valuable since the chimpanzees are relatively unaffected by human-transmitted disease, such as polio, which are reported from Gombe (Teleki, Hunt, and Pfifferling, 1976).

Social interactions of primates are influenced by demographic factors (e.g., Altmann and Altmann, 1979). Behavioral differences have been noticed between groups and even within the same group at different times. Many such differences are influenced by demographic factors, such as the size and age-sex composition of the social group.

For example, the extent of competition for playmates, social partners or allies, and sexual partners may be affected by the number of individuals of the same age or sex class and by the operational sex ratio. These are often referred to as "demographic constraints" (e.g., Dunbar, 1988). Thus, an adult male chimpanzee may employ different mating strategies depending on the number of other adult males and estrous females in a party (see Chap. 5).

The sex-ratio in a group may also influence mother-infant relations. For example, since there are more sexually mature females than males in a population (see below), a mother may place a different amount of parental investment in male vs. female offspring. In particular, it is interesting that mothers maintain long-lasting relations with sons, but usually not with daughters (Trivers, 1972; Clutton-Brock and Albon, 1982).

Behavior differs across ages as well, and so reflects the life history of an individual. For example, a female chimpanzee may alter her mating strategies depending upon the time that has elapsed since she immigrated into a new group (see Chap. 5). Social differences observed between chimpanzees and most Old World monkeys can be explained to some extent by the different genealogical relationships within and between sexes. These, in turn, are caused by different patterns of migration resulting in patrilocal/patrilineal compared to matrilocal/matrilineal mating systems (Harcourt and Stewart, 1983).

Thus, without knowledge of demography, one cannot understand properly the social life of primates. This chapter serves to provide background information for the succeeding chapters.

II. METHODS

1. Subjects of Study

This chapter is based on data collected from observations on the chimpanzees of K and M Groups. In total, 228 chimpanzees have been identified. Of them 125 were female, 92 were male and 11 were of unknown sex (Table 3.1). Life history and demographic parameters have been obtained over 22 years of observation, beginning in 1966 when most chimpanzees of K Group were identified, and lasting to 1988.

For analysis of population dynamics, data obtained from M Group individuals during 1979–1988 were used. Nishida *et al.* (1985) provides similar data for K Group. In addition to the authors, data were collected by people who are listed in the Acknowledgments.

2. Estimates of Age

Of the individuals still under observation, only three females and two

Table 3.1 Number of individuals under observation (K Group and M Group).

Birth year	Certain	Uncertain	Total
Sex			
Female	47	78	125
Male	50	42	92
Unknown	10	1	11
Total	107	121	228

males born during the study period have reached reproductive maturity. Age estimates for individuals who were adolescent or younger in the early years of the study are likely to be reasonably accurate.

The age of a parous adult female was taken to be the estimated age of her oldest offspring plus 15 years (which is the average age of a female's first parturition at Mahale: Hiraiwa-Hasegawa, Hasegawa, and Nishida, 1984). This method was first used for Gombe's chimpanzees (Teleki, Hunt, and Pfifferling, 1976).

3. Definitions of Age-sex Class and Sex Ratio

The distribution of age-sex classes basically follows Goodall (1983). These are: infancy 0 – 4 years, juvenile 5 – 7 years, early adolescence 8 – 12 years (male) and 8–10 years (female), late adolescence 13–15 (male), 11–13/14 years (female), and adult 16 years or older (male) and 14/15 years or older (female). However, at Mahale, we use the category "old adult" for those individuals older than 34 instead of 32 years. In addition, infancy was divided into "older" (more than one year) and "younger" (less than one year) classes.

Sex ratio is here defined as the number of males divided by the number of females × 100.

The age-sex composition of M Group was determined at the end of each year. Those animals that survived each New Year's Eve were counted (Table 3.3).

4. Distinguishing Death from Emigration

Since there has been no record of strange adult or adolescent males entering K or M Group during the 22 years of study, adult or adolescent males that disappeared were regarded as having died rather than having emigrated. In some cases of male disappearance ($N=7$), the animals apparently were senile or were suffering from some disease when they were last seen. However, only one dead body was found, that of an adolescent male.

The disappearance of fully adult females poses a difficult problem. There have been many confirmed cases of transfer of adult females from K to M Group. However, these should be regarded as exceptional owing to the unusual disappearance of K Group's adult males. Thus, no adult female of K Group transferred to M Group before the number of K Group adult males decreased to 2 (Nishida *et al.*, 1985). All the adult females who disappeared were, therefore, regarded as dead. The presumption that parous females who disappeared had died was occasionally supported by the discovery of their juvenile offspring roaming alone ($N=3$).

In contrast, the disappearance of adolescent females should be regarded as emigration, since both K and M Groups have accepted many such strange females. No adolescent female was regarded as having died because none was seriously ill when last sighted. The one exception was MD, whose dead body was recovered.

5. Distinction of Orphans from Immigrants

M Group is a large-sized group and ranges over a large area. Some mothers and their dependent offspring are observed only once or twice a year (Hiraiwa-Hasegawa, Hasegawa, and Nishida, 1984). Therefore, if a strange young chimpanzee is suddenly seen, he or she may be an immigrant, a shy female's daughter, who upon reaching sexual maturity has begun to move with adult males, or an orphan of a shy female who has recently died.

The daughters of shy females were finally identified when they reached late adolescence and were sooner or later observed to associate with their mothers. If the strange females were young juveniles, they were regarded as orphans, since such small juvenile females almost always follow their mothers. If the strangers were in early adolescence, they were regarded as orphans if they associated at the outset with some of the resident chimpanzees, and as immigrants if they tended to be isolated from the resident chimpanzees (see Hayaki, 1988; Nishida, 1989b).

Four strange juvenile males appeared in M Group, one (AJ) in 1979, two (HT, HR) in 1981, and one (FX) in 1988. All of them were regarded as orphans, because between 1965 and 1983 (when K Group virtually became extinct) no juvenile males were observed to transfer between K and M Groups without being accompanied by their mothers (Kawanaka, 1981; Nishida *et al.*, 1985, Hiraiwa-Hasegawa and Hasegawa, 1988; Takahata and Takahata, 1989). Once, a juvenile male was described as having been "deserted by his mother" when she transferred alone from K to M Group (Hiraiwa-Hasegawa and Hasegawa, 1988). However, this may be regarded instead as the son refusing to emigrate along with his mother.

III. RESULTS

1. Population Structure

Chimpanzees of the Mahale Mountains live in unit-groups or communities of up to 100 individuals. Nishida (1968) reported that there were six unit-groups (subsequently called B, P, K, M, N, and L) in the study area. Although very few systematic studies on the distribution of unit-groups have been conducted since then, many short-term surveys have revised these early conclusions (Table 3.2, Fig. 3.1, Fig. 3.4).

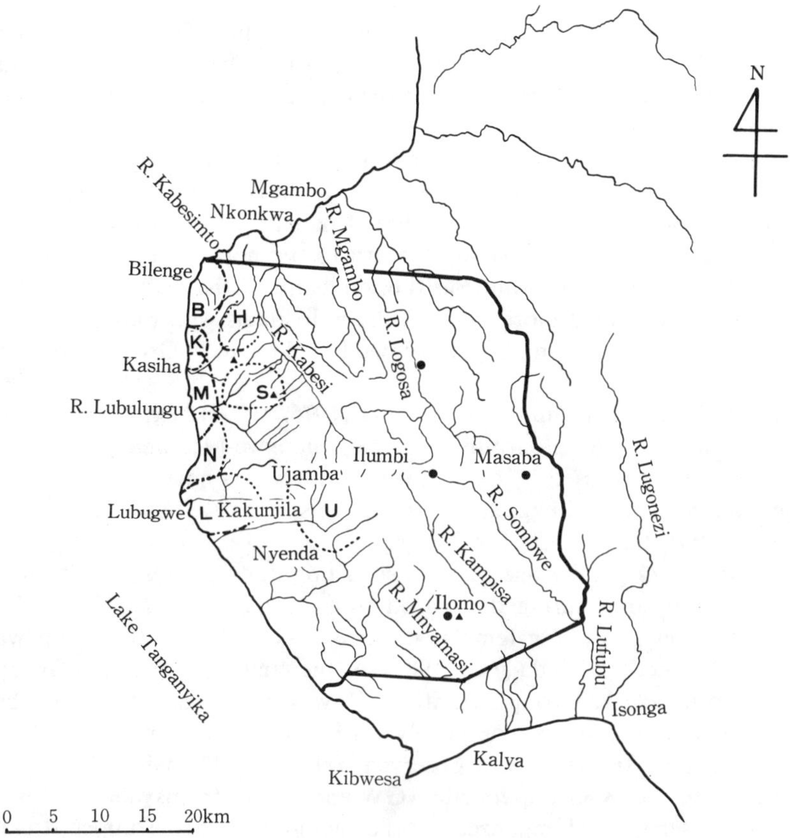

Fig. 3.1 The distribution of the chimpanzee unit-groups in the northern half of the Mahale Mountains. ● Location of unit-group.

A group census conducted by more than 10 people in 1975 concluded that B Group's range was considerably larger than had been initially realized. This resulted in "P Group" being included in B Group (Nishida and Norikoshi, 1975).

Nishida and Uehara (unpublished) heard an outburst of calls from the uppermost part of Kabesimto Valley in January 1974. Meanwhile, Nishida (1976b) found a well-trodden path used by chimpanzees which ran from the Mganga Valley, reached the Mahale main ridge and then descended to the Kasiha Valley. This suggests that some chimpanzees use both the eastern and western sides of the Mahale ridge. Subsequently, Itani and Norikoshi found a concentration of beds and food remnants at the head of the Kasiha Valley in 1976. This suggests that another unit-group, designated "H" Group, uses the mountain area to the east of the ranges of K and M Group (Norikoshi, 1976). In 1980, Hasegawa and Hiraiwa-Hasegawa (1980) actually observed the chimpanzees of H Group and counted up to 9 individuals on the eastern edge of the Mahale main ridge near Mt. Humo.

At least four observations in 1979 and 1981 strongly suggest that the area to the south of the lower Lubulungu River is used by both M Group and N Group. For example, in 1981 Takahata and Takasaki saw five adult males of M Group suddenly turn to the north in response to the sight of a chimpanzee group on the left bank of the Lubulungu River. The group contained more than 40 shy individuals who were unfamiliar to the observers, and who were assumed to belong to N Group (Takahata *et al.*, 1982).

In 1975, and again in 1979, Nishida (1975, 1979c) heard or observed the chimpanzees of L Group near the Lubugwe-Musuma junction. A well-trodden path used by chimpanzees, which runs north-south through the Kakunjila ridge, suggests a large range size, as well as a large group size (estimated at over 60) for this group.

K Group, originally one of the two major study groups, lost its adult males one by one during the period from 1970 to 1983. In 1978 –79, a mass transfer of cycling females occurred, and by 1983 K Group was almost "extinct" (Nishida *et al.*, 1985). At the same time, M and B Groups began to invade K Group's territory. However, even as late as October 1988, K Group was not completely extinct; three observers (K. and H., Takasaki and T. Hasegawa) observed a 16-year-old male, LI, moving with two former K Group females (GW and GM) in Kansyana.

The presence of chimpanzees in the region around the upper stream

Fig. 3.2 An adult female, GW, was lame of her left leg. (Photo by T. Nishida)

Fig. 3.3 WB, one of the oldest females, carry her last offspring under her belly. (Photo by T. Nishida)

of the Sansa Valley was first confirmed by Itani and Nishida who, in 1966, were near the summit of Mt. Nkungwe (Nishida, 1973a). In 1967 Nishida (1973a) found many fresh feces and food remnants on the Mahale main ridge from Mt. Sisaga to Mt. Mfitwa. In 1969, Nishida and Kawanaka (unpublished) observed chimpanzees in various parts of the Sansa and Sibindi Valleys. They also heard a subgroup moving from the uppermost edge of the Sibindi Valley northeastwards, crossing the Mahale ridge. It is likely that these chimpanzees form a single unit-group ("S" Group) judging from the direction of their movements.

Adding "U Group", which was encountered four times around the Upper Lumbye River (Nishida, 1975, unpublished), there were at least 8 unit-groups in the northern half of the Mahale Mountains in the 1970s (Table 3.2). Given an estimated average group size of 60, there was a total population of 480 individuals within an area of 500 km^2. Thus, the population density in this area was about one animal per km^2.

Nishida (unpublished) has confirmed the presence of at least four other unit-groups within the Mahale Mountains National Park (Fig. 3.1). Assuming an average group size of 60, the park harbors about 700 chimpanzees.

2. Age-sex Composition of M Group

M Group has been continuously monitored by at least one primatologist since 1975. Since a few shy resident females have never been habituated to human observers, and since young females move between unit-groups,

Table 3.2 Sizes and ranges of unit-groups in the northern half of the Mahale Mountains.

Group	Group size	Known (*) or estimated range (names of major valleys)	Range size (km^2)
B	60–80	* Lakeside to Mt. Pasagulu (Bilenge, Kabesimto, Kabwe)	25–30
K	10	* Lakeside (Myako, Nkala)	7
M	90	* Lakeside (Mpila, Kasiha, Ntale Lubulungu, Kakolomo)	21
N	40<	Lakeside to Mts. Mkulume and Mhela (Msoffwe)	
L	40 – 80	Lakeside (Lubugwe, Msuma, Ntala)	
H	20 – 40	Mahale ridge (Kasiha, Mahembe, Mganga)	
S	?	Mahale ridge (Mgalla)	
U	?	Mahale ridge (Msangabale)	

it has been almost impossible to determine the exact size of the group. Between 1980, when all chimpanzees were individually identified (Hiraiwa-Hasegawa, Hasegawa, and Nishida, 1984), and 1985, it appeared that M Group had not changed much in size. However, between 1985 and 1987, the number of members decreased by about 10% (Table 3.3). This was a result of deaths which were probably due to an epidemic which lasted over 2 years in the area (see below).

Over the course of 9 years, M Group averaged 10 adult males, 36 adult females, 9 adolescent males, 11 adolescent females, 4 juvenile males, 7 juvenile females, 8 infant males, and 11 infant females, for a total of 96 individuals (Table 3.3).

The overall sex ratio was female-biased in all 9 years in which the censuses were taken. In particular, mature females were three times as numerous as mature males. The number of "younger infants" did not differ by sex, but females outnumbered males among "older infants" in 7 of the 9 annual censuses. This suggests higher male mortality in the first year of life. This female bias was also present in the juvenile period. Although adolescence lasts longer in males than in females, females nevertheless outnumbered males in 5 of the 9 annual censuses. This was partly because adolescent female immigrants outnumbered emigrants (see below).

The number of adult individuals almost equals that of immatures,

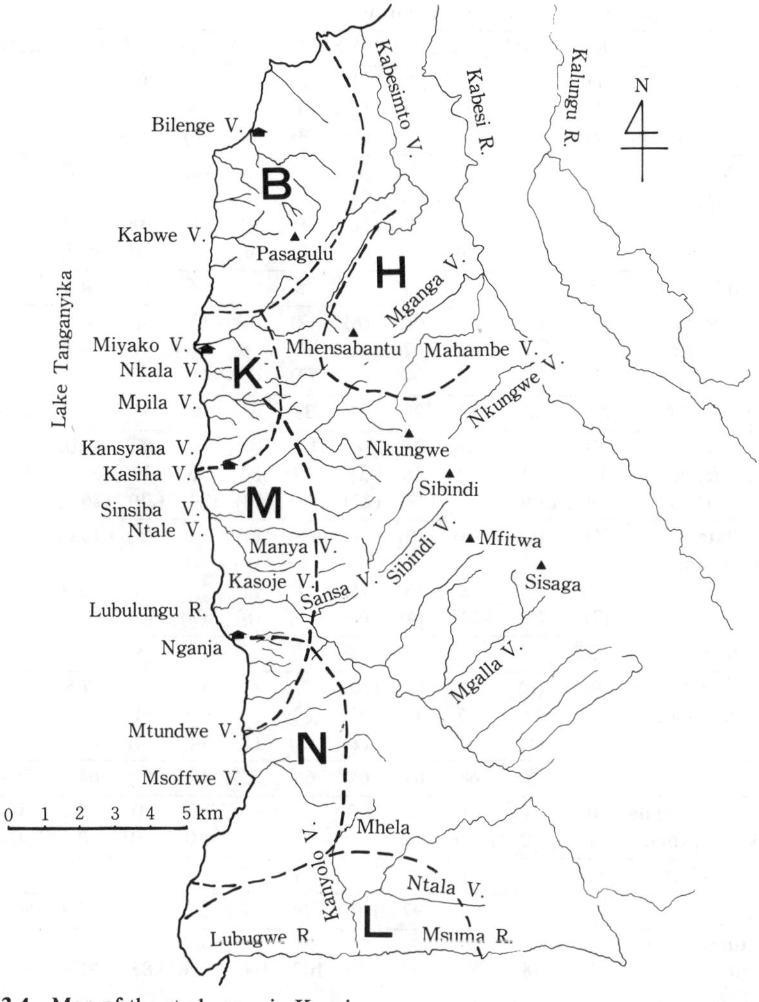

Fig. 3.4 Map of the study area in Kasoje.

although the latter exceeded the former in 7 of the 9 annual censuses. These numbers agree roughly with earlier reports on M Group (Hiraiwa-Hasegawa, Hasegawa, and Nishida, 1984) and K Group (Nishida *et al.*, 1985) and are similar to those reported by Goodall (1983) for the Kasakela community of Gombe (1972–1980) after fissioning.

One difference between Gombe and Mahale is that M Group contained a large number of "old" females, while Gombe had more "old" males than females in 8 of 12 census years (females outnumbered males in 2

Table 3.3 Age-sex composition of M Group.

Age/sex class	1980*	1981*	1982	1983	1984	1985	1986	1987	1988	Mean	Range
(old)	(1)	(1)	(1)	(2)	(2)	(2)	(1)	(0)	(0)	(1.1)	
Adult male	10	10	10	9	10	11	9	8	9	9.6	8 – 11
(prime)	(9)	(9)	(9)	(7)	(8)	(9)	(8)	(8)	(9)	(8.4)	
(late)	(1)	(2)	(3)	(2)	(2)	(2)	(4)	(4)	(4)	(2.7)	
Adolescent male	7	10	9	9	10	9	10	10	10	9.3	7 – 10
(early)	(6)	(8)	(6)	(7)	(8)	(7)	(6)	(6)	(6)	(6.7)	
Juvenile male	5	3	4	3	3	3	4	6	6	4.1	3 – 6
(older)	(4)	(5)	(6)	(3)	(6)	(7)	(6)	(3)	(4)	(4.9)	
Infant male	7	7	9	7	11	10	8	4	9	8.0	4 – 11
(younger)	(3)	(2)	(3)	(4)	(5)	(3)	(2)	(1)	(5)	(3.1)	
SUBTOTAL	29	30	32	28	34	33	31	28	34	31.0	29 – 34
(old)	(8)	(9)	(10)	(9)	(8)	(12)	(10)	(9)	(6)	(9.0)	
Adult female	34	39	37	36	35	39	34	33	35	35.8	33 – 39
(prime)	(26)	(30)	(27)	(27)	(27)	(27)	(24)	(24)	(29)	(26.8)	
(late)	(4)	(4)	(6)	(9)	(9)	(2)	(5)	(6)	(5)	(5.6)	
Adolescent female	11	9	8	13	15	11	11	13	7	10.9	7 – 15
(early)	(7)	(5)	(2)	(4)	(6)	(9)	(6)	(7)	(2)	(5.3)	
Juvenile female	3	9	8	9	5	5	3	9	9	6.7	3 – 9
(older)	(9)	(6)	(11)	(11)	(10)	(8)	(6)	(2)	(7)	(7.8)	
Infant female	13	11	15	11	12	10	7	6	11	10.7	6 – 15
(younger)	(4)	(5)	(4)	(0)	(2)	(2)	(1)	(4)	(4)	(2.9)	
SUBTOTAL	61	68	68	69	67	65	55	61	62	64.0	55 – 69
Sex unknown Juvenile	0	1	0	0	0	0	0	0	0	0.1	0 – 1
Sex unknown Infant	1	0	0	0	0	1	0	0	0	0.2	0 – 1
SUBTOTAL	1	1	0	0	0	1	0	0	0	0.3	0 – 1
Total	91	99	100	97	101	99	86	89	96	95.3	86–101
Mature/immature ratio	94	98	89	87	80	102	100	85	85	91	80–102
Overall sex ratio	48	44	47	41	51	51	56	46	55	48	41 – 56
Adult sex ratio	29	26	27	25	29	28	26	24	26	27	24 – 29
Adolescent s.r.	64	111	113	69	67	82	91	77	143	85	64–143
Juvenile s.r.	167	33	50	33	60	60	133	67	67	58	33–167
Infant (old) s.r.	44	83	55	27	60	88	100	200	57	59	27–200
Infant (young) s.r.	75	40	75	∞	250	150	200	25	125	108	25–∞

* Calculated from Hiraiwa-Hasegawa, Hasegawa, and Nishida (1984) and their unpublished data.

of 12 years, and the number of "old" individuals of each sex was equal in 2 of 12 years). It is possible that the ages of some Mahale females who lacked older offspring were overestimated from their appearance. When these females are excluded, however, there were still more females than males in the "old" age class.

3. Overall Demographic Trend of the Study Population

Populations increase by birth and immigration and decrease by death and emigration. Since almost all K Group females transferred to M Group in 1978–1983, data for the two groups were combined in order to estimate overall demographic trends of the study population (Table 3.4). During the 10-year study period, the number of births and deaths were exactly the same ($N=74$ each). When migration between K and M Groups was discounted, the number of immigrants ($N=14$) was also very similar to that of emigrants ($N=13$). Thus, with K Group and M Group combined, the study population maintained a state of virtual equilibrium, with an increase of only one chimpanzee in 10 years.

4. Patterns of Migration

1) Females that Transferred

From 1979 to 1988, at least 18 adolescent females immigrated into M Group (Table 3.5). Three out of the 18 were from K Group, and two came with their mothers (see below). During the same period, 9 fully adult females immigrated into M Group (Table 3.5). All of them came from K Group, which had lost all of its adult males by 1982 (Nishida *et al.*, 1985).

Table 3.4 Overall population trend during a 10-year period (K and M Groups combined).

Year	1979	1980	1981	1982	1983	1984	1985	1986	1987	1988	Total		
Birth													
Male	2	3	4	3	2	5	3	3	1	5	31		
Female	2	5	7	4	0	2	3	2	5	5	35	74	
Unknown	1	0	1	1	2	2	1	0	0	0	8		
													88
Immigration	?0	?0	2	2	3	1	1	1	2	2		14	
Emigration	?0	?0	0	2	1	0	1	2	2	5		13	
													87
Death													
Male	3	1	3	5	6	0	5	4	4	2	33		
Female	5	1	3	1	4	2	3	11	5	1	36	74	
Unknown	1	0	0	0	3	0	0	1	0	0	5		
Balance	−4	+6	+8	+2	−7	+8	−1	−12	−3	+4			+1

Table 3.5 Immigration into and emigration out of M Group.

Year	No. of females immigrating into M Group	Natal transfer	Multiple transfer	No. of females emigrating from M Group	Natal transfer	Multiple transfer
1979	6	0	6 (6)	0	0	0
1980	0	0	0	0	0	0
1981	3	3 (1)	0	0	0	0
1982	2	2	0	1	1	0
1983	5	3	2 (2)	2	2	0
1984	1	1	0	0	0	0
1985	1	1	0	1	1	0
1986	2	2	0	2	2	?2
1987	5	4 (2)	1 (1)	2	2	0
1988	2	2	0	5	5	0
Total	27	18 (3)	9 (9)	13	13	?2

Note: Parentheses indicate immigrants from K Group.

On the other hand, 13 females born in M Group emigrated during late adolescence (Table 3.5) and probably transferred into one of the unmonitored neighboring unit-groups. Two multiparous females (IA and OO), who exhibited no physical deterioration due to age, disappeared with 2 or 3 offspring each around 1986. It is possible that they transferred, but since they had not often been observed because of shyness, they are assumed to have died (see above).

Two females of M Group (TY and FB) did not emigrate, but remained and produced multiple offspring within their natal group (Hiraiwa-Hasegawa, Hasegawa, and Nishida, 1984).

In sum, excluding 9 parous immigrants and their 2 dependent daughters from K Group, M Group received 16 adolescent females from the outside and lost at least 13 adolescent females to emigration. Thus, emigration and immigration of females were more or less balanced during the 10-year period.

2) Females that Failed to be Integrated into their New Group

During the 10 year period, there have been more than 3 adolescent females that immigrated into M Group, stayed for a few weeks, months, or years, and disappeared again. For example an early adolescent female, NP, immigrated into M Group in 1981, stayed for 3 years and then disappeared. She was healthy when she was last seen. A late adolescent female, "TM-no. 2", was identified in M Group in December 1983 and disappeared within one month. A few strange adolescent females were occasionally observed and disappeared again before they were identified individually.

These individuals may be adolescent females who failed in their transfer to M Group and who returned to their natal groups or transferred once again to another new group.

5. Female Reproductive Life

1) Physical Development and Sexual Maturity

a) Sexual Swelling

The first tiny swellings of the sex skin appear around 8 years of age. These gradually become larger until the first full-sized swelling appears at around 10–11 years when mating with adult males occurs. Six females were observed from birth to sexual maturity; all showed their first sexual swelling and most were first observed to mate when they were 10 years old (Table 3.6).

Menarche has been observed only once at Mahale. K. Takasaki (unpublished) observed BU having menstrual blood on October 19, 1988. BU was estimated to be 11 years old at the time. At Gombe, menarche occurs between one and six months after the first full-sized swelling (Goodall, 1986).

b) Age at Transfer

Twenty females (12 from K and 8 from M Group) whose age was known or whose birth year can be presumed fairly accurately have been observed to emigrate from their natal group. Some of the K Group females immigrated into M Group (Table 3.6), which means that both processes of leaving the natal group and entering a new group were observed. The average (mean $\pm$ sd) age at emigration was 11.0 ± 1.5 years (range = 8–14, $N = 20$).

Of seven females that transferred whose age was known, SF first temporarily immigrated to M Group with her mother (WN) when she was only 5 years old. At the age of 9, SF immigrated alone to M Group, her mother having probably died in the interim. Another female, AS, following her mother WL, immigrated to M Group at the age of 8. If these two females are excluded, the average age at transfer in females of known age was 11.2 ± 1.2 years (range = 10–13, $N = 5$).

Additional evidence, besides these estimated ages of immigrants, indicates that transfer usually occurs after menarche. Thus, most immigrants to K or M Group exhibited full-sized sexual swellings. There were clear exceptions, however. For example, the K Group female CH did not exhibit maximum swelling while in K Group in August of 1968, but trans-

Table 3.6 Age at sexual maturity and natal transfer in females.

Name	Presumed or known (*) birth year	First observed maximum swelling	First observed copulation	Last obs. in natal group	First obs. in new group	Presumed or known (*) age at transfer
[Emigrants from K Group]						
TA	1957	Aug. 67	–	>Aug. 67	Oct. 68	11
CH	1958	Nov. 68	Sep. 69	Aug. 68	Oct. 68	10
SA	1960	Aug. 69	Aug. 70	Nov. 72	Mar. 73	12
ND	1963	<Aug. 73	Sep. 73	Aug. 72	Oct. 72	9
WH	1963	Feb. 73	Jun. 74	Aug. 75	–	12
TT	1964	Jun. 75	Oct. 75	Jan. 78	–	14
SI	1965	Jul. 77	Jul. 77	Jan. 78	–	13
MD	1966	Dec. 77		Sep. 77	Oct. 77	11
MW	*1967	Jul. 77	Aug. 77	Dec. 77	–	*10
NK	*1970	Sep. 80	Nov. 80	Aug. 81	Aug. 81	*11
SF	*1978	not 88	virgin	–	Jun. 87	*9
AS	*1979	Apr. 89	virgin	–	Nov. 87	*8
[Emigrants from M Group]						
EB	1970		–	Jan. 82	–	
MR	1970	Feb. 81	Oct. 81	Oct. 82	–	12
IB	1971	>May 82	–	Oct. 83	–	12
MM	1971	Aug. 81		Jan. 85	–	14
II	1971	Aug. 81		Feb. 87	–	16
LS	*1975	Sep. 85	Nov. 85	Jun. 88	–	*13
PA	*1975	Sep. 85	Sep. 85	Dec. 87	–	*12
DB	1975	<Aug. 87	<Aug. 87	Jun. 88	–	13
IC	1976	–	–	Feb. 86	–	10
CB	1976	–	–	? 86	–	10
AN	1977	Sep. 87		Nov. 88	–	11
SH	*1978	Oct. 88		Nov. 88	–	*10
WC	1978			Apr. 88	–	10
[Immigrants to K Group, from unknown natal group]						
WI	1956	Jun. 69	Oct. 69	–	Jun. 69	13
WL	1959	Jul. 71	Jul. 71	–	Jul. 71	12
WD	1961	Oct. 72	Oct. 72	–	Oct. 72	11
GW	1962	Oct. 72	Oct. 72	–	Oct. 72	10
WP	1962	Nov. 72	Nov. 72	–	Nov. 72	10
GM	1966	Jul. 76	Aug. 76	–	Dec. 75	9
[Immigrants to M Group, from unknown natal group]						
PL	1971	Aug. 81	Dec. 81	–	Aug. 81	10
NP	1973	Dec. 83		–	Aug. 81	8
TM	1972	Jul. 82		–	Jan. 82	10
PU	1970	Feb. 82		–	Feb. 82	12
OP	1971	Jul. 83	Oct. 83	–	Mar. 83	12

New unit group	Notes 2tr. = secondary transfer ret = return	Date of first parturition	Age at first parturition	Years after immigration to first birth	Years from transfer to disap./ death(*) or to present
M	? 2tr.	–	–	–	?
M	ret. to K	Dec. 74	16	–	1
M	die Sep. 86	Oct. 73	13	0.9	*14.2
M		May 75	12	2.8	16.8
?		–	–	–	–
?	peripheral	–	–	–	–
?		–	–	–	–
M	die Feb. 79	–	–	–	–
?		–	–	–	–
M		no	–	–	–
M	accomp. mo.	no	–	–	–
M	accomp. mo.	no	–	–	–
?	peripheral	–	–	–	–
?		–	–	–	–
?	peripheral	–	–	–	–
?		–	–	–	–
?		–	–	–	–
?		–	–	–	–
?		–	–	–	–
?	peripheral	–	–	–	–
?	peripheral	–	–	–	–
?	peripheral	–	–	–	–
?		–	–	–	–
?		–	–	–	–
?		–	–	–	–
K	2tr. ? Sep. 85	Feb. 72	16	2.7	*6.3
K	2tr. M Sep. 87	Jan. 79	20	7.5	*16.2
K	2tr. M Jul. 83	Jun. 74	13	1.7	*10.8
K	2tr. M Sep. 79	barren	–	–	*6.9
K	2tr. M May 79	barren	–	–	*6.5
K	2tr. M Oct. 79	Apr. 83	17	[–]	*3.8
M	die Dec. 86	Nov. 84	13	3.3	*5.3
M	disap. Jul. 84	–	–	–	*2.9
M	die Oct. 87	Sep. 84	12	2.7	*3.6
M		no	–	–	7.3
M		Jun. 84	13	1.3	6.3

Table 3.6 (continued).

PI	1972	Oct. 83	Dec. 83	–	Sep. 83	11
VL	1972	Oct. 83	Oct. 83	–	Oct. 83	11
KK	1973				? Sep. 85	12
BT	1973			–	Feb. 86	13
PT	?1970			–	Aug. 86	
MN	1974			–	Oct. 84	10
JN	1974			–	Jun. 87	13
[M Group females, from unknown natal groups]						
WE	1956			–		
WO	1958			–		
FT	1963			–		
IK	1965			–		
FB	1967			–		
TY	1967			–		

ferred to M Group in September. She was not observed mating with M Group males at least from September to early November. GM, who immigrated to K Group in December 1975, was small in body size, had a tiny swelling, and did not mate until August 1976. She was estimated to be 9 years old. NP, who immigrated to M Group, was even smaller and was estimated to be only 8 years old.

c) Adolescent Sterility

There is a period of adolescent sterility in the female between menarche and the first conception which results in a live birth (Goodall, 1986). At Mahale, young females did not give birth for a long time after immigration.

Thirteen immigrant females produced their first live birth 0.9 to 7.5 years after immigration ($\bar{x}=2.9$, sd$=1.6$, median$=2.7$). Three immigrant females did not give birth even after 6.5, 6.9, and 7.3 years had elapsed since their transfer to M Group (Table 3.6).

One female (GM) immigrated into K Group, stayed there without giving birth for 3.8 years, and retransferred to M Group during the "mass emigration" period. She produced her first offspring 3.5 years after the retransfer.

d) Age at First Parturition

Despite 22 years of study, no female whose age is known exactly has yet been observed to give birth. This is because almost all females born in K Group disappeared, probably to migrate into unmonitored unit-groups. Those females of known age who immigrated to M Group have not yet given birth, although one of them (NK) is 18 years old (Table 3.6). All the females of known age who were born in M Group disap-

M		Jan. 87	15	3.3	5.7
M	die Jul. 86	Dec. 85	13	2.2	2.8
M	disap. May 86	Nov. 85	15	?	*>0.7
M		Apr. 89	16	3.2	3.3
M		–		–	3.8
M		Apr. 89	15	4.5	4.7
M		Nov. 88	14	1.4	2.0
		Dec. 70	16		
		May 75	15		
		Apr. 76	13		
		Jan. 81	16		
		Dec. 81	14		
		Jan. 82	15		

peared at adolescence. Therefore, age at first parturition is known relatively precisely only for three K Group-born females who were juveniles when first seen, and who transferred to M Group. They gave birth when they were about 12, 13, and 16 years old respectively.

Along with these three, the first parturition was documented for 22 females whose ages were estimated when they were juvenile or adolescent (at the start of the study or when they first transferred into K or M Group). The estimated age at first parturition ranged from 12 to 20 years ($\bar{x}=14.6$, sd$=1.8$, median$=15$). Hiraiwa-Hasegawa, Hasegawa, and Nishida (1984) gave a similar median (15), range (12–20), and mean (15.0).

2) Female Estorus Cycle

Females show an estrous cycle of 27–39 days. They experience stages of tumescence, maximum swelling, detumescence, and no swelling (Hasegawa and Hiraiwa-Hasegawa, 1983).

Females cease to show maximum sexual swelling a few months after conception. However, the time when a female stops swelling varies considerably from individual to individual. A few females were observed to show maximum swelling and to mate with adult males up to 3 months before parturition.

3) Gestation Period

Captive studies indicate that the gestation period of chimpanzees is about 8 lunar months, or 227.6±4 days (range=203–240, $N=8$; Martin, Graham, and Gould, 1978). In some pregnancies, periods of sexual skin swelling occur at irregular intervals; most frequently during the first trimester of pregnancy both in the wild (Hiraiwa-Hasegawa, 1986) and in captivity (Fairman *et al.*, 1981). Pregnant females near term display an unusual resting posture, lying on the belly on the ground or on branches (Nishida, unpublished; Takasaki, unpublished). Most females near term also tend to separate from core subgroups, living alone (with their offspring if any) for a few weeks to months. Hiraiwa-Hasegawa (pers. comm.) refers to this period as "maternity leave." This varies, however, as a high-ranking female (WO) and some other resident females never did take "maternity leave."

4) Lactational Amenorrhea and Postpartum Anovulatory Estrus

Lactating females typically do not show swelling until their infants are weaned, or (as occurs far more often) until their infants die (see below). Again, this tendency is variable, as lactating females occasionally show maximum swelling even as soon as one month after parturition (Takasaki *et al.*, 1986). The adaptive significance, if any, of such swelling is unknown. Rather such early resumption of postpartum cycling is characteristic of young adult females (e.g., PI, OP), and they tend to lose their unweaned infants (OP, twice).

5) Menopause and Longevity

It is often said that few animals experience post-reproductive life in the natural environment (Comfort, 1956). However, there is evidence that wild female chimpanzees experience post-reproductive life, and that some show menopause. Female chimpanzees typically produce their last offspring in their late 30s to early 40s. There are three measures that indicate post-reproductive life (Table 3.7), as follows.

First, at least 5 of 11 old females did not give birth nor show any sign of pregnancy during a period from the death of their last infant offspring to their own death. This period varied from 2.5 to 9.5 years (median = 4.2). Second, seven of these females did not give birth or show any sign of pregnancy during the prolonged period from the resumption of their last estrous cycle to their death. This period varied from 3 to 9 years (median = 3.6). Finally, at least five of these females completely ceased to show estrous swelling in their last few years before death. The most remarkable example is WB who was presumed to be born before 1943. She reared at least two daughters to sexual maturity. She gave birth to her last offspring (male) in 1974 (Fig. 3.3, p. 69). This infant died in 1977. WB showed estrus swelling until 1983, but from 1983 until September, 1986, when she was last seen, she did not exhibit any swelling. NG, who is estimated to be over 45 years old, has not shown signs of estrus for at least 3 years. Both NG and WB were in good health during their anestrous periods.

6) Sterile Females

Some females did not show any signs of pregnancy during a consider-

Table 3.7 Female post-reproductive life and menopause.

Initial	(A) Year of birth	(B) Last parturition	(C) Age at last parturition (years)	(D) Death of last offspring	(E) Beginning of last estrus	(F) Last observed estrus	(G) Year of death or last obs.	(G–D) Min. sterile years	(G–E) Min. sterile years	(G–F) Non-cycling years
WM	<1936	72	36	survive	Oct. 78	Oct. 80	Oct. 81	n.a.	3y 0 m	1y 0 m
WW	<1936	May 77	41	Dec. 78	–	–	Dec. 78	0	–	–
GA	<1940	Dec. 79	39	Nov. 83	Feb. 84	Jan. 88	Jan. 88	4y 2 m	3y 11 m	0
GP	<1940	79	39	survive	Aug. 82	Sep. 84	Aug. 85	n.a.	3y 0 m	0y 11 m
NG	<1940	?	?	?	?	Jan. 86	alive	–	–	>3y 5 m
WT	<1942	Jul. 74	32	Mar. 77	no cycling		Sep. 79	2y 6 m	–	2y 6 m
WB	<1943	May 74	31	May 77	Sep. 77	Dec. 83	Sep. 86	9y 4 m	9y 0 m	2y 9 m
WN	<1945	Jul. 78	33	survive	Nov. 83	Sep. 84	? Jun. 87	n.a.	3y 7 m	2y 9 m
UM	<1945	Dec. 80	35	Feb. 81	Aug. 81	Sep. 86	Feb. 87	6y 0 m	5y 6 m	0y 5 m
WS	c1946	Jan. 81	35	Apr. 81	Jul. 81	Aug. 84	Sep. 84	3y 5 m	3y 2 m	0y 1 m
LN	<1950	?73	23	survive	–	Jan. 89	alive	–	–	–

able period of observation, although they continuously exhibited estrous cycling (Table 3.6). Two examples are GW, who is now about 27 years old, and who has been observed for more than 16 years and WP who had been observed for more than 12 years, until she disappeared in 1984 (at about 24 years old). Neither of these females showed evidence of ill health or unusual copulatory behavior which might have prevented insemination, such as that seen in Gombe's sterile female, Gigi (Goodall, 1986).

6. Male Reproductive Life

Males copulate (show intromission and pelvic thrusting) with estrous females from early infancy. Male infants do not select estrous females, and copulate even with their mothers when they resume postpartum estrus. Usually, such incestuous mating stops after the male reaches adolescence (see Chap. 1).

Three males of known age, TB, LL, and NS, were observed to ejaculate for the first time when they were 9 years old, 9 years 3 months, and 10 years old respectively (Hayaki and Nishida, unpublished). There have been no cases of senile males ceasing to mate with estrous females.

7. Birth Rate

1) Birth Interval when the Previous Offspring is Living

Birth interval consists of (i) the period of lactational amenorrhea, (ii) the period from the resumption of estrus to conception, and (iii) the gestation period (Table 3.8).

The period of lactational amenorrhea varies considerably between individuals and for the same individual. The weaning of the infant is accompanied by the resumption of estrous cycling but females do not necessarily become pregnant immediately.

Goodall (1983) stated that, compared to primiparous mothers, multiparous mothers typically exhibit longer periods of lactational amenorrhea after giving birth, but have fewer periods of sexual swelling before conceiving. Our sample size was too small to test this hypothesis. However, the age of the mother was not correlated with the period of lactational amenorrhea ($r_s = 0.14$, $N = 18$, $p > 0.05$), the period from the resumption of estrus to the next birth ($r_s = 0.14$, $N = 18$, $p > 0.05$), or the birth interval itself ($r_s = 0.12$, $N = 19$, $p > 0.05$).

The birth interval was significantly correlated with the period of lactational amenorrhea ($r_s = 0.60$, $N = 18$, $p < 0.05$), but not with the period from the resumption of estrus to the next birth ($r_s = 0.39$, $N = 18$, $p > 0.05$). The period of lactational amenorrhea was not correlated with the period

Table 3.8 Birth interval.

Initials of mother/infant	Age of mother at birth of previous infant	Sex of previous infant	Period between two births	Period of lactational amenorrhea (months)	Period from resumption of estrus to birth (months)	Birth interval (months)	Notes
CH/KB	16	M	Dec. 74–Dec. 80	57	15	72	PRIM
WA/LL	18	M	May 72–May 79	64	20	84	PRIM
SA/NS	13	M	Oct. 73–May 79	<55	12	<67	
SO/TB	23	M	Oct. 76–Jan. 82	44	19	63	
WD/MA	16	M	Sep. 77–Jul. 83	30	40	70	
WO/BB	23	M	Nov. 81–Jul. 87	<59	9	<68	
ND/NC	19	M	Nov. 82–May 88	49	17	66	
BO/BG	29	M	Mar. 81–Jul. 88	66	22	88	OLD
WX/AL	21	M	Jan. 82–Sep. 88	57	23	80	
IK/IW	17	M	May 82–Feb. 88	53	16	69	
WO/PA	17	F	May 75–Nov. 81	51	26	77	PRIM
WE/LS	19	F	May 75–Jul. 81	47	27	74	
TY/TZ	15	F	Jan. 82–Jan. 88	56	16	72	PRIM
SA/TL	20	F	May 80– 85	53	<10	<63	
FT/PN	17	F	Mar. 80–Jun. 86	59	16	75	
WL/AS	20	F	Jan. 79–Dec. 85			83	PRIM
SL/SV	27	F	Dec. 82–May 87	33	20	53	
FB/FD	14	F	Dec. 81–Jul. 87	58	9	67	
SO/AB	29	F	Jan. 82–Jun. 88	67	10	77	OLD

PRIM: primiparous mother, OLD: old mother.

from the resumption of estrus to the next birth ($r_s = -0.44, N = 18, p > 0.05$).

The median period of lactational amenorrhea was 55.5 months ($\bar{x} = 53.2$, sd = 9.7, range = 30–67, $N = 18$). The period of lactational amenorrhea was as long when the previous offspring was male (median = 57, $\bar{x} = 53.4$, sd = 9.9, $N = 10$) as when it was female (median = 54.5, $\bar{x} = 53.0$, sd = 9.4, range = 33 – 67, $N = 8$). The difference in median length was not significant (Mann Whitney U-test, $p > 0.05$).

The median number of months from the resumption of estrus to the next birth was 16.5 ($\bar{x} = 18.2$, sd = 7.6, range = 9–40 months, $N = 18$). This period was not significantly different when the previous offspring was male (median = 18, $\bar{x} = 19.3$, sd = 8.0, $N = 10$) than when it was female (median = 16, $\bar{x} = 16.8$, sd = 6.6, $N = 8$) (Mann Whitney U-test, $p > 0.05$).

The median birth interval was 72 months ($\bar{x} = 72$, sd = 8.2, range = 53–88, $N = 19$). Again, this was not significantly different when the previous offspring was male (median = 69.5, $\bar{x} = 72.7$, sd = 7.9, range = 63–88, $N = 10$) than when it was female (median = 74, $\bar{x} = 71.2$, sd = 8.5, range = 53–83, $N = 9$; Mann Whitney U-test, $p > 0.05$).

2) Birth Interval when the Previous Offspring Died

When females lost their infants, they resumed cycling within a few months ($\bar{x} = 62.6$, sd = 42.0 days, median = 55, range = 0–187, $N = 29$). The number of days from the death of the infant to the resumption of estrus was not correlated with either the mother's age ($r_s = 0.32, N = 29, p > 0.05$), or the infant's age ($r_s = -0.18, N = 29, p > 0.05$).

These females usually gave birth to their next offspring about 12 months after resuming estrus ($\bar{x} = 12.9$, sd = 7.6, range = 7– 41.5 months, $N = 23$). This was significantly earlier than when the previous offspring was alive (Mann Whitney U-test, $z = -2.91, p < 0.01$). The period from the resumption of estrus to the birth of the next infant was not correlated with either the mother's age ($r_s = -0.16, N = 23, p > 0.05$) or the infant's age at death ($r_s = -0.18, N = 23, p > 0.05$).

In this context, the mean birth interval (i.e., the number of months from the death of the previous infant to the next birth) was 14.4 months (sd = 7.6, median = 12, range = 7– 44 months, $N = 25$) (Table 3.9). This birth interval was correlated significantly with both the period from the resumption of estrus to the next birth ($r_s = 0.94, N = 23, p < 0.01$) and the period from the infant's death to the mother's resumption of estrus ($r_s = 0.46, N = 23, p < 0.05$). The birth interval was not correlated with either the mother's age ($r_s = -0.12, N = 25, p > 0.05$) or the age at death of the offspring ($r_s = -0.29, N = 25, p > 0.05$), nor was it related to the previous offspring's sex (Mann Whitney U-test, $p > 0.05$).

Table 3.9 Interbirth interval and influence of infant's death on maternal reproductive physiology.

Mother's and infant's initials		Infant's age (month) at death	Mother's age (year) at infant's death	Month at infant's death	Month at the onset of resumption of estrus	Days from infant's death to resumption of estrus	Months from resumption of estrus to next birth	Months from infant's death to next birth
[Son]								
WB	LB	36	34	May 77	Sep. 77	102	–	Menopause
CH	CU	3	26	Dec. 83	Jan. 84	23	12	13
WN	KO	8	32	Aug. 77	Oct. 77	52	9	11
WA	–	14	17	Jul. 71	Aug. 71	32	10	11
WA	–	1	25	Jun. 79	Sep. 79	83	16	19
WA	–	0	27	Dec. 81	Feb. 82	55	8	10
WS	NY	23	30	Dec. 75	Feb. 76	67	13	15
WD	LK	19	15	Jan. 76	Jun. 76	147	15	20
WD	–	3	22	Jul. 83	Aug. 83	49	9.5	11
FT	–	10	24	Apr. 87	Jun. 87	46	8.5	10
SL	RY	46	27	May 82	May 82	0	7	7
UM	–	1	36	Feb. 81	Aug. 81	187	–	Menopause
TM	–	10	13	Jul. 85	Aug. 85	33	12	13
ND	HM	2	14	Jan. 77	–	–	–	12
OP	–	13	14	Jul. 85	?Aug. 85	?24	9	10
[Daughter]								
ND	AM	19	18	Oct. 81	Oct. 81	24	12	13
CH	–	1	23	Feb. 81	Mar. 81	36	30	31
ND	SN	13	16	Feb. 79	May 79	76	10.5	13
WA	LC	12	29	Oct. 83	Dec. 83	55	8	10
WS	NN	2	31	May 77	Aug. 77	73	41.5	44
WS	WY	3	35	Apr. 81	Jul. 81	95	–	Menopause
WD	DL	3	25	Jul. 86	Sep. 86	42	10.5	12
FT	CM	25	14	May 79	Jun. 79	30	9	10
GM	–	10	22	Nov. 87	Mar. 88	103		
GA	GB	47	44	Nov. 83	Feb. 84	89	–	Menopause
IK	–	4	16	May 81	May 81	3	12	12
SA	–	0	19	May 79	Sep. 79	110	8.5	12
ND	–	1	12	Jun. 75	Aug. 75	70	14.5	17
GK	–	0	22	Feb. 83	May 83	99	–	Sterile
FT	–	0	13	May 76	–	–	–	14
GM	–	2	17	Jun. 83	–	11	10.5	11

8. Mortality

1) Mortality Factors

Mortality factors may be divided into senility, the death of the mother, aggression, disease, and predation.

(i) Senility: Some chimpanzees died at an advanced age. For example, KL, a male who in 1973 already showed signs of aging such as his white-haired lumbar region and his bent back, lived for another 13 years.

(ii) Death of the mother: Infants dependent on their mothers for milk cannot survive their mothers' death. Infants are vulnerable to this risk until they are about four years of age.

(iii) Aggression: Infants born to females whose core areas were in the overlapping area between two unit-groups were often killed and eaten by adult males (see Chap. 1). At least 9 males and one female were killed directly or died of injuries inflicted by conspecifics during the 22 year study period. Moreover, some other infants died from injuries which were unlikely to have been inflicted by other chimpanzees.

Adult males were occasionally severely injured by other males of the same unit-group. Thus, most, if not all, past-prime males had some body part mutilated, usually ears, lips, nose, toes, or fingers. However, there has been no clear case where such aggression has led to the death of an adult individual.

(iv) Disease: Probably the most important illnesses that wild chimpanzees suffer are respiratory diseases such as pneumonia and bronchitis. Many chimpanzees were seen or heard to cough deeply and to discharge thick mucous from the nose particularly in the dry season. A long period of rain may have a debilitating effect on sick individuals. In fact, one important distinction between infants and juveniles is that the latter can stand a period of long, heavy rain sitting alone without being hugged by the mother.

In 1977, influenza, probably transmitted by humans, attacked the K Group chimpanzees. Those suffering from the disease could not pant-hoot even if they tried to do so (Uehara and Nyundo, 1983). Uehara gave aspirin to sick chimpanzees by putting it in bananas. Even so, a few infants disappeared during this period.

Chimpanzees also often suffer from diarrhea. Diarrhea is the rule, rather than a notable disorder, after eating particular types of fruits such as *Pycnanthus angolensis* and *Toddalia asiatica*. Vomiting was occasionally observed after eating unripe fruits such as *Voacanga lutescence*. Huffman and Seifu (1989) described a case of gastro-intestinal disease, which apparently gave an adult female pain when defecating. She also showed signs of lethargy and lack of appetite.

Chimpanzees of Mahale also suffer from intestinal parasites; 65.4% (100/153) of fecal samples from Mahale chimpanzees had at least one kind of intestinal parasite (Kawabata, unpublished). However, the extent to which parasite load contributes to the mortality of Mahale's chimpanzees is unknown.

In 1985–87, an epidemic disease that research assistants called "AIDS-like" appeared to strike M Group. A prime adult female, WE, progressively lost weight and her 4-year-old infant died in July 1985. WE herself was last seen in September of that year (Hayaki, pers. comm.). Nishida observed that VL, a primiparous female with a small infant, was very thin in February 1986 and disappeared with her infant after July of that year. LJ, a young adult male who had just attained the second-highest rank lost weight and declined to the lowest rank of the adult males (Masui, unpublished). He recovered only after one year. Another adult male, LU, similarly lost weight but not to the extent of LJ.

A primiparous female, PL, became very thin and was last seen in January 1987 (Bunengwa, pers. comm.) after her two-year-old infant had died in December 1986. MN and JN, both young adult females, were thin and not sexually cycling in the first half of 1987 (Takasaki, unpublished). A young adult male (SM) also disappeared sometime after July 1986. He apparently suffered from a congenital heart disease, for he had long been observed to breathe heavily after even light exercise (for a chimpanzee) such as climbing a hill.

Three other shy middle-aged females disappeared during the same period although no symptoms of disease were observed. SA disappeared with her infant after September 1986. IA and OO (who had a newborn infant) were last seen in November 1986 and have not been sighted since. Both SA and IA had 6-year-old daughters (TL and ID) who began to move with core subgroups in October 1986 and September 1987, respectively. The 8-year-old male, UU, disappeared almost simultaneously with his mother OO.

In total, 6 lactating females, with their dependent offspring, and an adult male disappeared in this 2-year period (in addition to old males and females (KL, KI, WB, WN) who might have died of old age, and many newborn infants who were likely to have died in any year). Since some of these individuals were rarely observed, it was not known whether all these prime chimpanzees died of the same cause. However, it is unlikely that all of their deaths were mere coincidence.

An adult female (CH) suffered from goiter when she was about 20 years old (Uehara, unpublished). Another prime female (HA) exhibited a similar swollen throat.

(v) Predation: At Mahale, chimpanzees are not hunted by humans, nor is there any record of chimpanzees being preyed upon by carnivores. However, Tsukahara (pers. comm., June 1989) twice recently found something resembling chimpanzee hair in the feces of lions that were staying in the study area.

(vi) Falling: In 1980, when she was about 18 years old, an adult female (GW) was found to have injured her left leg (Fig. 3.2, p. 68). At the age of 4, a male (TB) became lame because of a dislocated hip joint. It is likely that these injuries were caused by falls from a tree. Chimpanzees have been observed to fall from a tree all the way to the ground (e.g., Nishida, 1983a). More often, they managed to save themselves by grabbing branches as they fell (Nishida and others, unpublished). These observations suggest that falling sometimes caused the death of individuals, as was the case in Gombe (Goodall, 1986).

(vii) Unknown: Many chimpanzees disappeared without the exact reason being known. The disappearance of two adolescent males (LL and HT) is particularly puzzling because both were in good health when last seen. Another adolescent male (KE) probably died of disease. These examples might suggest a high mortality rate among adolescent males. It is not known whether the stress documented in adolescent males of other species—for example, in bonnet macaques (Dittus, 1977)—was a factor in the deaths of these male chimpanzees.

During the study period, 107 chimpanzees have been born and monitored throughout infancy. Of these, 56 infants (52%) died before weaning, 26 died of disease (46%), 10 died because of aggression (18%), 5 deaths were due to the death of the mother (9%), and 15 died of unknown causes (27%) (Table 3.10). Two died of disease after weaning, and 49 are still alive.

One hundred-twenty chimpanzees were born before the study period began or grew up without being closely monitored. Seventy-four are still alive, 22 died of known causes, and 24 of unknown causes.

2) Age-specific Mortality Rate

Mortality rate was highest (0.283) in the first year of life. It decreased

Table 3.10 Mortality factors.

Cause of death	Born from 1979 onward	Born in and before 1978
Still alive	49	74
Senility	0	16
Disease	28	6
Aggression	10	2
Mother's death	5	0
Unknown	15	24

Table 3.11 Age-specific infant/juvenile mortality.

Age at death (months)	No. of deaths	No. of individuals	Mortality
0=<12	26	92	0.283
12=<24	16	61	0.262
24=<36	4	44	0.091
36=<48	5	38	0.132
48=<60	1	31	0.032
60=<72	0	30	0
72=<84	1	24	0.042
Male			
0=<12	12	44	0.272
12=<24	10	31	0.323
24=<36	1	21	0.048
36=<48	3	18	0.167
48=<60	0	14	0
60=<72	0	14	0
72=<84	0	11	0
Female			
0=<12	7	40	0.175
12=<24	5	29	0.172
24=<36	3	23	0.130
36=<48	2	20	0.100
48=<60	1	17	0.059
60=<72	0	16	0
72=<84	1	13	0.077
Sex unknown			
0=<12	7	8	0.875
12=<24	1	1	1

to 0.262 in the second year, and to less than 0.1 thereafter (Table 3. 11). Overall, 53% (34/64) of chimpanzees died before weaning by the sixth year of life.

3) Infant Mortality and Mother's Age

Infant mortality was correlated with the age of the mothers. Infant mortality was highest when the mother was younger than 15 years of age, lowest when she was between 16 and 30 years of age, and increased again in older mothers (Table 3.12). These differences were significant ($\chi^2=8.7$, df$=2$, $p<0.05$). (In this calculation, some mothers were used twice as samples, but only in different age categories. This pooling procedure can be justified since it should bias the test only in the direction of favoring the null hypothesis.)

9. Factors Influencing the Sex Ratio

1) Sex Ratio at Birth

The sex ratio at birth for Mahale Chimpanzees was virtually one-to-one. Thus, of the infants born in K Group after 1966, and in M Group after 1975, 54 were male and 55 were female. The sex of the firstborn infant was male in 9 cases, and female in 7 cases, and unknown in 4 cases.

2) Sex-biased Mortality

Although the difference is not statistically significant ($\chi^2=2.1$, df=1, $0.2>p>0.1$), mortality was higher in male (21/39=0.54) than in female (10/32=0.31) infants (Table 3.12). This trend has already been reported elsewhere (Hiraiwa-Hasegawa, Hasegawa, and Nishida, 1984). Although the rate is somewhat inflated because all the victims of infanticide ($N=6$) were males, male infant mortality is still higher even if such cases are excluded (15/33=0.45).

10. Female Lifetime Reproductive Potential

Theoretically, if a female chimpanzee produced her first infant at 13 years of age and then gave birth every 5.5 years, she could produce 5–6 infants before her death (Goodall, 1983; Tutin, 1980).

No individual has been followed through her complete reproductive life. However, several females have died at an advanced age leaving a number of known and suspected offspring. Using data collected on 8 mothers at Gombe, Tutin (1980) calculated that the median number of births for these females was 3 (range 1–5), and the median number of

Table 3.12 Infant mortality and maternal age.

Mother's age	Male infants	Female infants	Sex unknown	Total	
	a/b	a/b	a/b	a/b	%
<15	6/ 7	3/ 6	3/ 3	12/16	75.0
16–20	3/10	2/10	4/ 4	9/24	37.5
21–25	5/ 9	1/ 4	1/ 1	7/14	50.0
26–30	3/ 7	1/ 5	0/ 0	4/12	33.3
31–35	4/ 5	2/ 3	1/ 1	7/ 9	60.0
36<	0/ 1	1/ 4	1/ 1	2/ 6	
Total	21/39	10/32	10/10	41/81	50.6

a=number died in infancy.
b=sample size.

Table 3.13 Female lifetime reproductive potential.

Initials	No. of weaned sons	No. of weaned daughters	No. of deliv-eries	No. of sons	No. of daughters	Unknown
WB	0	2	>3	1	2	0
WN	0	4	>5	1	4	0
SA	1	1	4	1	2	1
WT	0	2	>3	0	3	0
WM	1	1	>2	1	1	0
WW	2	0	>4	3	0	1
WG	0	1	>1	0	1	0
WS	0	1	5	3	2	0
WE	0	1	3	1	2	0
*SO	2	1	5	3	1	1
*BO	2	2	5	2	3	0
NG	1	0	>1	1	0	0
*WA	1	0	8	6	1	1
Totals (median)	10 (1)	16 (1)	>49	23	22	4

*Still nursing an infant in 1988.

offspring that survived to reproductive age was 2 (range 1–3). A similar result was obtained from 11 females of K and M Groups at Mahale (Table 3.13). The median number of offspring surviving to adolescence was 2 (mean = 2.2, range = 1–4).

11. Seasonal Factors Influencing Reproduction

1) First Estrus after Lactational Amenorrhea

Females usually resume full estrous cycles after a period of lactational amenorrhea (not including the resumption of estrus after the death of an infant) during the late dry season (a year is divided into 4 seasons: February-April, May-July, August-October, and September-December). Half of the females sampled (18/36) resumed cycling in September-October. Some females were used two or three times in the sample. (Table 3.14a). Data from these individuals are used more than once in the analysis only when the seasons in which they resumed postpartum estrus were different (Table 3.14b). Even so, the seasonal difference is still highly significant ($\chi^2 = 23.2$, df = 3, $p < 0.001$), with the major concentration occurring during the late dry season.

When females resumed full cycling after infants died before being weaned, the peak is again in August-October, although the seasonal difference is not significant (Table 3.14b: $\chi^2 = 4.5$, df = 3, $p > 0.05$).

Table 3.14a Monthly distribution of first postpartum estrus, births, and deaths.

		First estrus after infant was weaned		died		Total	Birth		Death	
Late rainy season	Feb.	2		4			4		3	
	Mar.	2	5	3	8	13	5	13	1	7
	Apr.	1		1			4		3	
Early dry season	May.	1		4			14		7	
	Jun.	1	2	3	10	12	7	31	3	18
	Jul.	0		3			10		8	
Late dry season	Aug.	5		7			3		2	
	Sep.	11	23	6	17	40	4	10	1	7
	Oct.	7		4			3		4	
Early rainy season	Nov.	3		3			7		3	
	Dec.	2	6	2	6	12	11	31	5	11
	Jan.	1		1			13		3	
Total			36		41	77		85		43

2) Seasonal Distribution of Birth

Eighty-five instances where the month of birth was known were recorded during the 22-year study period. The seasonal distribution of births is apparently not random (Table 3.14a); however, some mothers who repeatedly lost infants contributed to the data as often as six times. For statistical analysis (McGrew and Webster, 1989), data from these mothers were reused only when the birth seasons were different (Table

Table 3.14b Seasonal distribution of first postpartum estrus and births (data modified., see text).

	First estrus after infant was weaned	died	Total	Birth
Late rainy season	5	6	11	9
Early dry season	2	8	10	21
Late dry season	20	13	33	7
Early rainy season	6	6	12	24
Total	33	33	66	61

3.14b). The seasonal difference is significant ($\chi^2 = 14.2$, df=3, $p < 0.01$), and is bimodal in distribution, with peaks in May and January.

3) Seasonal Influences on Mortality

During the 22-year study period there were 43 cases when the month of death of infants (72 months old or younger) was known (Table 3.14a). Deaths tended to occur in the early dry season ($\chi^2 = 7.56$, df=3, $0.05 < p < 0.1$).

IV. DISCUSSION

1. Comparison of Reproductive Profiles

Reproductive profiles appear to differ between the Mahale and Gombe populations. According to Goodall (1983), only two adolescent females are thought to have emigrated permanently from their study community during a 15-year period. This means that most known females gave birth in their natal community.

Perhaps related to this, the average age at first parturition was lower (13 years according to Goodall (1983)) than at Mahale. Although the precise age at which females first produce an offspring is unknown at Mahale, the long interval between immigration and parturition makes it unlikely that females give birth as early as at Gombe. Female chimpanzees of Mahale thus experience a longer period of adolescent sterility.

It is likely that transfer is stressful to individuals (Freeland, 1976). New female immigrants are often attacked by resident females, and such hostility may cause them to leave their new group (Nishida, 1989b). Some adolescent immigrants stayed at M Group for only a few weeks, months, or even years and then disappeared again. Stress might cause new immigrants to suffer more from disease because of immunological disruption. For example, MD was in good health when she transferred to M Group from K Group in 1977, but she died 15 months later (Uehara, 1981). Possibly her death was related to the stressful social environment she encountered.

Thus, the differences in reproductive profiles between Gombe and Mahale are likely to be caused by differences in the level of stress related to contrasting life-history patterns of females. Thus, the females of Mahale experience inter-group transfer before reproducing, while most of Gombe's females reproduce in their natal group.

2. Sexual Selection

Current evolutionary theory predicts that parents should invest more heavily in individual sons than daughters, but that they should adjust the sex ratio of their progeny so as to divide their total investment equally between offspring of both sexes (Trivers, 1972; Clutton-Brock and Albon, 1982).

In captive studies, no significant differences were found between male and female offspring in birth weights (Gavan, 1953) or in the duration of gestation (Peacock and Rogers, 1959). The sex ratio at birth is also one-to-one. These facts suggest that the intra-uterine costs of producing sons and daughters are similar.

If the cost of lactation differs according to the sex of the offspring, the interval between births should vary accordingly. Although no difference between the sexes exists in birth weight, female chimpanzees are heavier than their male counterparts when they are between 2.5 and 10 years of age (Gavan, 1953). Therefore, it might be expected that the cost of lactation for raising daughters is greater than that for raising sons (see Chap. 13). However, the present study showed no difference in birth interval regardless of whether a son or a daughter had been weaned.

It is also unlikely that post-weaning maternal investment differs much between sons and daughters (see Chap. 13). Both sons and daughters share their mothers' ranges until adolescence, hence there are no obvious differences in rearing costs. Daughters emigrate in late adolescence, while sons do not; however, since mothers and their late adolescent sons usually do not stay together, sons exact no important cost from their mothers. In addition, they have few opportunities to exchange benefits, such as helping each other in social conflicts.

Probably marked sex-biased differences in parental investment did not evolve in chimpanzees, although more data are needed (see Chap. 13).

3. Menopause

Although fragmentary evidence of menopause in free-living primates has been collected (mangabey: Waser, 1978; rhesus macaque: Tilford, 1981; Japanese macaque: Okayasu, unpublished; chimpanzee: Goodall, 1983), this study clearly demonstrated that wild female chimpanzees may experience this phenomenon.

No definite menopause has been observed in any captive female (Graham, 1979); the oldest mother in captivity conceived her last infant at 41 years and 8 months (Martin, 1981). Therefore, menopause in nature may partly be the product of seasonal stresses in the environment, particularly changes in the food supply.

The adaptive significance of human menopause was discussed by Mayer (1982), following Hamilton (1966), who thought that post-reproductive life-spans may evolve when the old animal still benefits its younger kin.

However, the evolutionary advantage of menopause in female chimpanzees is puzzling, since they rarely, if ever, care for younger relatives such as grandchildren or nieces/nephews. Because their daughters emigrate, old females have no opportunity to care for young kin unless they can recognize the offspring of their sons or transfer with their sisters to the same unit-group. As a matter of fact, aged females typically live a lonely life, although some of them are often followed by unrelated (at least through their daughters) female orphans (Hayaki, 1988).

Gaulin (1980) asserted that higher female parental investment is, by itself, a sufficient explanation for female menopause: "if female reproductive success has been more limited by the ability to expend reproductive effort as parental investment than has male reproductive success, menopause would have been more likely to evolve in females." This argument is a more likely explanation linking human with chimpanzee menopause. It follows from this that menopause in the human female has not specifically evolved in our species, but is part of a very old mammalian pattern shaped by sexual selection.

4. Seasonality of Reproduction

Goodall (1983) hit the mark when she stated that "the seasonal factors influencing the female estrous cycle provide a good subject for future research." Female chimpanzees resumed postpartum estrus after weaning their infants significantly more often in the late dry season. At Gombe, the peak of sexual swelling is also September (Goodall, 1983). This is the season when food supply begins to increase at Mahale. For example, the body weights of chimpanzees increased from the early to the late dry season on 5 of the 6 occasions when such data were available (Uehara and Nishida, 1987). Thus, it is likely that the first post-partum estrus is triggered by changes in food abundance. The fact that females whose unweaned infants died failed to show this seasonal trend indicates that they do not "wait" until September if their physiological conditions are ready.

The present study found that the chimpanzees of Mahale have two birth peaks in a year, in the early dry season and in the early rainy season. The first may be due to females resuming post-partum estrus in September (i.e., September plus an 8 lunar month gestation period). However, the early rainy season birth peak is puzzling, since there was no corresponding increase in the resumption of post-partum estrus. It remains to be seen why the estrus patterning is not reflected in the birth patterning.

A recent study on a human population shows that births among nomadic Turkana women are highly seasonal in distribution, with more than half of all births occurring between March and June. Leslie and Fry

(1989) suggest that this is a result of the highest rate of conception occurring during the early dry season, when food supply is at its best and women attain their peak nutritional state. However, quantitative study of seasonal changes in the food supply of chimpanzees is necessary in order to assess whether the underlying reason for birth concentration is similar in both species.

V. SUMMARY

1. Population of chimpanzees in the Mahale Mountains National Park is at least 600.

2. The overall sex ratio of M Group is always female-biased and there are three times as many mature females as mature males.

3. The study population (data of M Group and K Group combined) maintained zero net growth ($R_0 = 1$) during a period of 10 years. During this time, the population experienced 74 births, 74 deaths, 14 immigrations, and 13 emigrations.

4. M Group "exports" and "imports" nulliparous females to and from other groups.

5. Some nulliparous females failed to be integrated into M Group and left.

6. Females showed their first maximal swelling and mated at around 10 years of age.

7. Most natal females emigrated at around 11 years of age, probably at the time of menarche.

8. Immigrant females did not give birth until after about 3 years of residence. This did not include three females who have not given birth after 6.5, 6.9, and 7.3 years.

9. It was hypothesized that the difference in the age of first parturition between Gombe and Mahale was caused by the difference in the proportion of immigrant females in the samples.

10. The median age of first parturition is estimated to be 15 years, compared to 13 years at Gombe.

11. In contrast with Gombe, there were more females than males in the "old adult" category (>34 years).

12. Post-reproductive life in female chimpanzees is not unusual, with some individuals ceasing their estrous cycling in their last few years before death.

13. The median interbirth interval was 72 months when the previous offspring was alive.

14. The interbirth interval did not differ if the previous live offspring was male or female.

15. Most infant deaths occurred in the first 24 months, with 30% of all infants dying in the first year. Fifty-seven percent of chimpanzees died before the sixth year of life.

16. Most infant deaths were due to disease.

17. Infant mortality was lowest among prime mothers (16–30), highest among young mothers (< 16 years), and intermediate among old mothers (> 30 years).

18. The sex ratio at birth was unity.

19. Mortality was higher in male than female infants, although this was not statistically significant.

20. The lifetime reproductive output of a female, in terms of the median number of offspring surviving to adolescence, was 2.

21. After lactational amenorrhea, females usually resumed full estrus cycling in the late dry season, particularly in September. This seasonal difference was highly significant.

22. The seasonal distribution of births was not random, but was bimodally distributed with peaks in May and January. However, the January peak cannot be explained from the currently available data.

Acknowledgements

We are grateful to K. Norikoshi, K. Kawanaka, S. Uehara, M. Hiraiwa-Hasegawa, T. Hasegawa, H. Hayaki, K. Masui, M.A. Huffman, K. Takasaki, S. Kobayashi, K. Hunt, R. Olomi, T. Torigoe, T. Tsukahara, R. Nyundo, M. Seifu, J. Katensi, M. Bunengwa, H. Bunengwa, and many others of the Mahale Mountains Wildlife Research Centre for cooperation in collecting demographic data. We thank T. Hasegawa, H. Hayaki, M. Hiraiwa-Hasegawa, J. Itani, K. Norikoshi, K. Takasaki, and S. Uehara for the permission to use their unpublished data. We are much indebted to W.C. McGrew for critical comments and revision of the English. This chapter also benefited from constructive comments by T. Hasegawa and S. Uehara. Research was mainly supported by grants under the Monbusho International Scientific Research Program (#56041018, 57043014, 58041025, 59043022, 60041020, 61043017, 62041021, and 63043017 to T.N.), Fellowship for Research Abroad (February 1987—February 1989) of the Japan Society for the Promotion of Science to H.T., and the Japan International Cooperation Agency to Y.T.

4

Sex Differences in Ranging Patterns

Toshikazu HASEGAWA

Department of Psychology, Faculty of Letters, Teikyo University

I. INTRODUCTION

Unlike most large gregarious primates, chimpanzees characteristically associate and travel in temporary subgroups (Nishida, 1968; Goodall, 1968). This fission-fusion pattern, however, does not occur at random. First, temporary subgroups, or parties, are formed within stable unit-groups, or communities, whose members share a common home range (Nishida, 1968; Goodall, 1983). Second, there exist distinct sex differences in associate patterns; adult males are consistently more sociable than adult females (Nishida and Hiraiwa-Hasegawa, 1987; Wrangham, 1986; Chap. 6). The third point is concerned with sex differences in ranging patterns, which is the theme of this study.

So far, detailed reports on sex differences in ranging behavior are available only about Gombe chimpanzees (Wrangham, 1975, 1979b; Wrangham and Smuts, 1980; Goodall, 1986). Wrangham (1979b) reported that males range widely and evenly over the community range, while the core area of different females is smaller than that of males and scattered evenly throughout the community range. He interpreted female dispersion as being due to strong female-female competition for feeding ranges.

In Mahale, however, comparative studies on this subject have been scant, partly because most Japanese researchers have been more interested in sociological aspects of group dynamics rather than in behavioral ecology of individual chimpanzees. Although Uehara (1981) and Kawanaka (1984) revealed that males and females share a common home range if annual ranges of both sexes are compared, they did not analyze how individuals use areas within the group range.

How individuals disperse or distribute in their habitat is one of the most basic and important questions to be answered when we consider sexual and life-history strategies of any species. In this paper, I will examine the ranging patterns of individual male and female chimpanzees of M Group in Mahale, and briefly discuss what kinds of factors determine dispersion of chimpanzees.

II. METHODS

Observations for this study were made on the chimpanzees of M Group during the periods from July 1980 to June 1981, and from August 1981 to May 1982. On encountering a subgroup of chimpanzees (see below), I followed them as long as possible, namely, until I lost them or until they made a bed up in trees to sleep. I recorded their travel routes on the map and checked the membership at least every 30 minutes.

As the membership of a subgroup changed temporally and spatially, a subgroup was defined operationally in terms of time and space for the purpose of quantitative analysis. That is, I called a subgroup a set of individuals observed within a particular quadrate of 400 square meters drawn on the map (Fig. 4.1) during the same 30-minute period. For example, when I continued to observe individuals A, B, and C within the same quadrate for 60 minutes, I counted them as two different subgroups even though the place and membership was the same. Also, when the same set of individuals was observed to move from one quadrate to another within 30 minutes, I counted them as two subgroups.

During the study period, a total of 1843 subgroups were recorded. Almost all the members of the subgroups were identified individually. In this paper, I analyze the data of 11 males and 23 females who were late adolescent or older (for age definition, see Chap. 3) and were observed more than 100 times in different subgroups. I excluded the data collected in one quadrate where the Kansyana camp locates, because chimpanzees were usually fed when they visited the camp during 1980–1981 periods for weighing their body size and for habituation of shy individuals. Although I also fed them small pieces of sugarcane or bananas in addition to the camp feedings at the beginning of observation in order to check their daily attendance records, the effect on their ranging behavior was regarded to be minimal.

Chimpanzees of Mahale utilize a thin strip of area stretching from north to south along the shore of Lake Tanganyika. I divided the study area into 6 zones and named them N2, N1, C, S1, S2, and S3, respectively, from north to south (Fig. 4.1). The vegetation and topology of the study area is described in Chapter 1.

The total number of subgroups observed in each zone was not equally distributed (N2=108, N1=325, C=849, S1=338, S2=170, S3=53).

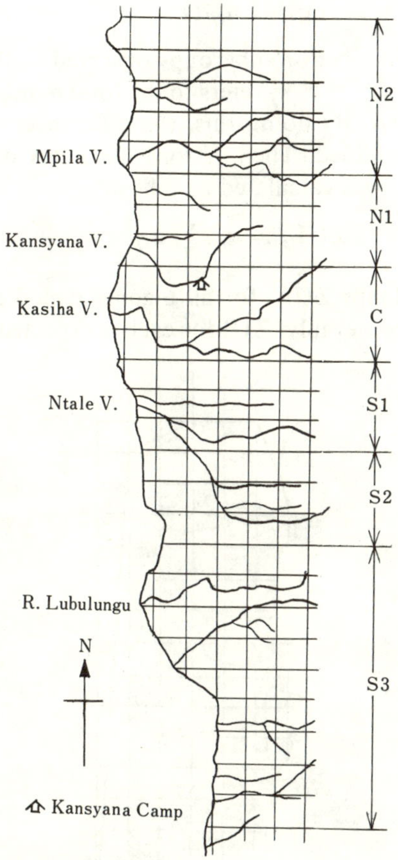

Fig. 4.1 Study area.
The group range of M Group was sectioned by every 400 m and divided into 6 zones.

These differences were partly due to sampling bias. For instance, it was easier to find chimpanzees if they traveled nearer to the camp. Also there were individual differences in observability: some chimpanzees were better habituated than others and they were observed more often. In order to correct these observational biases, raw data were processed as follows.

1) Zone Use Index: ZUI

In order to compare the ranging patterns of individuals, a zone use index (ZUI) was calculated for each individual, correcting the observational bias between zones and between individuals. First, the observation rate of an individual (i) at zone (j), q_{ij}, was calculated.

$$q_{ij} = n_{ij}/N_j,$$

where n_{ij} refers to the number of subgroups observed in the zone (j) that included the individual (i), and N_j refers to the total number of subgroups observed in the zone (j). By doing this, the differences in observability between zones were modified. Then, a zone use index of the individual for the zone (j), ZUI_{ij}, was calculated as follow.

$$ZUI_{ij} = q_{ij}/\sum_j q_{ij}$$

The sum of an individual's ZUI_{ij} for all 6 zones would make 1. If each individual used each zone evenly, ZUI for each is expected to make 0.167.

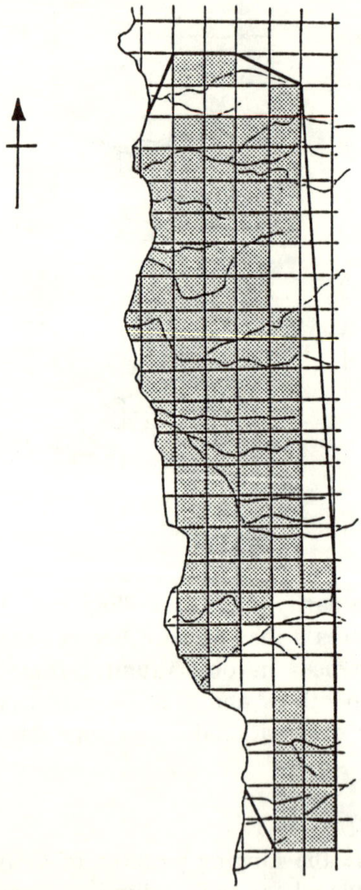

Fig. 4.2 Group range of M Group.
The members of M Group were observed in 90 quadrates. The group range (inside of bold lines) was 19.4 km².

In the following sections, an individual is said to use a zone preferentially when he/she showed one or more zones with ZUIs less than $0.167-0.067$ or more than $0.167+0.067$. (The figure 0.067 is put operationally and does not have statistical meanings.)

2) Home Range Similarity: HRS

In order to clarify the degree of overlap between ranges of 2 individuals, a home range similarity, HRS_{ab}, was calculated, using ZUIs of 2 individuals, a and b.

$$HRS_{ab} = \frac{\sum_j (ZUI_{ja} \cdot ZUI_{jb})}{\sqrt{\sum_j (ZUI_{ja})^2} \sqrt{\sum_j (ZUI_{jb})^2}}$$

III. RESULTS

1. Sex Differences in the Area of Individual Home Range

During the period of observation, M Group utilized a total of 90 quadrates, the area of 19.4 km^2 (Fig. 4.2). Figure 4.3 describes cumulative numbers of different quadrates used by 6 males and 6 females by the time of Nth observation of subgroups. The curves show the mean for

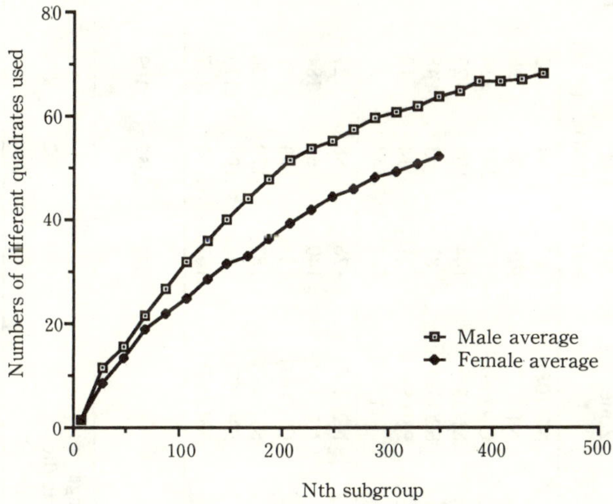

Fig. 4.3 Sex differences in the area of individual home range.
Cumulative numbers of quadrates visited by 6 males (NT, LU, MU, RA, KI, LJ) and 6 females (GW, CH, FT, WP, SO, WO) were plotted by every 20 × Nth observation of subgroups.

Table 4.1 Observation frequency and zone use indices of M Group males from July 1980 to May 1982.

	Zone	NT (P,H)*	LU (P,M)	MU (P,M)	RA (O,M)	KI (O,H)	LJ (Y,L)	DE (Y,L)	KZ (Y,L)	BA (P,H)	SM (Y,L)	KL (O,M)
	N2(108)	45	38	22	32	20	34	19	27	20	18	7
Obser-	N1(325)	108	98	96	72	75	65	70	60	67	57	27
vation	C (849)	267	227	234	210	195	195	195	150	169	106	56
Frequency	S1(338)	144	108	123	110	97	72	68	78	64	39	17
(n)	S2(170)	82	63	48	58	74	44	49	50	44	18	15
	S3(53)	21	14	10	11	16	12	11	10	7	6	1
Total	(1843)	667	548	533	493	477	422	412	375	371	244	123
	N2(.167)	.176	.188	.127	.181	.111	.218	.133	.189	.158	.208	.175
Zone	N1(.167)	.140	.161	.184	.135	.138	.139	.163	.139	.176	.219	.224
Use	C (.167)	.133	.143	.171	.151	.138	.159	.174	.133	.170	.156	.178
Index	S1(.167)	.180	.170	.226	.199	.172	.148	.153	.174	.162	.144	.135
(ZUI)	S2(.167)	.204	.198	.175	.208	.261+	.179	.219	.222	.221	.132	.238+
	S3(.167)	.167	.141	.117	.127	.181	.157	.157	.142	.113	.141	.051−
Maximal difference in ZUIs		.071	.057	.109	.082	.150	.080	.085	.089	.108	.087	.187
Home range similarity to the group range		.989	.992	.976	.983	.960	.987	.987	.982	.982	.979	.937

(−: ZUI<.167−.067=.100, +: ZUI>.167+.067=.234)

* Categories of males O: Old, P: Prime, Y: Young, H: High-rank, M: Middle-rank, L: Low-rank.

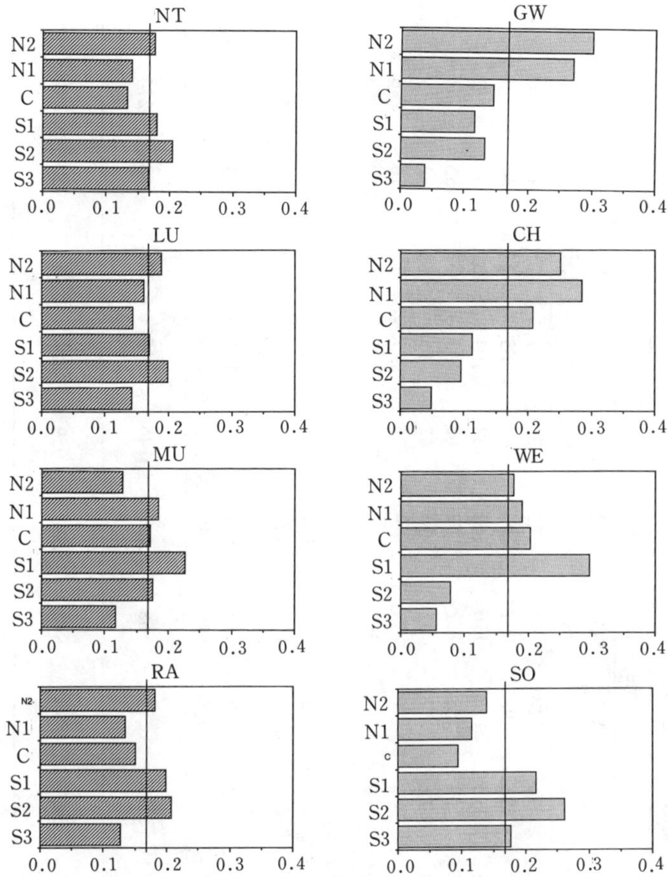

Fig. 4.4 Profiles of ZUI for selected individuals.
Males (left) used each zone evenly, while females (right) did preferentially.
The vertical line indicates. 0.167 of ZUI which is expected if individuals used each zone evenly.

males and females. The number of different quadrates used by individuals increased as the number of observation of subgroups increased. The area used by females included fewer numbers of quadrates than those used by males, being approximately 80% of male home ranges. That is, in general, males travel a wider area than females.

2. Sex Differences in Zone Use

Tables 4.1 and Table 4.2 indicate the observation frequencies (n_{ij}) and zone use indices (ZUI_{ij}) for each zone of 11 males and 22 females. Figure 4.4 shows the profiles of ZUIs of each zone for selected individuals. As

Table 4.2 Observation frequency and zone use indices of M Group females from July 1980 to May 1982.

		Name of females										
	Zone	GW (I,E)*	CH (I,E)	FT (L)	WP (I.E)	SO (E)	WO (E)	IK (E)	ND (L)	WA (I,E)	AD (L)	SA (L)
	N2(108)	48	32	24	28	21	25	25	17	25	18	10
Obser-	N1(325)	130	109	95	98	52	66	47	65	48	44	73
vation	C (849)	183	208	195	168	110	167	133	194	130	103	132
frequency	S1 (338)	58	45	39	51	102	57	68	33	51	55	33
(n)	S2 (170)	33	19	11	17	62	30	47	15	33	32	5
	S3 (53)	3	3	6	1	13	9	6	1	12	16	6
Total	(1843)	455	416	370	363	360	354	326	325	299	268	259
	N2(.167)	.300+	.251+	.214	.252+	.139	.202	.206	.199	.210	.155	.130
Zone	N1(.167)	.270+	.285+	.282+	.293+	.115	.177	.129	.253+	.134	.126	.315+
use	C (.167)	.145	.208	.221	.192	.093–	.172	.139	.289+	.139	.113	.218
index	S1(.167)	.116	.113	.111	.147	.216	.147	.179	.123	.137	.151	.137
(ZUI)	S2(.167)	.131	.095–	.062–	.097–	.261+	.154	.246+	.112	.176	.175	.041–
	S3(.167)	.038–	.048–	.109	.018–	.176	.148	.101	.024–	.205	.281+	.159
Maximal difference in ZUIs		.262	.237	.219	.275	.168	.055	.145	.265	.076	.168	.274
Home range similarity to the group range		.878	.887	.907	.874	.943	.993	.959	.880	.981	.950	.892

Table 4.2 (continued).

	Zone	WE (L)	SL (L)	GM (I,E)	GK (E)	PL (I,A)	NG	TI (E)	GP (L)	BO (L)	UM	LN	MR
		Name of females											
	N2	13	7	33	18	12	3	13	4	9	11	2	3
Obser-	N1	42	40	60	53	32	23	45	13	21	34	19	18
vation	C	117	128	83	69	100	90	44	77	67	67	55	34
frequency	S1	68	34	30	32	37	42	29	48	35	23	28	28
(n)	S2	9	10	10	30	13	27	28	16	18	8	12	23
	S3	2	1	0	8	2	7	4	3	6	1	1	2
Total		251	220	216	210	196	192	163	161	156	144	117	108
	N2	.177	.125	.415+	.200	.202	.045–	.189	.080–	.152	.243+	.059	.073–
Zone	N1	.190	.238+	.251+	.196	.179	.114	.217	.087–	.118	.249+	.186	.146
use	C	.203	.292+	.133	.098–	.214	.171	.081–	.197	.144	.188	.206	.106
index	S1	.296+	.195	.121	.114	.199	.201	.135	.308+	.188	.162	.264+	.219
(ZUI)	S2	.078–	.114	.080–	.212	.139	.256+	.259+	.204	.193	.112	.225	.357+
	S3	.056–	.036–	.000–	.181	.068–	.213	.119	.123	.206	.045–	.060–	.100
Maximal difference in ZUIs		.241	.255	.415	.114	.145	.211	.177	.228	.088	.204	.205	.284
Home range similarity to the group range		.898	.891	.779	.966	.957	.923	.939	.902	.902	.918	.903	.864

(– : ZUI<.167–067=.100, + : ZUI>.167+.067=.234)

* Categories of females L: Lactating mother for longer than 10 months, E: Sexually active for longer than 6 months, A: Adolescent females, I: Immigrant females who transferred to M Group from 1978 on.

is shown in these tables and figures, males tended to use each zone more evenly than females, and females were likely to use particular zones more preferentially than males. For 9 out of 11 males, ZUIs for all zones ranged within the criteria mentioned above (i.e., between 0.167±0.067). On the other hand, 20 out of 23 females showed one or more zones with ZUIs outside the criteria.

The ranges of ZUIs (i.e., the differences between the maximal and minimal ZUIs) were significantly greater in females than in males (Mann-Whitney U-test, N1=11, N2=23, z=3.41, p<0.01) (Tables 4.1, 4.2; Fig.

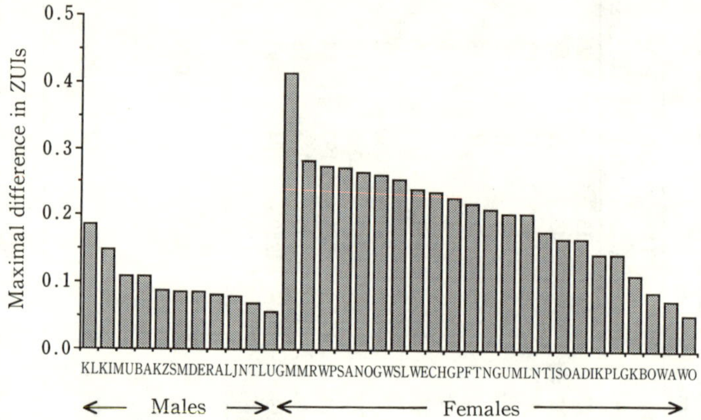

Fig. 4.5 Sex differences in maximal range in ZUIs.
The indices are lower in males than in females, which indicates that the males visited each zone more evenly.

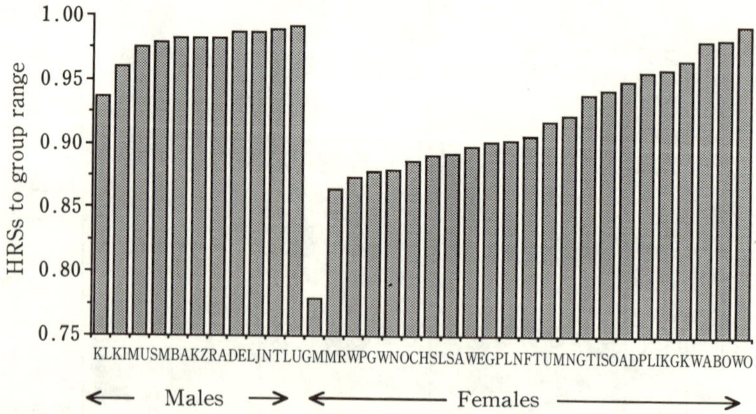

Fig. 4.6 Sex differences in HRSs to group range.
The indices are almost equal to one in males, which indicates that home range of each male is almost identical to the group range. On the other hand, females use each zone more preferentially.

4.5), which also shows that males used each zone more evenly than females.

In addition, when individual HRSs to the group range were calculated (Tables 4.1, 4.2; Fig. 4.6), the indices were greater in males than in females (Mann-Whitney U-test, $N1=11$, $N2=23$, $z=3.805$, $p<0.01$). The indices were almost equal to 1 in males, which indicates that a home range of each male was almost identical to the group range. On the other hand, females used each zone more preferentially.

Although the females used zones preferentially, their preferences were not concentrated to the same particular zones. The female ranges shifted individually. Figure 4.7 depicts a dendrogram drawn from the HRSs of 34 individuals. In this figure, 4 distinct categories of individuals can be identified: northern females, north to central females, southern females, and males and widely-ranging females.

3. Life-history Differences in Female Ranging Patterns

There was little difference in individual ranging patterns among males, except for 2 old males, KL and KI, who showed some degree of preference in a particular zone. On the other hand, there were considerable differences among individual females, which can be attributed to several life-history factors described below.

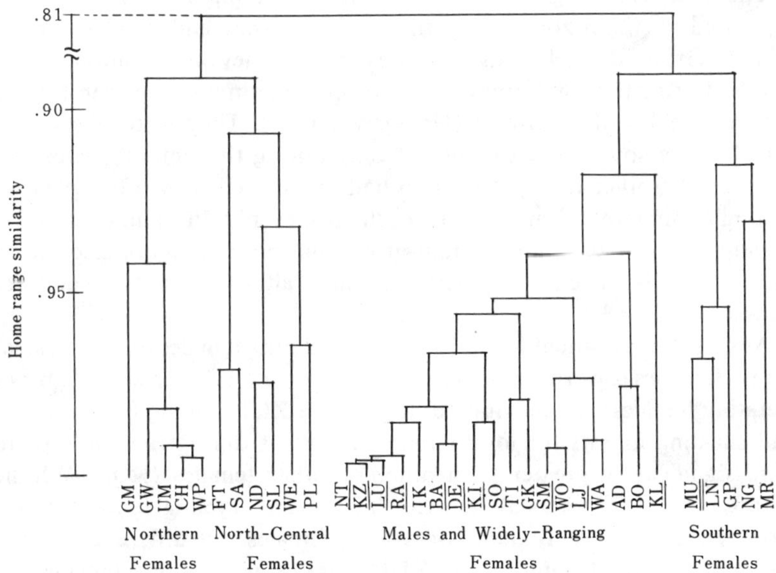

Fig. 4.7 Dendrogram of HRSs for 11 males and 23 females. Males' names are underlined.

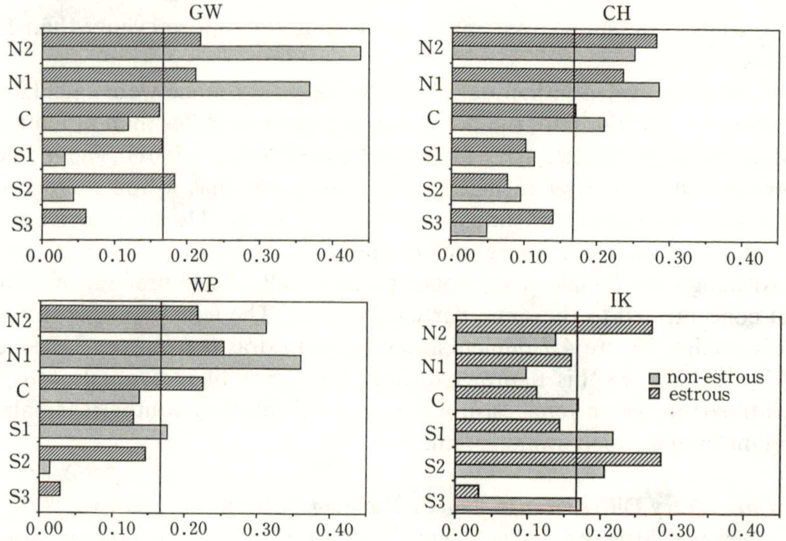

Fig. 4.8 Comparison of female ZUIs between when they were in estrus and in non-estrus periods. Immigrant females (GW, WP, CH) ranged northern zone preferentially when they were not in estrus and moved to the south when they were in estrus. A resident female, IK, used the group range evenly regardless of her estrous condition.

The 4 newly-immigrated females from K Group, GM, WP, GW, and CH, used northern zones preferentially where they had used to range in their K Group days. During the study period, they were still on the way of transferring from K Group to M Group, sometimes being seen together with the males of K Group (Hasegawa, 1989). They were observed to travel in the southern areas almost only during the time of estrus (Fig. 4.8). On the other hand, WA, who had transferred to M Group earliest among 6 immigrant females, used each zone evenly. PL, a late adolescent immigrant from an unknown unit-group (but not K Group), also did not show strong persistence in particular zones, although she was rarely observed in zone S3.

Among adult resident females ($n=16$), lactating females exhibited stronger preferences toward particular zones and used a smaller area than non-lactating females: the maximal differences in ZUIs among 8 females who had suckling infants for longer than 10 months during the study period were significantly greater than among 8 other females (Mann-Whitney U-test, $U=11.5$, $p<0.05$). On the other hand, cycling females ranged more widely and evenly than non-cycling females: the difference between the maximum and minimum in ZUIs were smaller among females who had estrous cycles for longer than 6 months than among other females (Mann-Whitney U-test, $N1=5$, $N2=11$, $U=5.5$, $p<0.05$).

IV. DISCUSSION

It remained untested whether sex differences in the ranging patterns observed in Gombe (Wrangham, 1975, 1979b; Wrangham and Smuts, 1980; Goodall, 1986) are the norm among chimpanzees in general. Previous reports on Mahale chimpanzees have not clarified this point in detail. The analysis in this paper revealed that in Mahale also there are sex differences in the ranging patterns, with males traveling widely through the whole communal area and females using land more preferentially. These results mostly agree with the reports on Gombe chimpanzees, though some differences seem to exist (see below).

Uehara (1981) argued that the differences among female ranging patterns in Mahale would be the reflection of the process of female transfer; newly-immigrated females use the overlapping area between the two unit-groups in the early stage of transfer but eventually would come to range as widely as the males. The effects of transfer on the female ranging patterns were also detected in this study. However, this study also demonstrated that the ranging patterns of individual females differed widely beyond the influence of transfer.

The extent to which each female confines herself in her core area seemed to differ between Gombe and Mahale. Although the methods of observation and calculation are not directly comparable in these two study sites, it seems that the sex differences are more clearly expressed in Gombe than in Mahale. According to Wranghams (1979b) data, for example, the mean area covered in four days by 4 females was about 54% of that of 5 males. Figure 4.3 of this study, however, showed that on average females used about 80% of males' area. Wrangham (1986) also estimated a female's stable core area, where 80% of observations of each individual occurred over at least eight days, to be less than 20% of the community's range. On the other hand, a core area of a Mahale female is thought to be much wider (Table 4.2, Fig. 4.4). Especially, a female's core area became wider and its boundary was more or less ambiguous when seasonal staple food was abundant. For instance, during the best time for *Saba florida*, many females were often observed to travel together all over the group range.

These differences between Gombe and Mahale may be due to the differences in vegetation between the two sites: riverine forests spread more widely and develop more densely in Mahale than in Gombe (Collins and McGrew, 1988). Therefore I speculate that food patches are bigger in average size in Mahale and thus can support larger feeding groups than in Gombe (Wrangham, 1986). For instance, a total of 53 individuals (including 13 adult males and 16 adult females) were observed to feed

Fig.4.9 Group travelling. An adult male(BA) is play-biting a juvenile's finger during his walk. (Photo by T. Hasegawa)

Fig. 4.10 A Tanzanian assistant follows chimpanzees on a research trail. (Photo by T. Hasegawa)

together on a single tree (*Zanha golungensis*) in Mahale (December 12, 1988; Hasegawa, Hayaki, and Takasaki, unpublished), and such a big number of individuals at a single patch has not been reported in Gombe. Thus females of Mahale may utilize the majority of food patches in the whole unit-group's range together when food is abundant, which weakens the sex differences in the ranging patterns. All these speculations need more detailed empirical data on ranging behavior of females in relation to vegetation and food distribution in Mahale.

Wrangham (1979b) extended his observation in Gombe to argue that the "community" concept was not applicable to females who typically live in isolated core areas and lead a solitary life, but only to males who share and communally defend the community range. He observed that female chimpanzees, who relied heavily upon fruits of patchy distribution for their staple food, cannot form a large-sized group but have to isolate themselves from each other. On the other hand, for males, communal defense by closely related individuals of the area that covers several females' core areas would be the most efficient way to secure their mates (Wrangham, 1979b).

His "male-only community" hypothesis was opposed by several researchers who argued that the chimpanzee social group is really a bisexual group. Response to Wrangham's idea was based mainly on the point of whether females as well as males have identity toward one particular unit-group. Many argued against Wrangham's idea on the ground that (1) females seldom associate with males from different unit-groups at the same time unless they are estrous, (2) females sometimes range widely together with males, leaving their core area far behind, (3) resident females direct aggression toward newly-immigrated females, and (4) female ranging is not only influenced by food distribution, but also by the availability of mates (Pusey, 1980; Uehara, 1981; Goodall, 1983; Kawanaka, 1984; Nishida *et al.*, 1985).

My observations also did not support Wrangham's explanation. Indeed, females' ranging patterns seem to be more directly influenced by food availability than males' are, as Wrangham discussed. However, females in Mahale were not as solitary as the Gombe females he described. Mahale females were often observed to travel together with other members of a unit-group and to interact with each other actively. I think that Wrangham stressed too much the feeding competition among females, so he underestimated the benefits of group living for females, such as cooperative nursing, getting grooming partners, finding suitable mates, avoiding predators, and so on.

V. SUMMARY

1. Two year's data on ranging patterns of 11 males and 23 females of M Group were analyzed.

2. Males traveled a wider area and used the unit-group's range more evenly than females. Females, on the other hand, used land more preferentially.

3. Newly-immigrated females from K Group visited new areas only when they were in estrus. Among resident females, lactating mothers

showed stronger preferences toward particular places and used a smaller area than non-lactating females, and sexually cycling females traveled more widely and evenly than non-cycling females.

4. It seems that the sex differences in ranging patterns are more clearly expressed in Gombe than in Mahale. This may be due to differences in vegetation between the two sites.

Acknowledgements

I would like to express my gratitude to M. Hiraiwa-Hasegawa for her assistance in data processing and for her valuable comments and M. Shimada and H. Matsuda for their advice on data analysis. I also thank T. Nishida for his comments on the draft of this paper. This research was supported by Japan International Cooperation Agency.

5

Sperm Competition and Mating Behavior

Toshikazu HASEGAWA* and Mariko HIRAIWA-HASEGAWA**

* *Department of Psychology, Faculty of Letters, Teikyo University*
** *Department of Anthropology, Faculty of Science, University of Tokyo*

I. INTRODUCTION

It has been well documented among primates and other mammals that testes size, after controlling for the effect of body size, is related to the mating system. Males belonging to species with multi-male mating systems possess larger testes than those belonging to species with single-male systems (Harcourt *et al.*, 1981; Kenagy and Trombulak, 1986; Møller, 1988). Male chimpanzees have unusually large testes relative to body size (Short, 1979; Harcourt *et al.*, 1981, Fig. 5.1). In addition, females show conspicuous genital swellings when they are in estrus. Chimpanzees live in multi-male, multi-female groups, and females usually copulate with more than one male during each estrous period (Goodall, 1968; Tutin and McGinnis, 1981; Nishida and Hiraiwa-Hasegawa, 1987). These characteristics strongly imply the presence of intense sperm competition in chimpanzee mating (Harvey and Harcourt, 1984).

Mating patterns observed in Mahale have been analyzed elsewhere (Hasegawa and Hiraiwa-Hasegawa, 1983; Takasaki, 1985; Hasegawa, 1989; Chap. 6). It has been demonstrated that there are two distinct patterns of mating among chimpanzees: opportunistic mating (which is non-competitive and temporary) and restrictive mating (which is a continuous sexual relationship between a particular pair, and which includes possessiveness and consortship) (Tutin, 1979; Hasegawa and Hiraiwa-Hasegawa, 1983).

In this chapter, we will analyze the factors that affect the rate of copulation for males and females. The rate of conception and the intensity of sperm competition in the chimpanzee mating system will also be discussed.

II. METHODS

1. Subjects

Observations were made on the chimpanzees of M Group at Mahale. The data were collected by both authors during the period from July, 1980 to June, 1981, and from August, 1981 to May, 1982. Additional observations were made by the first author during the period from December, 1979 to June, 1980, and from August to December, 1988. Total hours of observation amounted to about 1,000 hours in 1980–1982 and 180 hours in 1988. During the 1980 –1982 research, M Group consisted of 12 adult males, about 40 adult females, and about 55 immature individuals, under 12 years old (Hiraiwa-Hasegawa, Hasegawa, and Nishida, 1984).

2. Observation Methods

In 1980–1982, data were collected using the "focal subgroup" sampling method. In this method, a subgroup of chimpanzees, once encountered, was followed as long as possible and all sexual activities and interactions which could be seen were recorded. Five hundred and seventy one copulations (701 intromissions) were recorded by this method. In 1988, the focal animal sampling method was adopted for 3 adult females. In most cases, the data collected included the time, place, names of the mating pair, patterns of courtship behavior, duration of intromission (in seconds), copulatory vocalizations, occurrence of ejaculation, behavior of other individuals toward the mating pair, mating pattern, and interactions between the male and female after copulation.

3. Definition of Terms

Copulation: Following Tutin (1979), copulation was defined as an interaction between a female and a male which includes at least one intromission. Successive intromissions which occurred within 10 minutes of each other were considered to be the same copulation.

Opportunistic mating: Non-competitive and temporary mating in which the sexual interactions between the pair did not continue over one hour. On such occasions, an estrous female may mate with all the males in the group.

Possessiveness: Continuous mating in which a male establishes a special relationship with a receptive female for a period of time longer than one hour and often prevents lower-ranking males from copulating with her. This type of mating can take place in the presence of other group members.

Consortship: A long-term relationship between a male and a female who travel together away from other members of the group and maintain an exclusive mating relationship.

Male age: The age of each individual was known or estimated (Hiraiwa-Hasegawa, Hasegawa, and Nishida, 1984). In this paper, males under 12 years old were called immature, and those older than and equal to 13 years old, i.e., late adolescent and adult as defined by Goodall (1983), were called adult males. Adult males were further divided into 3 classes; young (13–19-years-old), middle-aged (20–29-years-old), and old (over 30 years old).

Male rank: Using agonistic interactions, adult males were classified into 4 rank categories: the alpha male (N=1), high-ranking (N=3), middle-ranking (N=4) and low-ranking (N=4).

Female age: Receptive females were divided into 2 age classes: those over 16 years old (N=23) and those younger and equal to 15 years old (N=5). The older group contained 21 parous and 2 nulliparous individuals. There were no receptive and parous females under 15 years old during the study period.

Female life history: We classified the females of M Group into 2 categories according to their tenure in the group. The first, "residents" (N=18), consists of 15 older females and 3 younger females who have associated with the males of M Group since before 1977. The second, "immigrants" (N=10), consists of 8 older females and 2 younger females who have transferred into M Group after 1977.

Female estrous stages: The reproductive profile of the females of Mahale has been described in Hasegawa and Hiraiwa-Hasegawa (1983). With the last day of maximal tumescence (which is the expected date of ovulation) set as Day 0, the female receptive phase was divided into 3 stages:

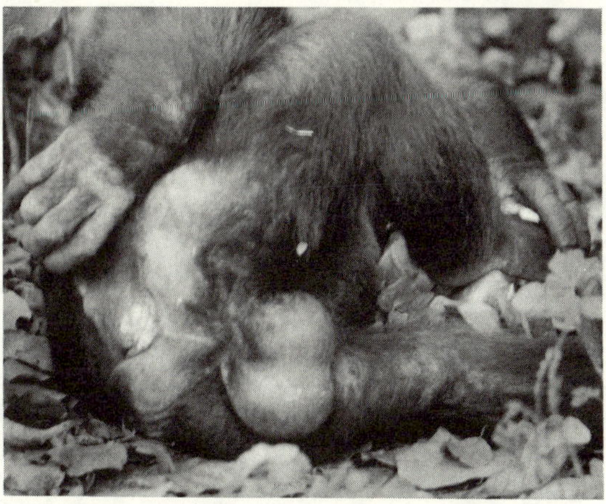

Fig. 5.1 Testes of a 15-year-old male (AJ). (Photo by T. Hasegawa)

stage I (up to Day −8), stage II (Day −7 to Day −4), and stage III (Day −3 to Day 0).

Relative copulation rate (RCR): RCR was defined as the number of copulations observed for an individual divided by the sum of observation hours for subgroups in which that individual was included. For females, observation hours were restricted to their period of maximal tumescence. Since it is likely that some copulations in the subgroup were not observed, the RCR is thought to underestimate the actual hourly copulation rate.

III. RESULTS

1. Diurnal Variation of Copulation

The number of copulations per hour had two daily peaks, one early in the morning, and one just before noon. Figure 5.2 describes temporal fluctuation in the hourly rate of copulation per estrous female in a subgroup. The rate was highest between 7:00 and 8:00 am, and next highest between 11:00 and 12:00. It declined during the afternoon and reached its lowest level in the evening. These observations mostly agree with those recorded from Gombe (Goodall, 1968; Tutin, 1979) and in captivity (Allen, 1981).

2. The RCR for Males

Overall, the relative copulation rate was negatively correlated with male age ($r=-0.50$, $df=18$, $p<0.05$) (Fig. 5.3). This was also the case when the data were restricted to adult males only ($r=-0.77$, $df=9$, $p<0.05$).

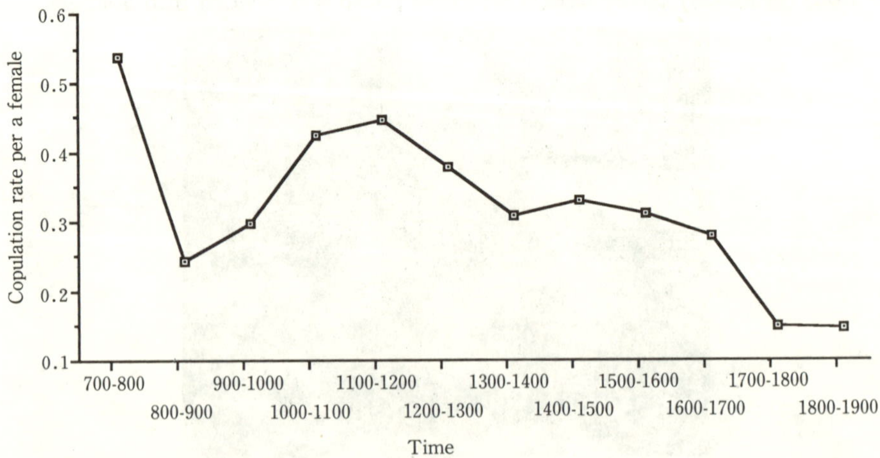

Fig. 5.2 Daily fluctuation of copulation rate.

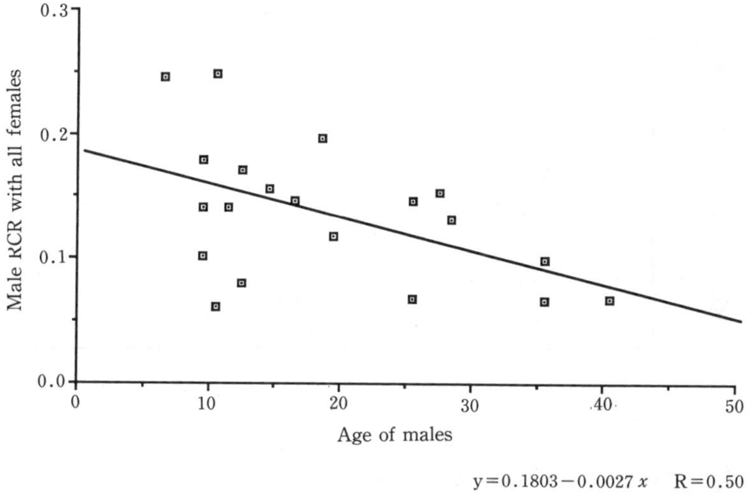

Fig. 5.3 Male RCR with all females.
Male RCRs with all females were plotted as a function of male age. Males (N=20) over 6 years old were observed for more than 50 subgroup-hours. Known or estimated ages were as of 1982.

However, there was no significant correlation between the RCR and the rank of adult males (Kendall's $\tau=0.236$, N=11, n.s.).

However, all copulations are not equal in terms of likelihood of insemination. The probability of conception varies according to female age, life history, and estrous stage. At Mahale, young immigrant females experience more estrous cycles before conception than do older females (see Chap. 3). The probability of conception is also thought to be the lowest during stage I. Therefore, copulations with such females are much less likely to result in conception. Figure 5.4 shows the RCRs of males for copulations with younger, immigrant females in any estrous stage and with other categories of females in stage I of estrus. Immature males were frequently responsible for the copulations with these less fertile females, while males of reproductive age performed fewer and fewer such unproductive copulations as they got older.

On the other hand, male RCR with females of estrous stage III peaked in middle age (Fig. 5.5). Thus, males in their prime engaged in "productive" copulations more frequently than younger and older males. Among females in stage III of estrus, the probability of conception is higher for older, resident females. Figure 5.6 shows the RCR of males for copulations with these most fertile females. With the exception of the very high

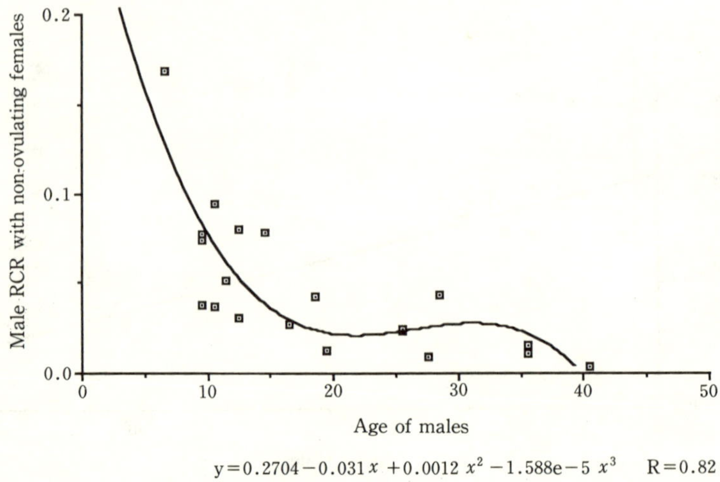

Fig. 5.4 Male RCR with less fertile females.
Male RCRs with younger, immigrant females in any stage of estrus and with other categories of females in stage I were plotted as a function of male age.

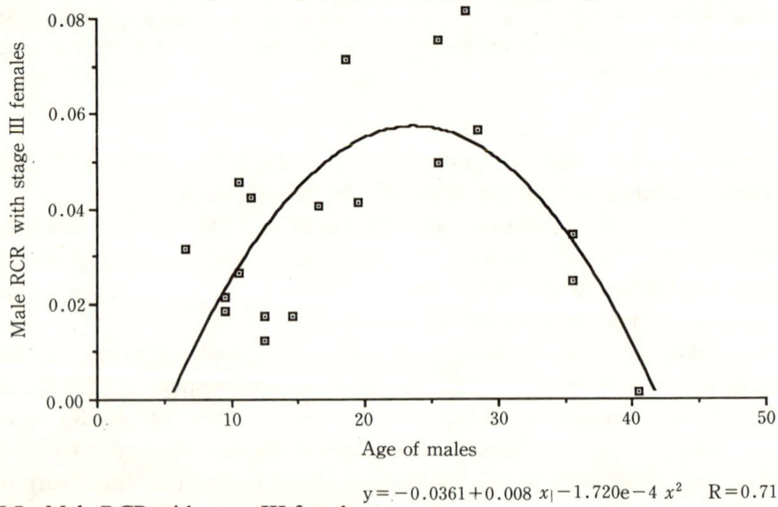

Fig. 5.5 Male RCR with stage III females.

value for the alpha male, there was no correlation between the rate of copulation and male age.

3. The Rates of Copulation for Females

1) Relative Copulation Rate

Female RCR, like male RCR, was negatively correlated with female

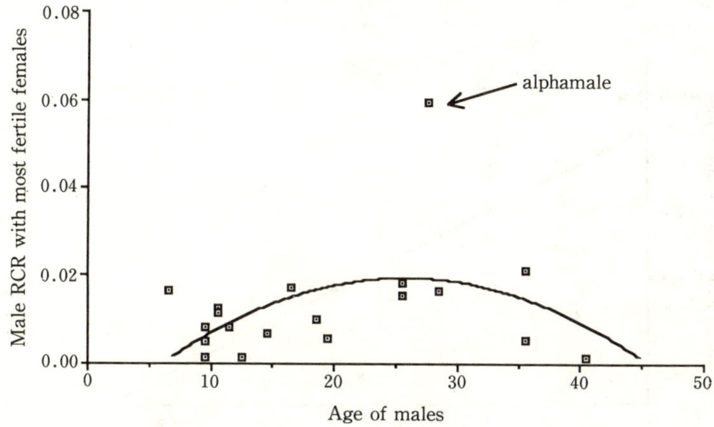

$y = -0.0139 + 0.0025x - 5.004e{-5}x^2$ $R = 0.43$

Fig. 5.6 Male RCR with the most fertile females.
Male RCRs with older, resident females in stage III were plotted.

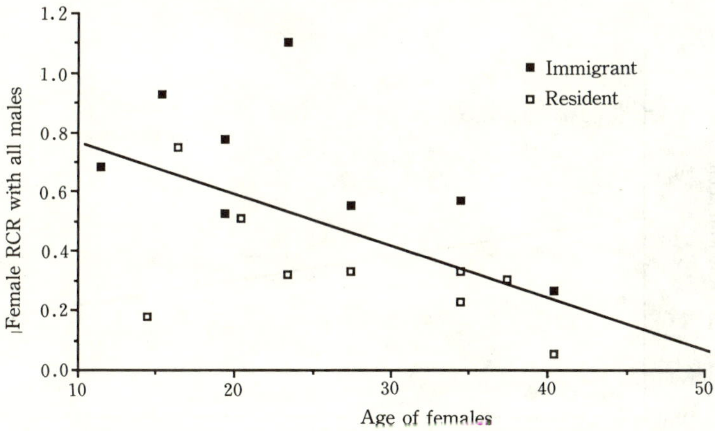

$y = 0.9132 - 0.0175x$ $R = 0.59$

Fig. 5.7 Female RCR with all males.
Female RCRs with all males were plotted as a function of female age. The females (N=18) were observed for more than 20 subgroup-hours when they were in full swelling. Known or estimated ages were as of 1982.

age ($r = -0.59$, df=16, $p < 0.05$) (Fig. 5.7). When copulations with immature males were considered alone, the negative correlation became clearer ($r = -0.67$, df=16, $p < 0.01$) (Fig. 5.8). On the other hand, the RCR for copulations with adult males was not correlated with female age ($r = -0.27$, df=16, n.s.) (Fig. 5.9). However, Fig. 5.10 shows that

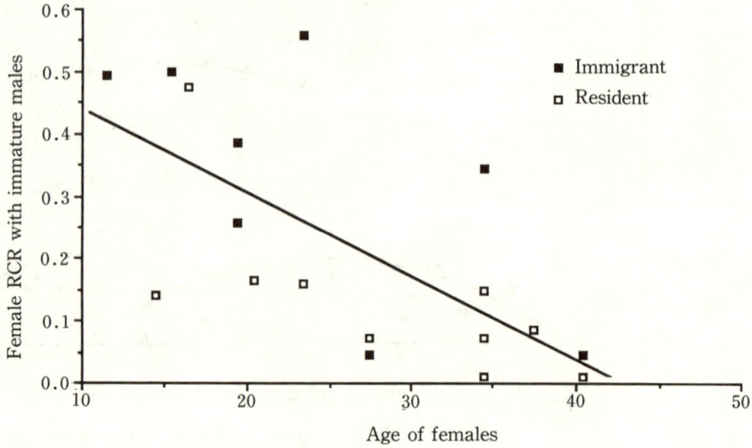

$y=0.558-0.0134x$ $R=0.67$

Fig. 5.8 Female RCR with immature males.

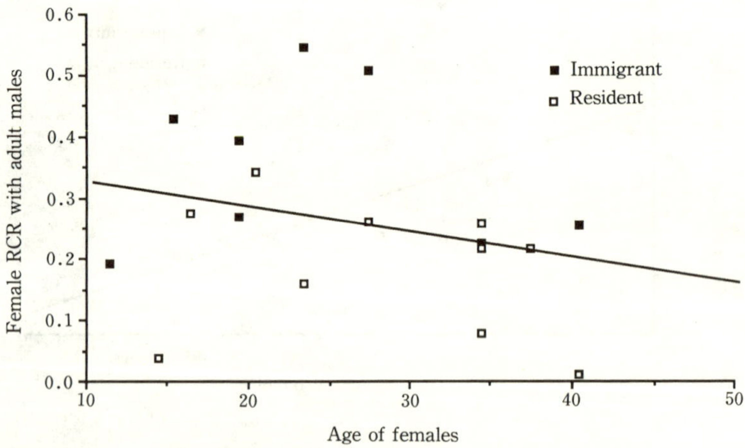

$y=0.3552-0.0041x$ $R=0.27$

Fig. 5.9 Female RCR with adult males.

the rate of copulation varied with female age when copulations with middle or high-ranking males were considered. Thus, females in their prime copulated with higher-ranking males more frequently than did young and old females.

RCRs of immigrant females were significantly higher than those of

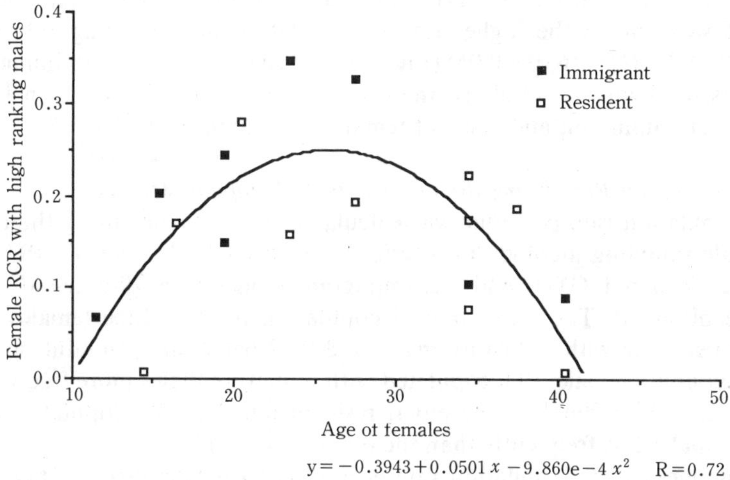

Fig. 5.10 Female RCR with middle- and high-ranking males.

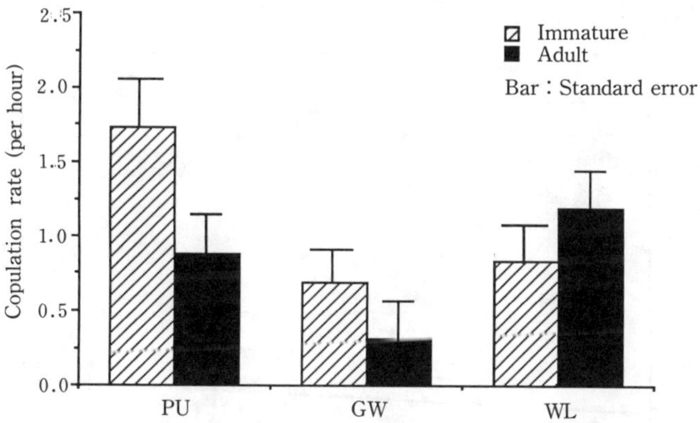

Fig. 5.11 Hourly copulation rate for 3 females.
PU was a resident female, 18 years old in 1988. GW was a resident, 26 years old. WL was a new immigrant female, 29 years old. The data were collected by focal sampling method.

resident females (Mann-Whitney $U=10$, $N1=8$, $N2=10$, $p<0.01$) (Fig. 5.7). When the effect of age was taken into account, 6 out of 8 immigrant females showed RCRs which were higher than expected, while only 2 out of 10 resident females did so ($\chi^2=5.45$, df=1, $p<0.05$). This difference was due to the higher rate of copulation of the immigrants with adult males (U=16, $p<0.05$) (Fig. 5.9). When copulations with immature males were considered alone, there was no significant difference in RCR between immigrant and resident females (U=20.6, n.s.) (Fig. 5.8).

2) Copulation Rate Using the Focal Female Sampling Method

Copulation rate per hour was calculated for 3 females using the focal female sampling method. Two resident females (the 18-year-old PU and the 26-year-old GW), and one immigrant female (the 29-year-old WL) were observed. The mean rate of copulation for the three females was 1.09 per hour with immature males and 0.79 per hour with adult males. The younger female, PU, copulated with immature males more frequently than the older females. The older, resident female, GW, copulated with adult males less frequently than the others (Fig. 5.11).

Changes in the copulation rate according to female estrous stage are illustrated in Fig. 5.12 and Fig. 5.13. The rate of copulation with immature males was not significantly correlated with female estrous stages ($r=0.16$, df=33, n.s.) (Fig. 5.12). However, the rate of copulation with adult males increased significantly as ovulation approached ($r=0.55$, df=33, $p<0.01$) (Fig. 5.13).

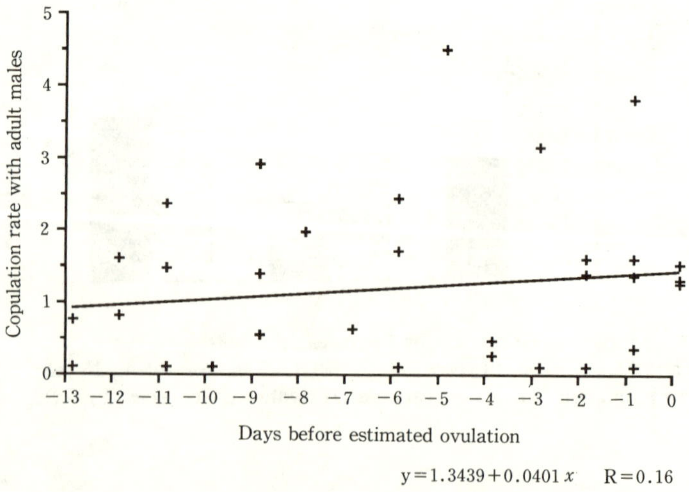

Fig. 5.12 Copulation rate with immature males as a function of female estrous stage. Data from PU, GW, and WL were merged. Each point was based on more than or equal to 80 minutes of observation.

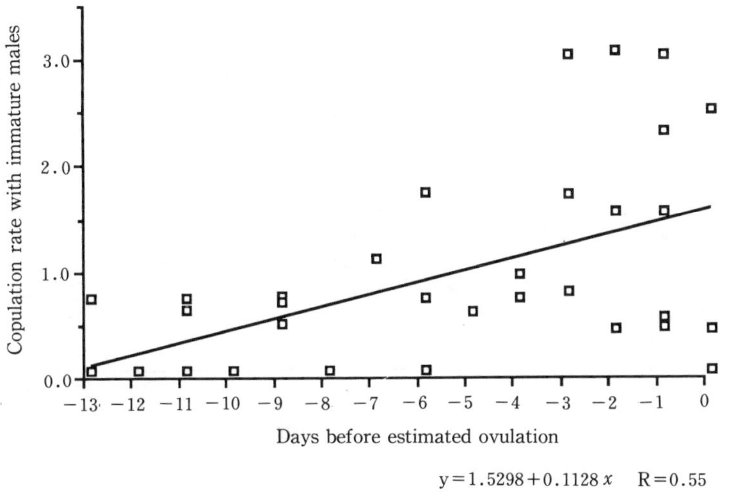

Fig. 5.13 Copulation rate with adult males as a function of female estrous stage.

4. Duration of Intromission

Duration of intromission ranged from 3 seconds to 72 seconds for all males. Those lasting longer than 20 seconds were performed only by immature males. Intromission by adult males lasted from 3 to 17 seconds. When those cases which exceeded 20 seconds were excluded, the mean duration of intromission was negatively correlated with male age ($r = -0.52$, $df = 19$, $p < 0.05$) (Fig. 5.14). The mean duration of intromission for 10 adult males (8.21 seconds, $SD = 0.71$) was significantly shorter than that for 11 immature males (9.56 seconds, $SD = 1.42$) ($t = 2.58$, $p < 0.05$).

The mean duration of intromission for opportunistic mating by adult males (8.27 seconds, $SD = 2.73$, $N = 255$) did not differ significantly from that for restrictive mating (9.04 seconds, $SD = 2.24$, $N = 25$) ($t = 1.367$, $df = 278$, n.s.). Furthermore, there was no significant difference in the mean duration of intromission by 10 adult males according to female estrous stage (stage I: 8.46 seconds, $SD = 2.57$; stage II: 8.05 seconds, $SD = 1.03$; stage III: 8.50 seconds, $SD = 1.16$).

5. Female Copulatory Vocalization

During intromission, females frequently emit a characteristic vocalization called the "copulatory squeak" (Tutin, 1979). Younger females emitted copulatory squeaks more often than older females ($\chi^2 = 27.52$, $df = 1$, $p < 0.01$). Out of 86 intromissions involving younger females, 68

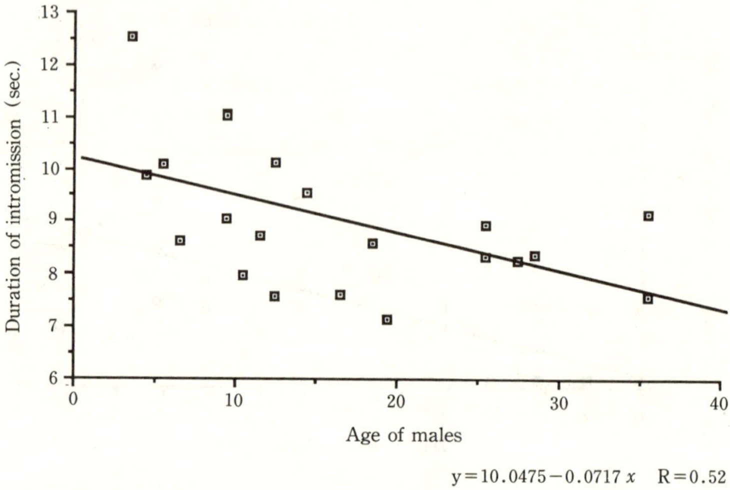

Fig. 5.14 Duration of intromission as a function of male age.
The males (10 adults and 11 immature) were observed at least 9 times when intromission was less than 20 seconds. Known or estimated ages were as of 1982.

(79%) were accompanied by the copulatory squeak, while 48% (240 out of 495) of intromissions involving older females were accompanied by this vocalization.

Immigrants vocalized more often than did residents ($\chi^2=8.14$, df=1, $p<0.01$). Immigrants emitted the copulatory squeak in 57% (234/412) of intromissions, while residents did so in 44% (74/169). Further analyses on the copulatory squeak at Mahale has been reported by Hasegawa (1989).

6. Reactions by Other Individuals toward the Copulating Pair

Most copulations took place in the presence of other individuals. Of the 780 cases of intromission or attempted intromission, overt reactions by other individuals toward the copulating pair were observed in 134 cases (17%).

1) Reactions by Adult Males

Conspicuous reactions by adult males were observed in only 17 cases. Characteristically, adult males were indifferent toward the copulating pair. Of the 17 cases of conspicuous reaction, 10 involved physical intervention in the copulation (hitting, kicking, chasing), 6 involved threats (branch-shaking, barking, charging display), and one involved staring. All the physical intervention was done by males who ranked higher than the copulating male. Aggression was directed mainly toward the copulating males rather than the females. In only one case did the intervening male copulate with the female after the invention. In addition, no pair whose copulation was disrupted was seen to restart mating later in the same day. Thus, the goal of intervention by an adult male is probably to prevent lower-ranking males from mating rather than to copulate with the females himself.

2) Reactions by Adult Females

Overt reactions by adult females were observed in 14 cases. Of these, only one involved physical intervention. An estrous female, SO, rushed in and hit a copulating male when he started to copulate with another female. Her action prevented intromission. The other 13 cases consisted

Table 5.1 Conception and mating patterns.

Name of mother	Estimated date of conception	Mating patterns observed or inferred during the period of conception			Group-setting (non-consort) conception
		Opportunistic	Possessiveness	Consortship	
1. FT	E. July-E. Aug. '79	Yes	No	No	Yes
2. ND	M. July-M. Aug. '79	Yes	No	No	Yes
3. SA	E. Sep.-E. Oct. '79	?	?	Possible	?
4. CH	L. Apr.-L. May '80	Yes	No	No	Yes
5. IK	E. May-E. June'80	Yes	?	No	Yes
6. WS	May-June '80	Yes	?	?	?
7. BO	Jul.-Aug. '80	Yes	?	?	?
8. WO	M. Mar.-M. Apr. '81	Probably No	Yes	No	Yes
9. WA	M. Apr.-M. May '81	Yes	?	?	?
10. FB	Mar.-May '81	?	?	Yes	?
11. WX	E. May-E. June '81	No	No	Possible	No
12. TY	L. May-L. June '81	Yes	?	?	?
13. SO	L. May-L. June '81	?	?	No	Yes
14. TI	E. July-E. Aug. '81	Yes	?	No	Yes
15. IK	E. Sep.-E. Oct. '81	Probably No	Yes	No	Yes
16. WA	L. Feb.-L. Mar. '82	Yes	?	No	Yes
17. ND	E. Mar.-E. Apr. '82	?	?	No	Yes
18. SL	M. Apr.-M. May '82	?	Yes	No	Yes

E: early in the month, M: middle in the month, L: latter part of the month
Yes: evidence of mating pattern observed.
Possible: indirect evidence of mating pattern.
?: no evidence of mating pattern observed, but pattern cannot be eliminated.
No: direct or indirect evidence against mating pattern.

of vocalization (pant-grunting, barking), non-aggressive contact with the copulating pair, and sexual contact with the copulating male. Unlike male intervention, female reaction seemed mainly non-aggressive.

7. Mating Patterns and Conception

Table 5.1 shows the pattern of mating at the time of 18 conceptions; the probable date of conception could be traced back from the date of birth of the offspring. The mating pattern at the time of conception was identified exactly in only 5 cases (no. 1, 2, 4, 8, and 15). In addition, consortship was strongly suspected for one case (no. 11). Of these 6 cases, 3 were opportunistic mating, 2 were possessiveness, and 1 was consortship. The possibility of a female copulating with more than one male can be excluded only in consortship. Thus, 5 out of 6 conceptions took place under group-setting conditions where sperm competition was likely.

Of the 18 conceptions, opportunistic matings were observed at least once during the conception cycle in 10 cases, and were absent in only 3 cases. On the other hand, consortship was confirmed in only one case, and positively rejected in 11 cases by the daily roll book.

IV. DISCUSSION

It has been demonstrated that chimpanzee reproduction is characterized by intense sperm competition (Short, 1979). However, there has been little analysis of the system of opportunistic mating, which is, by far, the norm among chimpanzees. Restrictive mating can be viewed as tactics to decrease the intensity of sperm competition by guarding a highly fertile female, but mate guarding behavior, especially consortship, is less successful in terms of insemination than opportunistic mating at Mahale.

This analysis has revealed that younger males copulate at a higher hourly rate than older ones. This may be a result of the higher sperm production rate in younger males. However, when sperm competition is present, it is also advantageous for males to select females who have a high probability of conceiving. When both the rate of copulation and the selection of highly fertile partners are considered, males in their prime have an advantage over both young and old males.

Although males in their prime tend to be higher in rank than young males, rank is not always highly correlated with male age. Therefore, the rank of a male might not be the primary factor affecting his reproductive success. However, in terms of the choice of a partner, the alpha male showed an exceptionally high rate of copulation with females who had the highest probability of conception (Fig. 5.6). This, together with his frequent possessiveness toward females near ovulation (Hasegawa and

Fig. 5.15 A 3-year-old male (CP) invites and mates with 18-year-old female (PU). (Photo by T. Hasegawa)

Hiraiwa-Hasegawa, 1983), suggests that alpha status confers especially high reproductive success.

This study also revealed that females in their prime copulated with higher-ranking males more frequently than did younger and older females. This may be the result of interactions between female choice for mate and male preference for the more fertile females. In order to clarify this point, further investigation will be necessary concerning pre-copulatory interactions between both sexes.

Females copulated with adult males more frequently as the date of ovulation approached. This was probably a result of increased solicitation by females as ovulation neared, although the data were insufficient to demonstrate this conclusively. Females, therefore, seemed to elicit sperm competition deliberately. If there is considerable variation among males in sperm production rate or in the viability of sperm, it would be advantageous for a female to induce intense sperm competition.

Tutin (1979), in emphasizing the importance of consortship in chimpanzee mating behavior, suggested that opportunistic mating might result in a low probability of conception. Tutin reported that 7 or more out of 14 conceptions which were observed from 1966 to 1975 occurred in consortship. Recent analysis by Goodall (1986), however, revealed that, of 18 conceptions from 1976 to 1983 at Gombe, 55.5% (N=10) resulted from copulations which took place in the group setting, 33% (N=6) in consortship, and 11% (N=2) in neighboring group. In contrast, the mating in group setting seems a more predominant pattern at Mahale. In M Group, more than 80% of conceptions where the mating pattern was known occurred in the group setting.

This difference may be due to differences in female ranging patterns between Gombe and Mahale (see Chap. 4). At Gombe, females tend to be more solitary, with individual core area overlapping less than at Mahale. Therefore, it would be easier for Gombe males to isolate one female from other group members for consortship. At Mahale, where females tend to be more gregarious, it would be more difficult for a male to separate a female from the group.

From the focal sampling of estrous females, it can be estimated that a middle-aged resident female copulates with adult males about 3.6 times a day during her period of maximal tumescence (Fig. 5.11). A middle-aged resident female typically experiences 3 cycles of maximal tumescence lasting an average of 12.5 days until she becomes pregnant (Hasegawa, unpublished). Therefore, females of this class are estimated to copulate approximately 135 times before each conception. For younger or immigrant females, this figure will be higher, probably up to several hundred copulations. It seems unlikely that such a large number of copulations are actually required for one conception.

Several possible functions for multiple copulations have been proposed (Bertram, 1975; Eaton, 1978; Parker, 1984). Bertram (1975) has argued that, in lions, the female may devaluate each mating by soliciting many matings. Thus, the value of fighting another male within the pride for a given copulation becomes negligible. This explanation might also be applicable to chimpanzees, which, like lions, have a social system which includes a group of cooperative males linked by kinship.

In chimpanzees, ovulation is supposed to occur on the last day of maximal tumescence. There is also no evidence that sperm precedence is advantageous. Therefore, the last male to copulate with a female is assumed to have the best chance of insemination. If so, the male may prefer to mate with a female who has already mated, creating a situation which will increase the number of multiple copulations.

V. SUMMARY

1. Mating behavior of male and female chimpanzees of Mahale was analyzed in terms of rate of copulation, duration of ejaculation, rate of conception, and the reaction by other individuals toward a copulating pair.

2. Male age was negatively correlated with relative rate of copulation with all females. However, males in their prime exhibited the highest rate of copulation with females near the time of ovulation. Although male rank was not correlated with copulation rate in general, the alpha male mated with the most fertile females (i.e., parous, resident, and ovulating females) at an exceptionally high rate.

3. Female age also was negatively correlated with copulation rate. Females tended to copulate with adult males more frequently as the day of ovulation approached. Immigrant females, after the effect of age was accounted for, copulated more frequently than resident females.

4. The duration of intromission by adult males did not differ according to the type of mating or the estrous stages of partner females. In general, other adult members of the group were indifferent toward the copulating pair.

5. At Mahale, mating in the group setting was responsible for approximately 80% of the conceptions in M Group.

Acknowledgements

We would like to express our gratitude to Dr. T. Nishida for valuable comments on the draft of this paper. This study was financially supported by Japan International Cooperation Agency and by grants under the Monbusho International Scientific Research Program (#404120 to T. Nishida and #63041078 to T. Kano).

6

Adult Males' Social Relations with Adult Females

Yukio TAKAHATA

Department of Zoology, Faculty of Science, Kyoto University

I. INTRODUCTION

In a unit-group (or community) of chimpanzees, adult males and females are rather alienated from one another (Nishida, 1979b; Goodall, 1986). Wrangham (1979b) reported that, in Gombe National Park, estrous female chimpanzees mingled with bisexual parties, but non-estrous females ranged independently and had only a few interactions with adult males. These relations may reflect chimpanzee's patrilineal social structure (Nishida and Kawanaka, 1972; Itani, 1985).

This study presents a preliminary description of the proximity and social interactions between adult male and female chimpanzees of the M Group, and analyzes how mature sons interact with their mothers, and how unrelated adult males and females interact with one another.

II. METHODS

The data were obtained from August 7 to December 31, 1981; this period corresponded to the seasons during which Mahale chimpanzees form large-sized parties because of the abundance of fruits (Nishida, 1968).

In 1981, M Group contained 12 males who were 13 or more years old (Hiraiwa-Hasegawa, Hasegawa, and Nishida, 1984). Of them, 10 males were observed by the focal-animal sampling method (Altmann, 1974) for about 20 hours (Table 6.1). An old male (KL) was excluded from the focal-animals, because he was too shy to be observed. The former alpha male (JG) was also excluded, because he was rarely observed, having few interactions with other M Group members.

There was a probable linear rank order among focal males as follows: NT > BA > KI > MU > LU > RA > DE > LJ > SM > KZ (see

Table 6.1 List of the focal males. The estimation of age was based on Hiraiwa-Hasegawa, Hasegawa, and Nishida (1984) and on their unpublished data.

Name	Ages (years)	Mother	Amount of observation (OU)
NT	26	NG	7,853
BA	27		7,761
KI	30–40		7,933
MU	24		7,041
LU	24		7,571
RA	30–40		7,050
DE	18		6,948
LJ	16		7,796
SM	15	GA	7,272
KZ	13	SO	6,719

Chap. 7). They also could be grouped into 3 status rankings, as follows: alpha male (NT), 5 non-alpha prime males (BA, KI, MU, LU, and RA), and 4 young adult males (DE, LJ, SM, and KZ). Of these males, only 3 (NT, SM, and KZ) had living mothers.

M Group also contained about 39 adult females and 10 adolescent females. This report generally follows Hiraiwa-Hasegawa, Hasegawa, and Nishida's (1984) description of the female age classes. These are adolescent females (8–14 years), young adult females (15–20 years), mid-aged females (21–40 years), and old-aged females (over 40 years). Females were also divided according to their sexual states; into: estrous females (in maximal tumescent phase of sexual swelling, see Hasegawa and Hiraiwa-Hasegawa, 1983), semi-estrous females (in tumescent and detumescent phases), and non-estrous females (no swelling).

Observation hours were divided into observation units (OU), each OU being 10 seconds. The incidence of proximity was measured by the number of OU in which females who were 10 or more years old stayed within 3 m of the focal male. Social interactions (approach, following, grooming, pant-grunt/bark, etc.) were also recorded. Data obtained while study animals fed on bait (sugarcane and banana) or meat were not included in the focal sampling data.

III. RESULTS

1. Dominance Relations with Females

Prime male chimpanzees dominated any adult female. Prime males seldom attacked females, but they unilaterally received from females the pant-grunts/barks (Table 6.2), which subordinates direct to dominants

Table 6.2 The observed numbers of approaches and pant-grunts/barks by females 10 or more years old. ( ): the observed number of pant-grunts/barks uttered by females when they approached focal males.

Name	Non-estrous females	Semi-estrous females	Estrous females
NT	23 (9)	2 (1)	14 (3)
BA	19 (2)	0 (0)	14 (1)
KI	13 (8)	0 (0)	14 (0)
MU	14 (6)	3 (1)	8 (1)
LU	9 (8)	1 (0)	12 (1)
RA	9 (2)	2 (0)	14 (0)
DE	7 (1)	2 (0)	21 (1)
LJ	9 (3)	1 (0)	13 (0)
SM	7 (2)	1 (0)	11 (0)
KZ	7 (0)	3 (0)	24 (0)
Total	117 (41;35%)	15 (2;13%)	145 (7;5%)

(Bygott, 1974). On the other hand, the youngest focal male (KZ), who was about 13 years old, received no pant-grunts/barks from females, and once pant-grunted at a mid-aged female.

KZ probably had not yet gained dominance over adult females. When there were no other adult males around, however, KZ sometimes charged or attacked the females, adolescents, and juveniles. At 12:52:40 on November 14, 1981, KZ slowly approached LL (adolescent male), WA (LL's mid-aged mother not in estrus), CH (mid-aged female in semi-estrus), WS (mid-aged female in estrus), and GM (young adult female in estrus). Seeing him, WS and GM fled. KZ began a charging-display, and rushed at LL, WA, and CH. The females barked "ua, ua" at KZ, and he retreated about 5 m. At 12:54:00, KZ charged them again, and slapped GM. Together the females threatened KZ, who performed another charging-display and ran away. As he left, he was showered with "uao, uao" from the females. In this case, females temporarily formed a coalition against KZ. In contrast, prime males rarely met with such a counter attack from females.

Female sexual states influence pant-grunts/barks. Non-estrous females pant-grunted/barked at the focal males in 41 cases (35%) of 117 approaches, semi-estrous females in 2 cases (13%) of 15 approaches, and estrous females in 7 cases (5%) of 145 approaches (Table 6.2). These results were significantly different from expected values ($\chi^2 = 40.2$, $p < 0.001$).

Non-estrous females often left the focal males immediately after performing greeting behaviors. Of the 41 approaches to focal males with pant-grunts/barks, non-estrous females left within 6 OU in 18 cases (44%).

This indicates that non-estrous females approached the males only for reconfirmation of their relationships.

2. Social Interactions between Mothers and Mature Sons

In the present study, mothers and mature sons tended to stay near each other, when the mothers were not in estrus or in semi-estrus, but not when the mothers were in estrus. Unfortunately, the data were inadequate for analysis, since of the 3 mothers, only SO (KZ's mid-aged mother) was well habituated. NG (NT's old-aged mother) and GA (SM's mid-aged

Table 6.3 KZ's proximity with his mother, SO.

	SO's sexual states		
	Non-estrous	Semi-estrous	Estrous
Amount of observation on KZ in OU	5,066	812	841
Total number of OU in which SO stayed within 3 m of KZ	315 (6.2%)	138 (17.0%)	12 (1.4%)

Table 6.4 Social interactions observed between mothers and their mature sons.

a. Not in estrus	Approach		Follow		Take initiative in parting	
	Mother	Son	Mother	Son	Mother	Son
NT	1	0	0	0	1	0
SM	0	1	0	1	1	0
KZ	2	2	2	4	1	4
Total	3	3	2	5	3	4
b. In semi-estrus	Approach		Follow		Take initiative in parting	
	Mother	Son	Mother	Son	Mother	Son
NT	0	0	0	0	0	0
SM	–	–	–	–	–	–
KZ	3	3	3	2	4	1
Total	3	3	3	2	4	1
c. In estrus	Approach		Follow		Take initiative in parting	
	Mother	Son	Mother	Son	Mother	Son
NT	0	0	0	0	0	0
SM	–	–	–	–	–	–
KZ	0	1	0	0	1	0
Total	0	1	0	0	1	0

mother) were too shy to be observed during the present study. As shown in Table 6.3, SO stayed within 3 m of KZ for 6.2% of the observation units when she was not in estrus, for 17.0% when she was in semi-estrus, and for 1.4% when she was in estrus.

While the mothers were not in estrus or in semi-estrus, they and their sons bilaterally approached, followed, and took the initiative in parting (Table 6.4a, b). Neither agonistic interaction, nor pant-grunt/bark, nor play was recorded between them.

While SO was not in estrus, KZ groomed his mother 6 times for 45 OU, and was groomed by her twice for 95 OU. While SO was in semi-estrus, KZ groomed her 6 times for 75 OU, and was groomed by her 5 times for 60 OU. Mutual grooming occurred 7 times while SO was in semi-estrus, and once while she was not in estrus.

While the mothers were in estrus, few interactions occurred between them and their sons (Table 6.4c). During the present study, 120 copulations were recorded in 70 dyads between 10 focal males and 14 females. No copulation was observed in 2 mother-son dyads (NT and NG, and SO and KZ; GA did not become estrous during the present study), though this is not significantly different from the expected value ($p = 0.496$, Fisher's exact probability test).

3. Social Interactions with Unrelated Females (I): Non-estrous Females

1) *Proximity*

Table 6.5 shows proximity relationships for each focal male with un-

Table 6.5 The mean number of unrelated females (10 or more years old) and other focal males who stayed within 3 m of each focal male per OU (OU: 10 seconds).

	Non-estrous females	Semi-estrous females	Estrous females	Focal males
NT	1.4×10^{-1}	0.6×10^{-1}	2.7×10^{-1}	3.2×10^{-1}
BA	1.9	0.1	1.0	6.4
KI	1.7	0.1	0.8	2.7
MU	0.8	0.01	0.8	4.1
LU	0.7	0.02	0.9	3.6
RA	0.8	0.02	0.8	4.0
DE	0.9	0.6	1.6	6.9
LJ	0.6	0.02	1.0	5.2
SM	0.6	0.01	0.8	0.6
KZ	0.5	0.1	1.2	2.4
Spearman rank coefficient	0.842 $p<0.01$	0.248 N.S.	-0.103 N.S.	0.236 N.S.

related females who were 10 or more years old, and with other focal males. With the exception of SM, focal males were more frequently surrounded by other focal males than by non-estrous females.

The mean number of non-estrous females who stayed around a focal male showed a significant correlation with the male's rank (Table 6.5). Also, male rank was significantly correlated with frequency of being approached by non-estrous females ($r_s = 0.964$, $p < 0.01$); non-estrous females chose to approach higher ranking males. On the other hand, there was no correlation between male rank and the frequency of male approaches to non-estrous females ($r_s = 0.370$, $p > 0.05$).

Only a few male-female dyads exhibited constant and frequent proximity. The individuals in the following 10 dyads stayed within 3 m of each other for 5% or more of the observation units in which the male was focal-observed and the female was not in estrus: NT and SO (6.9%), BA and PL (7.9%; adolescent female), KI and ND (7.8%; young adult female), KI and TI (6.2%; mid-aged female), KI and PL (5.6%), MU and CH (7.5%), DE and CH (5.3%), DE and WP (5.2%; young adult female), LJ and WP (5.4%), and SM and WE (5.0%; mid-aged female). Grooming was observed 49 times in 8 of these dyads; this represented 45% of the grooming episodes observed between focal males and non-estrous females. The females involved in these dyads, with the exception

Table 6.6 Social interactions observed between focal males and non-estrous females 10 or more years old.

a. Approach	Male approached	Female approached
Alpha male	9	22*
Non-alpha prime males	27	64**
Young adult males	12	28*
Total	48	114**
b. Follow	**Male followed**	**Female followed**
Alpha male	1	10*
Non-alpha prime males	14	28*
Young adult males	12	11
Total	27	49*
c. Take initiative in parting	**Male took**	**Female took**
Alpha male	13	16
Non-alpha prime males	29	61**
Young adult males	19	29
Total	61	106**

*: $p<0.05$; **: $p<0.01$, binomial test.

of WE, showed sexual cycling or were pregnant during the present study. For example, SO persistently associated with NT, even when she became pregnant and ceased to cycle in October 1981.

2) *Approach, Following, and Leaving Behaviors*

As shown in Table 6.6a, non-estrous females unilaterally approached focal males. Non-estrous females unilaterally followed both the alpha male (NT) and non-alpha prime males (Table 6.6b). Non-estrous females and young adult males followed each other. The initiative in parting was taken bilaterally when NT or young adult males were with non-estrous females (Table 6.6c). On the other hand, non-estrous females unilaterally left non-alpha prime males.

Of the 48 proximity bouts in which the focal males approached non-estrous females, the females left within 6 OU in 7 bouts (15%), and the males left within 6 OU in 5 bouts (10%; Fig. 6.1a, c, e). On the other hand, of the 114 proximity bouts in which females approached males, the females left within 6 OU in 37 cases (32%), and the males left within 6 OU in 12 cases (11%; Fig. 6.1b, d, f).

3) *Grooming and Play*

Compared with the frequent grooming interactions among focal males, grooming was rather infrequent between focal males and non-estrous females. During the focal-sampling, males had grooming interactions with non-estrous females 110 times for 1,224 OU, while grooming interactions were observed between focal males 389 times for 6,925 OU. Mutual grooming occurred in 6 dyads 11 times.

NT and non-estrous females bilaterally groomed each other; NT groomed 3 females 8 times for 60 OU, and was groomed by 2 females 9 times for 59 OU. On the other hand, non-alpha prime males and young adult males tended to groom non-estrous females more frequently than to be groomed by them. Thus, non-alpha prime males groomed 12 females 28 times for 443 OU, and were groomed by 15 females 23 times for 221 OU. Similarly, young adult males groomed 10 females 29 times for 289 OU, and were groomed by 5 females 13 times for 152 OU.

Play occurred infrequently between focal males and non-estrous females. During the focal-sampling, play was observed only twice; KI and TI played a finger-wrestling game for 4 OU, and RA played with AD (mid-aged female) and SH (AD's 3-year-old daughter) for 10 OU.

4. Social Interactions with Unrelated Females (II): Semi-estrous Females

The rank of focal males was not significantly correlated with their

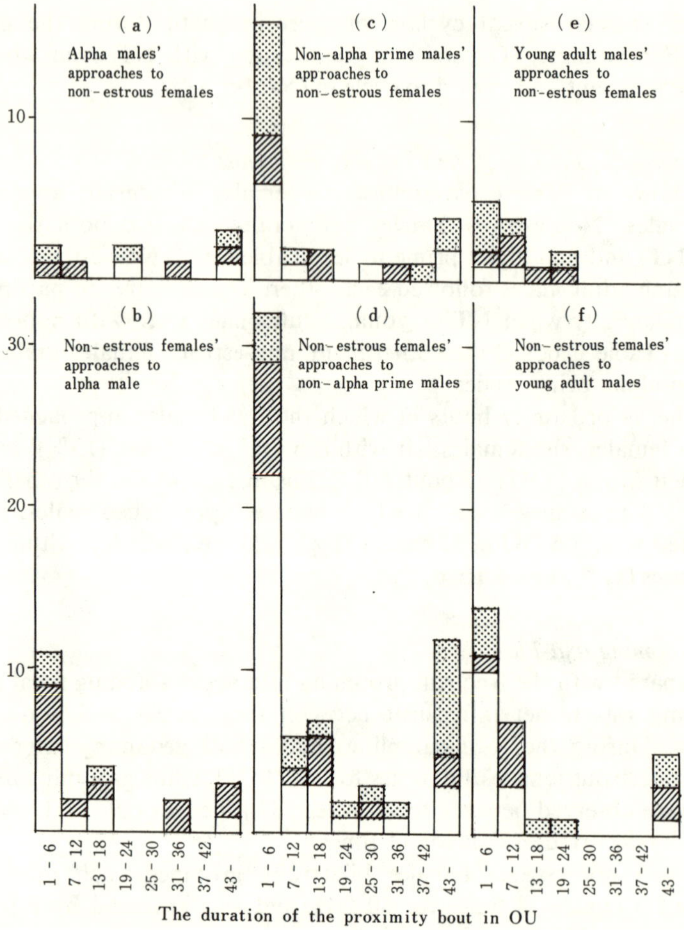

Fig. 6.1 The duration of the proximity bouts between focal males and non-estrous females. □: female leaves, ▨: male leaves, ▩: incomplete.

proximity relationships with semi-estrous females who were 10 or more years old (see Table 6.5).

Only a few interactions were recorded between focal males and semi-estrous females (Table 6.7). In general, focal males did not actively approach semi-estrous females, though the males occasionally inspected their vaginal openings.

Though there was no significant difference from the expected value, semi-estrous females tended to approach, follow, and leave focal males (Table 6.7). Play occurred twice; in both cases, DE wrestled with ND.

Prime males had few grooming interactions with semi-estrous females.

Table 6.7 Social interactions observed between focal males and semi-estrous females 10 or more years old.

a. Approach	Male approached	Female approached
Alpha male	1	2
Non-alpha prime males	4	6
Young adult males	1	4
Total	6	12
b. Follow	**Male followed**	**Female followed**
Alpha male	0	0
Non-alpha prime males	1	2
Young adult males	0	2
Total	1	4
c. Take initiative in parting	**Male took**	**Female took**
Alpha male	2	2
Non-alpha prime males	2	7
Young adult males	2	6
Total	6	15

NT groomed a female once for 10 OU, and was groomed by 2 females for a total of 10 OU. No non-alpha prime males had grooming interactions with semi-estrous females. On the other hand, young adult males groomed 4 females 6 times for 189 OU, and were groomed by 3 females 5 times for 39 OU. No mutual grooming occurred with semi-estrous females.

5. Social Interactions with Unrelated Females (III): Estrous Females

1) The Alpha Male

As shown in Table 6.5, male rank was not significantly correlated with proximity relationships of estrous females and focal males. However, estrous females most frequently stayed around the alpha male, NT, probably because NT often exhibited "possessive mating patterns" (Tutin, 1979); he paid close attention to some estrous females, and actively maintained close proximity to them.

For example, on August 24, 1981, NT "possessed" UM, a mid-aged female. At 14:15:00, he was reclining 2–3 m away from UM and LU. At 15:02:20, NT stopped grooming LU, and moved a little, sitting about 5 m away from the other 2 individuals. At 15:04:30, UM moved a little, and NT stared at her. UM then turned back to NT, barking "ugwag-wagwa". At 15:06:30, LU was about to approach UM. Immediately, NT charged them, perhaps to disturb mating. UM screamed and escaped to

Table 6.8 Social interactions observed between focal males and estrous females 10 or more years old.

a. Approach	Male approached	Female approached
Alpha male	12	14
Non-alpha prime males	25	62**
Young adult males	23	69**
Total	60	145**
b. Follow	**Male followed**	**Female followed**
Alpha male	4	13*
Non-alpha prime males	10	8
Young adult males	3	7
Total	17	28
c. Take initiative in parting	**Male took**	**Female took**
Alpha male	11	17
Non-alpha prime males	32	52*
Young adult males	30	63**
Total	73	132**

*: $p<0.05$; **: $p<0.01$, binomial test.

a tree, pant-barking, and LU ran away. UM then approached NT, pant-grunting, as a form of appeasement behavior. NT began moving, and UM followed him. Since NT was observed to copulate only 3 times during the focal-sampling, possessiveness may be a method for securing an estrous female and preventing her from copulating with other males, rather than being associated strictly with copulation.

Table 6.8a suggests that NT differed from subordinate males in that he positively approached estrous females. There was no significant difference in the duration of the proximity bouts in which NT approached non-estrous (Fig. 6.1 a) compared to estrous females (Fig. 6.2a). Similarly, there was no significant difference in the duration of the proximity bouts in which non-estrous (Fig. 6.1 b) compared to estrous females approached NT (Fig. 6.2b).

Estrous females unilaterally followed NT (Table 6.8b). On the other hand, NT and estrous females each left one another (Table 6.8c). NT groomed 4 estrous females 14 times for 157 OU, and was groomed by 5 females 10 times for 130 OU. NT's grooming relationships with estrous females was, thus, almost bilateral. Mutual grooming occurred 5 times with 3 females.

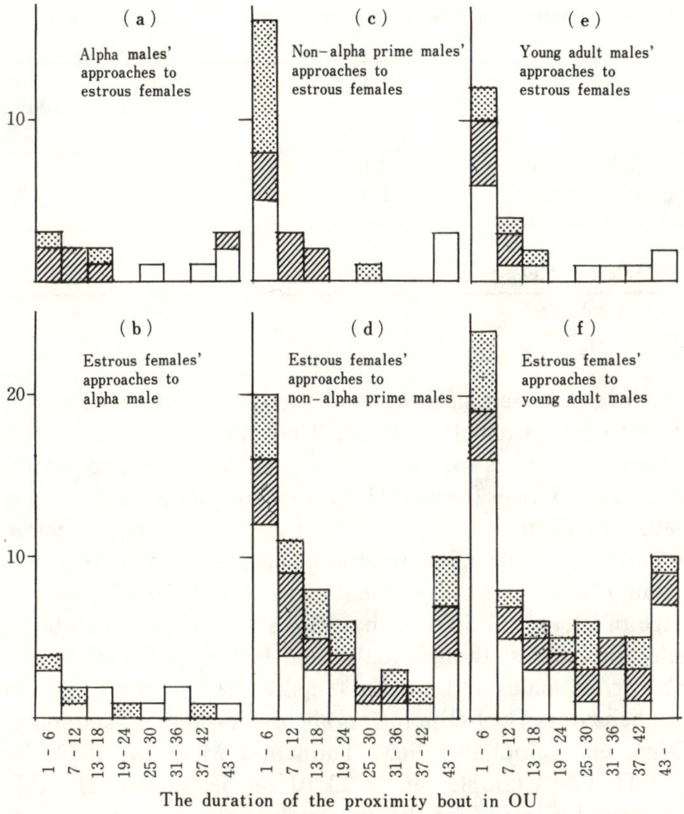

Fig 6.2 The duration of the proximity bouts between focal males and estrous females. □: female leaves, ▨: male leaves, ░: incomplete.

2) Non-alpha Prime Males

Estrous females unilaterally approached and left non-alpha prime males (Table 6.8a, c). Non-alpha prime males and estrous females followed each other (Table 6.8b). There was no significant difference in the duration of the proximity bouts in which non-alpha prime males approached non-estrous (Fig. 6.1c) compared to estrous females (Fig. 6.2c). On the other hand, there was a significant difference in the duration of the proximity bouts in which non-estrous (Fig. 6.1d) compared to estrous females approached non-alpha prime males (Fig. 6.2d; Mann-Whitney V-test, $z = 2.04$, $p < 0.05$). This was probably because non-estrous female approaches led to many short-term bouts.

Grooming was rather infrequent, compared to the amount seen with

Table 6.9 Social interactions observed during the proximity bouts in which copulation occurred.

	Alpha male (N=3)	Non-alpha prime males (15)	Young adult males (29)
Approach (Male/Female)	1/0	1/12*	10/19
Following (Male/Female)	0/3	1/1	0/5
Grooming (Male/Female)	0/0	3/2	13/5
Initiative in parting (Male/Female)	1/0	4/8	8/18

*: $p<0.01$, binomial test.

NT. In addition, non-alpha prime males tended to groom estrous females more than to be groomed by them. They were observed to groom 10 estrous females 10 times for 288 OU. In contrast, they were groomed by 3 estrous females 6 times for 68 OU. Mutual grooming occurred only once.

Non-alpha prime males usually showed an "opportunistic" mating pattern (Tutin, 1979). Table 6.9 shows the social interactions observed during the proximity bouts in which copulation occurred. Such bouts lasted for an average of 21.1 ± 17.5 OU. They also strongly tended to be initiated by female approaches, though males and females left each other. Both non-alpha prime males and estrous females occasionally displayed some courtship behaviors; leaf-clipping, branch/vine-shaking, stamping, cushion-making, etc. (Goodall, 1968; Tutin and McGrew, 1973; Nishida, 1980a, 1981). For example, at 13:23:10 on December 21, 1981, BA shook vines and stamped, at the approach of the mid-aged female, GK. GK then presented and copulated with him.

3) *Young Adult Males*

Estrous females unilaterally approached and left young adult males (Table 6.8 a, c). There was no significant difference in the duration of proximity bouts in which the males approached non-estrous (Fig. 6.1 e) compared to estrous females (Fig. 6.2 e). On the other hand, the proximity bouts in which estrous females approached the males (Fig. 6.2f) lasted significantly longer than those in which non-estrous females approached them (Fig. 6.1f; Mann-Whitney V-test, $z = 1.99$, $p < 0.05$).

Grooming was more frequent between young adult males and estrous females, compared to the amount seen with non-alpha prime males. Young adult males groomed 15 estrous females 46 times for 846 OU, and were groomed by 12 estrous females 37 times for 579 OU. Mutual grooming occurred 23 times in 7 dyads. DE was observed to play with GK twice.

Young adult males also showed opportunistic mating patterns. The proximity bouts in which copulation occurred lasted for an average of 31.6 ± 36.7 OU. In such bouts, young adult males and estrous females approached and left each other (Table 6.9). Young adult males occasionally used courtship behaviors. At 12:05:30 on November 14, 1981, KZ leaf-clipped, and stamped to WS. Then, KZ made a cushion on the ground. KZ leaf-clipped repeatedly and stamped again. At 12:08:50, WS approached and presented, and copulated with KZ. As soon as copulation was over, WS left KZ, and sat 8 m away from him.

As Hayaki (1985a) has pointed out for adolescent males, the sexual behaviors of young adult males tended to be affected by the presence of dominant males. For example, at 14:41:00 on December 30, 1981, NT, LU, KI, and LJ were sitting together. At 14:53:00, CH approached and sat near LJ, who began to groom her. Then, using an angle which was hidden from the sight of the dominant males, DE inspected her vaginal opening with the fingers of his right hand.

4) Male Rank and Copulation

Male rank was negatively correlated with the number of copulations observed during focal-sampling (Table 6.10). When *ad lib* and focal data were combined, neither the observed number of copulations nor that of mating partners was significantly correlated with male rank. However, as Hasegawa and Hiraiwa-Hasegawa (1983) and Takasaki (1985) have pointed out, higher ranking males, especially NT, tended to copulate with females closer to the expected date of ovulation (Table 6.11).

Table 6.10 Observed number of copulations by each focal male.

	Focal data	Focal and ad-lib data	
	No. of copulations	No. of copulations	No. of mating partners
NT	3	21	7
BA	4	8	7
KI	3	11	5
MU	3	6	3
LU	3	11	8
RA	2	6	4
DE	4	11	6
LJ	11	24	12
SM	4	8	7
KZ	13	22	11
Spearman rank coefficient	−0.615 ($p<0.05$)	−0.23 N.S.	−0.40 N.S.

Table 6.11 The observed number of copulations during each of the estrous stages of females. After Hasegawa and Hiraiwa-Hasegawa (1983), the last day of maximal tumescence is set as Day 0, and the swelling cycles are divided into 3 stages: Stage I (up to Day −8), Stage II (from Day −7 to Day −4), and Stage III (from Day −3 to Day 0). Ovulation is expected to occur on Day 0.

	Males				
	Alpha male	Non-alpha prime	Young adult	Adolescent	Juvenile
Stage I	5	12	25	13	14
Stage II	6	14	12	5	11
Stage III	9	12	22	8	8

IV. DISCUSSION

1. Mother-son Dyads

A strong tie between the mothers and their mature sons may be long-lasting, though there may be individual variations (Goodall, 1971; Nishida, 1979b; Hayaki, 1988). In the present study, KZ, who was about 13 years old frequently associated with his mother (SO), while SO was not in estrus or in semi-estrus. Neither agonistic interaction nor pant-grunt/bark was observed between them. NT often associated and groomed with his mother (NG) even in 1988 when he is believed to have been 33 years old (Hasegawa, Hayaki, and Takasaki, unpublished; also see Chap. 12).

In contrast, mothers had few interactions with their mature sons while they were in estrus. No copulations were observed in 2 mother-son dyads, though this is not significantly different from the expected value. Including adolescent and juvenile males, 181 copulations were recorded in 110 dyads between 23 males and 18 females during the present study. No copulations were observed in 5 mother-son dyads, though again this is not significantly different from the expected value ($p = 0.423$, Fisher's exact probability test). These data suggest that these mother-son dyads tended to avoid inbreeding, as reported in *Macaca* species (Sade, 1968; Missakian, 1973).

2. The Relationships with Unrelated Females

Though adult male chimpanzees frequently have symmetrical interactions with one another (see Chap. 7), such interactions do not always occur between unrelated adult males and females. Rather, their interactions and relationships are strongly influenced by dominance; usually, the subordinate females unilaterally approach the dominant males.

Female sexual receptivity also influences their interactions and relationships with unrelated adult males. Though non-estrous females actively approach and greet adult males, they often leave soon after without further social interactions (see Fig 6.1). In other words, they approach adult males only in order to greet and reconfirm their social relationships and to avoid their attacks and infanticides. Actually, as Nishida and Hiraiwa-Hasegawa (1985) reported, adult males sometimes severely attacked strange females and juveniles.

Most non-estrous females who frequently stayed near focal males were cycling or pregnant; non-cyclic females with dependent infants did not always stay around adult males. Of course, some females were psychologically dependent on the males for personal reasons (Nishida, 1979b); e.g., pregnancy, parturition, or recent immigration into the non-natal group. For example, SO's pregnancy might have influenced her to follow NT. PL, a young immigrant female, frequently associated with BA and KI. Such a case was also described by Nishida (1989b).

3. Mating Patterns

Estrous females frequently mingle with mixed parties, having various mating relations with the males (Fig. 6.3). Tutin (1979) grouped the mating patterns of the chimpanzees of Gombe National Park into opportunistic mating, possessiveness, and consortship. As Hasegawa and Hiraiwa-Hasegawa (1983) and Takasaki (1985) have pointed out, this last was scarcely observed in the M Group chimpanzees of Mahale.

During the present study, only the alpha-male showed a clear posses-

Fig. 6.3 A female in maximal tumescent phase of sexual swelling. (Photo by Y. Takahata)

sive mating pattern (Hasegawa and Hiraiwa-Hasegawa, 1983). This type of mating pattern seems to enable him to mate with the females around their expected date of ovulation, and to prevent them from mating with other males.

Both non-alpha prime males and young adult males showed opportunistic mating. Non-alpha prime males were usually approached by estrous females. Their copulation sequence tended to be initiated by females, and to be of short duration. As shown in Table 6.10, they also did not copulate frequently.

In contrast, young adult males responded more actively to estrous females, approaching and grooming them or engaging in courtship behaviors. Their efforts might be an attempt to compensate for their subordinate status, and may result in increased mating success.

V. SUMMARY

1. Chimpanzee mothers and their mature sons tended to interact when the mothers were not in estrus or in semi-estrus. While the mothers were in estrus, they had few interactions, and no copulation was observed.

2. Female sexual states influenced the relationships between adult males and unrelated females. Non-estrous females actively and unilaterally approached and greeted high-ranking males, but then often walked away without any further interaction. In particular, non-cyclic females with dependent infants did not always remain near adult males.

3. The alpha-male showed possessive mating patterns, and tended to mate with the females around their date of ovulation.

4. Non-alpha prime males showed opportunistic mating patterns. Their copulation tended to be initiated by females.

5. Young adult males also showed opportunistic mating patterns. They actively responded to estrous females by demonstrating various courtship behaviors, which might result in high mating success irrespective of their subordinate status.

Acknowledgements

This study was financially supported by the Monbusho International Scientific Research Program (#56041019 and 57043014 to T. Nishida).

7

Social Relationships among Adult Males

Yukio TAKAHATA

Department of Zoology, Faculty of Science, Kyoto University

I. INTRODUCTION

In a unit-group (or community) of chimpanzees, adult males form a "male cluster" or coalition network (Nishida, 1979b; de Waal, 1982; Goodall, 1986). The adult male chimpanzees of a unit-group cooperatively defend their range from those of the neighboring unit-groups. However, within the group, they compete for dominance rank (Goodall, 1971), or for priority of mating with estrous females (Nishida, 1983a; de Waal, 1986b), and form long-term or opportunistic mutually supportive coalitions with one another.

The chimpanzee unit-group is patrilineal, since males tend to remain within their natal groups and females tend to transfer into others (Nishida and Kawanaka, 1972; Itani, 1985). The adult males of a unit-group, therefore, should be closely related. As younger males grow up, they form close relationships with older males, and gradually enter into the coalition network of adults.

In this report, the social interactions and proximity relations between adult male chimpanzees of the M Group are analyzed as follows: (1) how dominance relationships are expressed, (2) how they interact, and (3) how young males establish themselves in the coalition network of adult males. The social function of male coalitions is also discussed, and the life history of male chimpanzees is described.

II. METHODS

This study is based on data obtained from August 7 to December 31, 1981. This period corresponded to the last half of the dry season and the early third of the rainy season, when the trees bear much fruit, and the chimpanzees form large parties (Nishida, 1968).

Table 7.1 List of the focal males. The estimation of age was based on Hiraiwa-Hasegawa, and Nishida (1984) and on their unpublished data.

Name	Ages (years)	Mother	Amount of observation (OU)
NT	26	NG	7,853
BA	27		7,761
KI	30–40		7,933
MU	24		7,041
LU	24		7,571
RA	30–40		7,050
DE	18		6,948
LJ	16		7,796
SM	15	GA	7,272
KZ	13	SO	6,719

Study subjects were 6 prime males (over 20 years old) and 4 young adult males (13–18 years old) of M Group (Table 7.1). For detailed description of the subjects of this study and the sampling method, see Chapter 6.

III. RESULTS

1. Rank Order among the Adult Males of M Group

1) *Agonistic Interactions*

It is difficult to confirm the rank order among male chimpanzees, since overt agonistic interactions are seldom observed. In contrast to the cercopithecoid species, the "peanut-test" is not definitive, because the subordinate, usually the old male, sometimes takes the bait in the presence of the dominant.

Table 7.2a shows the agonistic interactions observed among focal males. The 3 high-ranking males (NT, BA, and KI) unilaterally threatened or attacked other males, though no agonistic interactions were observed among them. Triadic interactions were recorded twice, and both cases were "winner support" and "actor alliance (de Waal, van Hoof, and Netto, 1976)."

Though direct attacks rarely occurred, focal males sometimes exhibited serious wounds. During the present study, MU, RA, and DE suffered from fresh injuries. For example, at 10:30:40 on October 21, 1981, DE's brow was torn and bleeding. DE repeatedly pant-grunted at LU and groomed him. When DE was again observed, at 13:33:40, his upper

Table 7.2a Agonistic interactions observed between focal males. Ad. lib. data were also included. ( ): The support of third chimpanzees observed in triadic interactions.

		Subordinate									
		NT	BA	KI	MU	LU	RA	DE	LJ	SM	KZ
Dominant	NT	—	0	0	3	2	1	0	1	0	2
	BA		—	0	0	0	1	1	1	0	1
	KI			—	2	2	1	1	2	1	0
	MU				—	1	0	0	0	1	0
	LU				(1)	—	0	0	0	0	1
	RA						—	0	0	0	0
	DE							—	(1)	0	0
	LJ								—	0	1
	SM									—	0
	KZ										—

Table 7.2b Pant-grunts/barks recorded when the focal males encountered.

		Utterer									
		NT	BA	KI	MU	LU	RA	DE	LJ	SM	KZ
Received	NT	—	3	2	7	3	2	2	4	5	4
	BA		—	0	0	0	0	0	0	0	1
	KI			—	0	0	0	0	0	1	0
	MU				—	0	0	0	0	1	0
	LU					—	0	0	0	0	1
	RA						—	2	1	0	2
	DE							—	0	1	4
	LJ								—	0	0
	SM									—	1
	KZ										—

Table 7.2c All observed cases of pant-grunts/barks.

		Utterer									
		NT	BA	KI	MU	LU	RA	DE	LJ	SM	KZ
Received	NT	—	14	15	15	24	6	6	12	11	8
	BA		—	0	1	5	0	2	0	0	14
	KI			—	1	0	0	1	3	2	3
	MU			1	—	1	0	0	1	3	3
	LU					—	0	3	1	0	5
	RA	1			5	1	—	2	1	1	3
	DE							—	0	2	7
	LJ								—	2	2
	SM									—	1
	KZ										—

lip was also cut. Subsequently, he closely followed and groomed BA and RA, as if he depended on them for protection.

2) Pant-grunt/bark

Pant-grunts/barks are the vocalizations which subordinates direct to dominants (Bygott, 1974, see also Chap. 9). As shown in Table 7. 2b, when focal males encountered one another, pant-grunts/barks were focused on the alpha male (NT). NT was thus recognized as a special individual by the others. Table 7.2c shows all observed cases of pant-grunts/barks. These data indicate a probable linear rank order among the focal males as follows: NT > BA > KI > MU > LU > RA > DE > LJ > SM > KZ. KL was probably ranked around LU and RA.

One problem with the above hierarchy is the dominance relationships among MU, LU, and RA. NT pant-grunted at RA once, MU 5 times, and LU once, and RA did not pant-grunt/bark at any other male except NT (Table 7.2c). However, it appears that RA who was 30–40 years old had been declining in rank owing to the senility, and it is possible that the pant-grunts/barks which he received from higher-ranking males might be regarded as appeasement vocalizations toward a declining male who had been dominant in the past.

Young adult males were subordinate to any prime male. Rank was positively correlated with age among the 4 young adult males. Thus, the older individuals dominated the younger.

3) Charging-display

Among focal males, 34 episodes of charging-display defined by Goodall (1968) were recorded (Table 7.3). Of these, stone-throwing or log-brandishing was observed in 11 episodes. The charging-display tended to occur in a stream bed, at the feeding site, and/or in the heavy rain. In particular, when they passed across a stream bed, males often burst out in displays and threw big stones into the water.

No positive correlation was found between male rank and the frequency of charging-displays (Table 7.3). Of the 15 episodes where 2 or more males were present, the most dominant started the display 9 times, and others 6 times. LU most frequently started and participated in charging-displays. In contrast, the lowest-ranking SM and KZ displayed only when no older male was near them.

As Goodall (1968) and de Waal (1982) have pointed out, charging-displays might subtly reflect male competition for rank. For example, at 12:21:50 on December 14, 1981, BA began a charging-display under the tree in which KI was sitting. BA climbed the tree, and continued the display near KI, shaking vines, but KI only watched him. At 12:24:40,

Table 7.3 Number of charging-display episodes observed among focal males.

	NT	BA	KI	MU	LU	RA	DE	LJ	SM	KZ	
No. of charging-display episodes	10	3	5	4	12	3	6	6	2	2	N.S.
No. of charging-display which each male started	4	2	3	1	7	1	1	3	2	2	N.S.

LU joined BA, and violently displayed around the tree. BA and LU displayed until 12:29:00, while KI kept silent. The impression given by these observations was that BA and LU intended to demonstrate against an older male, KI.

2. Association among Focal Males

1) Proximity Relations

Though the unit-group, as a whole, usually split into small parties which repeatedly fissioned and fused, adult males tended to assemble together as a "core" of the unit-group (Nishida, 1968, 1979b; Goodall, 1986). Table 7.4 and Fig. 7.1 show proximity relations between focal males. Some dyads maintained frequent proximity, while others did not. For example, the 2 highest-ranking males, NT and BA, scarcely associated with each other. NT maintained frequent proximity with LU, KI, and MU, while BA associated with DE, MU, and RA. Thus, it appears that there were 2 clusters, one centered around NT and one around BA. MU linked these 2 clusters.

Table 7.4 Proximity index of each dyad of focal males. The index between individuals A and B is calculated by the following formula: {DP(A,B)+DP(B,A)} ×100/{DT(A)+DT(B)}. DT(A) is the observation duration on individual A, and DP(A,B) is the total duration in which individual B was within 3 m of the focal animal A.

	NT	BA	KI	MU	LU	RA	DE	LJ	SM	KZ
NT	—	1.77	7.56	5.63	12.1	3.82	0.49	1.82	0.13	0.31
BA		—	3.76	10.2	3.36	5.24	13.1	6.23	1.68	6.81
KI			—	3.49	5.17	6.79	2.51	5.28	1.26	1.12
MU				—	11.5	2.95	5.65	7.03	0.01	0.04
LU					—	1.66	6.72	6.34	0.11	3.84
RA						—	10.9	9.32	0.07	4.91
DE							—	3.71	2.30	3.96
LJ								—	1.13	3.31
SM									—	0.08
KZ										—

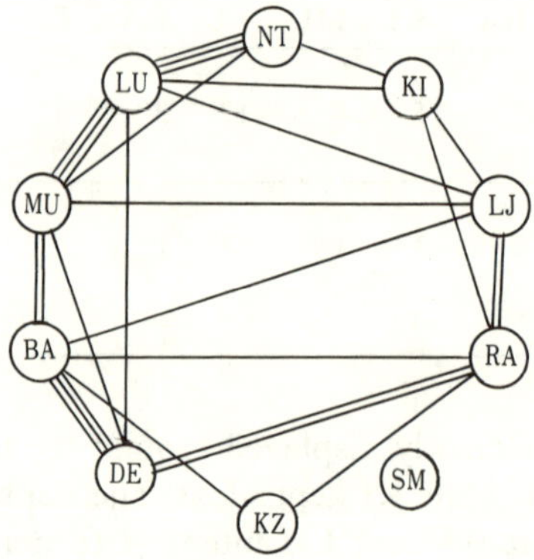

Fig. 7.1 Association sociogram among focal males. The sociogram was drawn from Table 7.4, according to the method of Kimura and Ihobe (1985). A line shows the proximity index between m (4.34) and $m+SD$ (7.84). A double line shows the index between $m+SD$ and $m+2SD$ (11.33). A triple line shows the index over $m+2SD$.

Young adult males rarely stayed near the alpha male. In particular, the youngest SM and KZ remained in peripheral positions. For them, this may have been a reflection of their stages of social development. Probably, they had just begun to be apart from their mothers and to associate with older males. On the other hand, LJ was probably trying to establish himself within the adult male coalition network, since he actively approached most non-alpha prime males. DE was the oldest of young adult males and he had already formed a stable relationship with BA. It should be noted that young adult males did not always associate with one another, and had respective relationships with prime males, as Kawanaka (1989) has pointed out.

While the young adult males were near the prime males, they often appeared tense, the slightest motion of prime males sometimes causing them to pant-grunt/bark. For example, when SM encountered NT, KI, and MU at 16:17:40 on November 26, 1981, he pant-grunted and approached KI. However, KI abruptly began moving. This frightened SM, who pant-barked and hurried away from the older males.

In other contexts, however, young adult males emotionally sought prime males' reassurance. At 15:07:50 on December 21, 1981, KI was sitting alone, when intense screaming took place in the bush. KZ came out, and rushed screaming up to KI. KI, with his hair erect, drew himself up and touched KZ's shoulders. KZ, as if encouraged, then turned around and rushed back into the bush. Ambivalent attitudes towards older males may be characteristic of the social state of young adult males.

2) Approach, Following, and Leaving Behaviors

There was a significant correlation between the proximity index (Table 7. 4) and the mean duration of a proximity bout (Table 7.5) ($r=0.318$, $p < 0.05$). There was also a significant correlation between the proximity index and the observed number of approach ($r = 0.801$, $p < 0.01$).

The alpha male was approached and followed by non-alpha prime males significantly more than the expected value (Table 7.6a, b). Besides pant-grunts/barks, various greeting behaviors, such as hand-stretching, embrace, open-mouth kiss, etc., were observed. For example, when KI approached NT at 13:25:00 on October 28, 1981, he pant-grunted at NT, and they embraced each other and open-mouth kissed. Then NT began to groom KI.

Proximity bouts between the alpha male and non-alpha prime males lasted for an average of 23.8 ± 34.6 OU. The alpha male tended to move first, and to be followed by non-alpha prime males, though this was not statistically significant (Table 7.6c). On the other hand, the initiative in parting was taken bilaterally (Table 7.6d). The approach/leaving index

Table 7.5 Mean duration of a proximity bout for each dyad in OU. ( ): standard deviation.

	NT	BA	KI	MU	LU	RA	DE	LJ	SM	KZ
NT	—	15.0	27.2	19.9	24.6	26.3	13.0	3.2	4.3	9.0
	(—)	(17.0)	(51.1)	(27.9)	(34.9)	(19.9)	(11.0)	(3.1)	(3.9)	(10.9)
BA		—	10.0	22.9	17.5	29.8	9.7	15.1	48.5	19.8
		(—)	(3.0)	(23.5)	(20.8)	(56.4)	(8.3)	(12.6)	(13.5)	(16.3)
KI			—	46.2	24.5	19.3	17.8	21.0	6.6	7.3
			(—)	(45.6)	(22.9)	(33.3)	(14.3)	(—)	(3.3)	(5.8)
MU				—	33.3	—	11.2	63.7	2.0	2.5
				(—)	(44.0)	(—)	(5.9)	(47.4)	(—)	(1.5)
LU					—	11.6	33.9	19.0	4.7	22.7
					(—)	(12.2)	(30.5)	(31.5)	(3.1)	(20.7)
RA						—	15.9	21.2	—	8.0
						(—)	(17.3)	(31.6)	(—)	(5.8)
DE							—	45.9	32.5	23.0
							(—)	(86.1)	(50.0)	(12.0)
LJ								—	17.0	4.0
								(—)	(19.7)	(0.8)
SM									—	11.0
									(—)	(—)
KZ										—
										(—)

Table 7.6 Social interactions between focal males.

a. Approach	Dominant approached	Subordinate approached
Between NT and non-alpha prime males	35	63**
Between non-alpha prime males	70	69
Between NT and young adult males	6	15
Between non-alpha prime males and young adult males	60	126**
Between young adult males	15	25
Total	186	298**
b. Follow	**Dominant followed**	**Subordinate followed**
Between NT and non-alpha prime males	14	27*
Between non-alpha prime males	36	41
Between NT and young adult males	2	1
Between non-alpha prime males and young adult males	25	90**
Between young adult males	4	12
Total	81	171**
c. Move first	**Dominant moved first (be followed)**	**Subordinate moved first (be followed)**
Between NT and non-alpha prime males	59 (12)	40 (6)
Between non-alpha prime males	75 (18)	72 (11)
Between NT and young adult males	10 (1)	12 (1)
Between non-alpha prime males and young adult males	134 (51)	65** (9)***
Between young males	20 (7)	14 (2)
Total	298 (89)	203**(29)***
d. Take initiative in parting	**Dominant took**	**Subordinate took**
Between NT and non-alpha prime males	53	44
Between non-alpha prime males	71	76
Between NT and young adult males	10	12
Between non-alpha prime males and young adult males	90	81
Between young adult males	15	16
Total	239	229

*: $p<0.05$; **: $p<0.01$, binomial test; ***: $p<0.001$, Chi-square test.

(Hinde, 1974) indicates that proximity was chiefly maintained by non-alpha prime males (Table 7.7).

In contrast, dominance relationships scarcely affected the approach, following, and leaving behaviors among non-alpha prime males. As shown in Table 7.6, dominants and subordinates bilaterally approached, followed, moved first, and took the initiative in parting. The approach/leaving index indicates that proximity was also bilaterally maintained (Table 7.7). Proximity bouts were rather long, lasting for an average of 27.1 ± 38.4 OU.

Young adult males unilaterally approached the alpha male, though this was not statistically significant (Table 7.6a). Following was infrequent (Table 7.6b), and the initiative in parting was taken bilaterally (Table 7.6 d). The approach/leaving index indicates that proximity was maintained by young adult males (Table 7.7). Proximity bouts tended to be short, lasting for only an average of 5.9 ± 7.8 OU, partly because young adult males soon left the alpha male without further social interactions. This attitude was not unreasonable, since the alpha male occasionally attacked the young adult or adolescent males who greeted him.

Young adult males unilaterally approached and followed non-alpha prime males (Table 7.6a, b). Non-alpha prime males tended to move first, and to be followed by young adult males (Table 7.6c). The initiative in leaving was taken bilaterally (Table 7.6d), and the approach/leaving index strongly indicates that proximity was maintained by the young adult males (Table 7.7).

Young adult males stayed near non-alpha prime males for longer periods than they did with the alpha male; proximity bouts lasted for an average of 17.8 ± 23.8 OU. Occasionally, prime males attacked the young adults. For example, when DE and KR (an adolescent male) rushed up to KI, pant-grunting, at 13:01:20 on December 6, 1981, KI charged them, chasing and hitting DE, who screamed and fled.

Table 7.7 Approach/leaving index (Hinde, 1974) between focal males.

Between NT and non-alpha prime males	$\{35/(35+63) - 53/(53+44)\} \times 100 = -18.9$
Between dominants and subordinates of non-alpha prime males	$\{70/(70+69) - 71/(71+76)\} \times 100 = 2.1$
Between NT and young adult males	$\{6/(6+15) - 10/(10+12)\} \times 100 = -16.9$
Between non-alpha prime males and young adult males	$\{60/(60+126) - 90/(90+81)\} \times 100 = -20.4$
Between dominants and subordinates of young adult males	$\{15/(15+25) - 15/(15+16)\} \times 100 = -10.9$

Among the young adult males themselves, dominance relations scarcely affected their approach and leaving behaviors (Table 7.6). The approach/leaving index indicates that proximity was maintained by the younger and more subordinate males (Table 7.7). A proximity bout lasted for an average of 27.5 ± 56.0 OU.

3) Leadership

In general, the dominant males tended to move first and to be followed by the subordinates (Table 7.6c). Within a chimpanzee unit-group, the dominant males (particularly the alpha male) thus took the leadership in ranging.

Occasionally, the follower rather than the leader determined the course of ranging. For example, at 13:07:00 on September 4, 1981, KI was following RA along a research trail. When they came to a trail junction, RA continued to move southward, while KI, ignoring RA, turned westward. RA continued about 2 m from the junction, and then turned around and followed KI. KI sat and let RA pass, then began to follow him again.

3. Grooming

1) Grooming Duration

Among prime males, the alpha male was the most active groomer. NT groomed other prime males for 757 OU, and was groomed by them for only 387 OU in total (Table 7.8). NT frequently had grooming relations with LU, RA, and KI, but infrequently with BA and MU. In contrast, NT had no grooming interactions with young adult males.

Among non-alpha prime males, dominants groomed subordinates for a total of 1,081 OU, and were groomed by them for a nearly symmetrical 854 OU (Table 7.8). Grooming frequently occurred between MU and LU, and between BA and RA. In contrast, no grooming occurred between 2 older males, KI and RA. Grooming frequency differed according to the individual. BA actively groomed other males, while KI and LU tended to be groomed by other males, rather than to groom them.

Young adult males unilaterally groomed prime males; they groomed non-alpha prime males for 2,829 OU, but were groomed by them only for 343 OU (Table 7.8). Similarly, among the young adult males, the younger subordinates groomed the older dominants for 611 OU, and were groomed by them for only 63 OU. DE, LJ, and KZ actively groomed the older males, but SM rarely had grooming interactions with any other male.

Table 7.8 Duration of grooming between focal males in OU. The upper row shows the dominant's grooming episodes, and the lower row shows the subordinate's grooming.

	NT	BA	KI	MU	LU	RA	DE	LJ	SM	KZ
NT	— —	54 59	57 104	9 0	470 70	167 154	0 0	0 0	0 0	0 0
BA		— —	164 49	111 6	16 0	231 124	81 290	0 102	0 0	1 207
KI			— —	38 193	73 86	0 0	0 47	0 359	0 7	0 15
MU				— —	340 253	54 68	0 13	0 67	0 0	0 0
LU					— —	54 75	14 275	4 391	0 0	14 213
RA						— —	69 186	109 181	0 0	51 476
DE							— —	49 99	0 13	14 212
LJ								— —	0 0	0 287
SM									— —	0 0
KZ										— —

2) Grooming Sessions

Like play bouts reported by Hayaki (1985b), grooming bouts frequently occurred in a series. Therefore, a grooming session is defined as a set of grooming bouts which are intermittently or reciprocally exchanged by 2 chimpanzees at intervals of 5 or less OU. In these grooming sessions, dominant and subordinate males bilaterally approached each other, and took the initiative in parting (Table 7.9a, b).

The highest-ranking NT and BA tended to groom other prime males first and last (Table 7.9c, d). This suggests that they actively took the initiative in grooming. Among the other prime males (KI, MU, LU, and RA), dominants and subordinates bilaterally groomed first and last. Young adult males strongly tended to groom the older males first and last.

At the start of a session, chimpanzees sometimes solicited grooming with gestural or vocal signals, though such signals were occasionally ignored. This is similar to behavior reported in Japanese macaques (Mori, 1975). Table 7.9e shows the conspicuous signals which were given to solicit grooming. It is notable that non-alpha prime males unilaterally signaled NT, and that NT responded to their requests. For example, at 11:43:10 on August 24, 1981, LU uttered "nguu" and sat in front of NT, with his back turned. NT immediately started to groom LU.

Table 7.9 Grooming sessions observed between focal males.

a. Approach	Dominant approached	Subordinate approached
Between NT and non-alpha prime males	8	13
Between BA and other non-alpha prime males	6	3
Between non-alpha prime males (excluding BA)	7	15
Between non-alpha prime males and young adult males	15	26
Between young adult males	5	7
b. Take initiative in parting	**Dominant took**	**Subordinate took**
Between NT and non-alpha prime males	9	8
Between BA and other non-alpha prime males	4	4
Between non-alpha prime males (excluding BA)	10	10
Between non-alpha prime males and young adult males	24	26
Between young adult males	3	2
c. First grooming	**Dominant first groomed**	**Subordinate first groomed**
Between NT and non-alpha prime males	23	9*
Between BA and other non-alpha prime males	9	2
Between non-alpha prime males (excluding BA)	16	14
Between non-alpha prime males and young adult males	7	77**
Between young adult males	0	15**
d. Last grooming	**Dominant groomed last**	**Subordinate groomed last**
Between NT and non-alpha prime males	20	10
Between BA and other non-alpha prime males	12	3*
Between non-alpha prime males (excluding BA)	11	19
Between non-alpha prime males and young adult males	7	87**
Between young adult males	0	11**

Table 7.9 (continued).

e. Gestural or vocal signal	Dominant sent	Subordinate sent
Between NT and non-alpha prime males	0	7*
Between BA and other non-alpha prime males	0	3
Between non-alpha prime males (excluding BA)	1	2
Between non-alpha prime males and young adult males	14	1**
Between young adult males	5	0

*; $p<0.05$; **; $p<0.01$, binomial test.

Non-alpha prime males also unilaterally signaled young adult males to elicit grooming, though in some cases, it was uncertain whether the prime male's motion was a gestural signal to solicit grooming, or whether the young adult male simply misunderstood the motion. When LJ was grooming KI at 10:48:10 on December 30, 1981, RA self-scratched. LJ stopped grooming KI and went behind RA, as if he had interpreted RA's self-scratching as a demand for grooming. However, RA closely approached KI and began mutual grooming with him. LJ then approached and began to groom RA.

3) Mutual Grooming

Mutual grooming frequently occurred between prime males, especially between MU and LU (Table 7.10). In the present study, 35 mutual grooming episodes were recorded between prime males, 13 episodes between

Table 7.10 Mutual grooming episodes observed among focal males.

	NT	BA	KI	MU	LU	RA	DE	LJ	SM	KZ
NT	—	1	0	0	3	6	0	0	0	0
BA		—	1	0	0	5	3	0	0	0
KI			—	2	2	0	0	0	0	0
MU				—	12	2	0	0	0	0
LU					—	1	1	1	0	1
RA						—	3	1	0	3
DE							—	2	0	0
LJ								—	0	0
SM									—	0
KZ										—

prime and young adult males, and only 2 episodes between young adult males.

While grooming prime males, young adult males sometimes tried to solicit mutual grooming. For example, at 13:18:40 on August 19, 1981, LJ was grooming KI. LJ then raised his left hand, and, in response, KI also raised his left hand. They then began a grooming hand clasp, a type of mutual grooming (McGrew and Tutin, 1978).

4. Play, Food-sharing, and Copulation

1) Play

Play is a symmetrical interaction (Hayaki, 1985b). Play most frequently occurred between prime and young adult males (Table 7.11).In this respect, play differed from another symmetrical interaction, mutual grooming. Usually, a male stretched out a hand to another male and solicited finger-wrestling. They gradually became excited and began wrestling or round-chasing.

At 14:17:40 on August 29, 1981, NT approached KI and lay down near him. At 14:34:30, KI uttered "nn", and opened his mouth and panted. Then he wrapped his left hand around NT's right hand, but NT hardly reacted. At 14:35:30, KI wrapped his hand around NT's hand again; they started finger-wrestling and gradually became excited. Play-panting, KI bit NT's jaw. At 14:38:40, KI uttered "nn", and raised his head. Interrupting the play, NT started to groom KI. At 14:39:40, KI panted and started to groom NT. At 14:47:10, KI play-panted, opening his mouth. NT also play-panted, and they started wrestling. Held down by KI, NT panted, and they bit each other.

2) Food-sharing

Meat-sharing occasionally occurred among the chimpanzees (Nishida,

Table 7.11 Play bouts observed among focal males.

	NT	BA	KI	MU	LU	RA	DE	LJ	SM	KZ
NT	—	0	3	0	0	0	0	0	0	0
BA		—	0	0	0	0	2	0	0	5
KI			—	0	0	0	0	0	0	0
MU				—	0	0	1	1	0	0
LU					—	0	0	0	0	0
RA						—	1	0	0	0
DE							—	0	0	1
LJ								—	0	0
SM									—	0
KZ										—

Uehara, and Nyundo, 1979; Kawanaka, 1982a). During predatory episodes, the possessors of the meat were tormented by persistent begging on the part of the other chimpanzees. This was true even of the alpha male. The possessor tended to be reluctant to share meat and tried to reject the beggars, but some beggers were allowed to eat directly from the prey as it was held by its possessor. If the prey was large, it was divided and shared among a few chimpanzees.

Meat transfer frequently occurred in peculiar combinations of males. This probably reflected their social relationships (Takahata, Hasegawa, and Nishida, 1984). For example, when NT possessed prey, BA, MU, and RA infrequently begged him for meat. In contrast, KI, LU, and KL gathered around NT, and were allowed to eat the meat held by him.

3) Copulation

The rank of a focal-male tended not to be significantly correlated with his observed number of copulations, or number of mating partners (Table 6.10 of Chap. 6). However, as Hasegawa and Hiraiwa-Hasegawa (1983) and Takasaki (1985) have pointed out, higher-ranking males tended to copulate with females around the expected date of ovulation (Table 6. 11 of Chap. 6).

The dominant male could easily interfere with the mating behavior of a subordinate. Therefore, subordinate males were sensitive to the presence of dominant individuals, as Hayaki (1985a) has pointed out with respect to the copulation of adolescent males. For example, at 11:31 on August 8, 1981, KI and IK (an estrous female) were with NT. As soon as NT moved away and was out of sight, IK rushed up to KI, who copulated with her twice.

5. Proximity and Social Interactions with Adolescent Males

Table 7.12 shows the number of OU in which adolescent males of 9–11 years old stayed within 3 m of the focal males. In general, adolescent males infrequently stayed near focal males. Most of them were still dependent on their mothers, and scarcely mingled with adult males. BE, for example, had a mother and 3 siblings, and had few interactions with focal males.

As shown in Table 7.13a, adolescent males unilaterally pant-grunted/barked to focal males. Focal males and adolescent males bilaterally approached and followed each other (Table 7.13b, c). On the other hand, adolescent males tended to take the initiative in parting (Table 7.13d). Adolescent males had infrequent grooming interactions with focal males. They were observed to groom BA, MU, RA, and KZ for 54 OU, and to be groomed by RA for 110 OU.

Table 7.12 Proximity relations with adolescent males. Figures show the number of OU in which the adolescent male stayed within 3 m of the focal males.

	Adolescent males							
	SU	KE	KR	YK	LL	BE	Unidentified	Total
NT	0	0	1	0	11	0		12
BA	7	11	0	89	93	0		200
KI	1	11	2	0	16	0		30
MU	0	0	29	0	12	0		41
LU	9	72	0	83	2	0		166
RA	64	68	104	0	154	0	62	452
DE	93	18	83	30	11	0		235
LJ	45	0	33	19	21	0		118
SM	0	0	32	18	0	0		50
KZ	138	0	41	52	3	0	2	236

Table 7.13 Social interactions observed between the focal males and adolescent males.

a. Pant-grunt/bark	The focal male panted	The adolescent panted
Alpha male	0	4
Non-alpha prime males	0	10*
Young adult males	0	0
Total	0	14*
b. Approach	**The focal male approached**	**The adolescent approached**
Alpha male	0	2
Non-alpha prime males	15	20
Young adult males	12	14
Total	27	36
c. Follow	**The focal male followed**	**The adolescent followed**
Alpha male	0	0
Non-alpha prime males	12	6
Young adult males	2	5
Total	14	11
d. Taking initiative in parting	**The focal male took**	**The adolescent took**
Alpha male	0	1
Non-alpha prime males	13	20
Young adult males	4	19*
Total	17	40*

*: $p<0.01$, binomial test.

Apparently, adolescent males chose to interact with the less-aggressive adult males. They most frequently stayed near RA, who was also persistently followed by a male juvenile orphan (AJ) during the present study. In contrast, adolescent males rarely remained near NT, who occasionally threatened or attacked adolescents who greeted him.

IV. DISCUSSION

1. Male Rank Order and Social Interactions

Based on the several criteria, it appears that there was a linear rank order among the M Group males during this study. Though the data are limited, chimpanzees generally seem to be aware of the dominance relationship between a pair of males. Focal males, with the exception of the youngest, KZ, apparently dominated over adult females and unilaterally received pant-grunts/barks from them (Hayaki, Huffman, and Nishida, 1989).

The data suggest that associations tended to be maintained by the subordinate, rather than by the dominant males. On the other hand, dominant males tended to take the leadership in ranging.

For anubis baboons, Rowell (1966) has pointed out that it should be the subordinate who cautiously observes and maintains the dominance relationship. Likewise, in the present study, dominance relationships tended to be expressed by the subordinate chimpanzees rather than by the dominants. In particular, pant-grunts/barks were focused on the alpha male (Nishida and Hiraiwa-Hasegawa, 1987). The alpha male responded in various ways to the greeting behaviors of subordinates. For example, he mutually embraced, kissed, groomed, ignored, or sometimes attacked. In a sense, the alpha male acts as a confirmer for the male relationships (see also Chap. 8).

Compared with the rigid male rank order within a BSU (Basic Social Unit; Itani, 1985) of cercopithecoid species, the rank order of prime male chimpanzees seems to be ambiguous. There remain not a few doubts about some of the dominance relationships among non-alpha prime males, because they seldom pant-grunted/barked at one another, and they mutually approached, followed, and groomed one another, irrespective of ranks.

In contrast, there was a definite distinction between prime and young adult males. Young adult males were subordinate to the older males, and unilaterally approached and followed them. Prime males plainly demanded grooming from young adult males, and mutual grooming was infrequent. Similarly, among young adult males themselves, there was a distinction between younger and older, the former being subordinate to the latter.

Of course, dominance relationships are not static, and males subtly compete for rank on the basis of everyday occurrences (Goodall, 1986). In the Arnhem chimpanzee colony, 3 males repeated "power takeovers and coalition changes" (de Waal, 1982, 1984), and 2 of them eventually brutally eliminated the strongest individual (de Waal, 1986b). In the present study, the wounds which some M Group males received suggests actual fighting, though the use of force was probably infrequent.

2. Coalition Network and Male Life History

Chimpanzee male relationships are characterized by high tolerance and intense competition (Nishida and Hiraiwa-Hasegawa, 1987). Against the males of neighboring unit-groups, they must unite, but within a unit-group, males also compete for rank or reproductive success. Males also try to "manipulate" social relationships through coalitions (de Waal, 1982; Nishida, 1981, 1983a; Goodall, 1986).

During the present study, stable coalitions were recognized between NT and LU, between MU and LU, and between BA and DE. In contrast, for some reason or other, the alpha male (NT) and the beta male (BA) seemed alienated; they had neither affinitive nor agonistic interactions. In the long term, however, these coalitions changed. Kawanaka (1989; see also Chap. 8) reported that in 1983, NT and BA continued to be alienated, but, in 1986, BA frequently had grooming interactions with NT. In January, 1987, KI disappeared from M Group and DE became the beta male. Then, BA and NT gradually formed a coalition (Takasaki and Hasegawa, unpublished).

Young adult males tended to be peripheral within the male association network. They assumed ambivalent attitudes toward the older males; they feared prime males, especially the alpha male, but were attracted to them. By frequent and persistent greeting behaviors, they may manifest their "presence" towards the older males (see Chap. 9). On the other hand, young adult males infrequently associated with one another, and no age-mate group was formed (Kawanaka, 1989).

As Hayaki (1988) has pointed out, late adolescent and young adult males are in the process of establishing their positions in the adult male cluster. They must become independent of their mothers and form coalitions with the older males. In this respect, KZ, the youngest focal male, was an example of an individual in transition from an adolescent to a young adult. During the present study, KZ actively approached older males and groomed them, but he received no pant-grunts/barks from females, and was observed to pant-grunt at an adult female once (see Chap. 6).

Male chimpanzees make their positions in the adult male cluster and reach high rank in their twenties. By 1984, for example, DE and LJ who were about 20 years old, gradually came to outrank MU and LU. In contrast, SM's case suggests the difficult situation of young adult males. In 1981, SM had no stable relationships with other focal males. In later years, he continued to be rather isolated from the others, and eventually disappeared in July 1986. In a sense, the young male chimpanzee must get through a "crisis of youth".

In Gombe, Goodall (1986) reported that males reach their zenith in the hierarchy between 20 and 26 years, but that rank gradually or suddenly drops in the thirties.

NT became the alpha male of M Group in 1979, when he was about 24 years old. Ten years later, at the age of 34, he still maintained his alpha status (Tsukahara, unpublished).

Regardless of their rank, old males seem to be treated differently from other males. In the present study, for example, KL was frequently permitted to take meat from prey held by NT. KL is believed to have been higher-ranking in former days (Nishida, 1979b), and his past relationships might affect his present social interactions.

3. Inequality and Equality Principles

Itani (1984, 1988), in discussing the origin of human "egalitarianism", points out that primates have 2 contrasting principles, the "inequality principle" and the "equality principle." Itani speculates that the inequality principle is the basic rule that exists *a priori*, based on the recognition of dominant-subordinate relationships. He further points out that co-existence based on the inequality principle is maintained by the self-restraint of the subordinate. For example, the linear rank order of female Japanese monkeys, based as it is on their nepotistic relationships, strongly reflects the inequality principle. On the other hand, the equality principle, which is the negation of the inequality principle, appears in some symmetrical interactions or relationships of primates, such as play, grooming, food-sharing, etc. These types of interactions are probably "supported by an implicit agreement for achieving peaceful co-existence among them" (Itani, 1988).

Symmetrical interactions observed among the M Group males may also reflect an incipient equality principle, or "egalitarianism", and may allow adult males to coexist easily within a unit-group. For example, non-alpha prime males interact with each other, avoiding the overt expression of rank differences. Mutual grooming and meat-sharing also frequently occur among them. Furthermore, the alpha male actively takes the initiative in grooming, and tends to groom subordinate males rather than to be groomed by them.

Male chimpanzees of a unit-group cannot recognize their paternal kinships. Thus, lacking established kinship connections, they must intentionally form complicated relationships in which they simultaneously associate and compete with one another through social interactions. Symmetrical interactions probably facilitate the formation of such relationships.

Of course, the inequality principle is apparent in social interactions between young adult and prime males, and in reproductive behavior. The high-ranking males tend to achieve a high mating success around the time of ovulation (Hasegawa and Hiraiwa-Hasegawa, 1983; Takasaki, 1985, see also Chap. 6).

4. Some Personalities of the Focal-males

It is impossible to understand the coalition network of chimpanzee males without mentioning their personalities, though personality is very difficult to discuss in quantitative terms (Bygott, 1979). Of course, personality is closely related to age and status.

NT must have a remarkable personality, which allowed him to dominate and overbear other chimpanzees (see also Chap. 8; Fig. 7.2). His long tenure as alpha male indicates his great physical and social abilities (though there was no data regarding his body weight, NT was the biggest of the M Group males). NT was the most active groomer, and often

Fig. 7.2 NT was rushing through a trail. (Photo by Y. Takahata)

Fig. 7.3 A submissive follower, LU. (Photo by Y. Takahata)

allowed older males (KI and KL) to get meat from prey which he held. However, NT had another side, in that he was one of the most aggressive individuals in attacking strangers (Nishida and Hiraiwa-Hasegawa, 1985).

Among non-alpha prime males, LU also has a unique personality (Fig. 7.3). He maintained intimate relationships with successive alpha males, and was accepted by them. From 1978 to 1979, when JG was the alpha male, LU usually associated with him (see Chap. 8). Kawanaka (1982a) reported that, when JG got a bushbuck fawn in 1978, KI, LU and BA formed "a meat-eating cluster", and LU got some meat. However, in 1981, he closely followed NT, and was frequently groomed by him. LU might be described as a "submissive follower", a term which has been coined for Japanese monkey females (Fujii, 1975). It is uncertain why LU has so often been accepted by alpha males. In any event, his close coalition with NT could not prevent his decline in rank after 1983.

On the other hand, RA was distinguished by his calmness and frequent interactions with young males. In the decline of his life (he probably died in May 1983), RA was peripheral within the prime male association network (see Fig. 7.1). Goodall (1986) also points out that, in Gombe National Park, some males may become more solitary during old age. Clearly, young adult or adolescent males chose RA as a target to associate and groom with. For example, from 1979 until his death, RA continued to take care of a male orphan (AJ).

Among young adult males, SM seemed to be unskilled in interacting with other males, and could not form any stable relationship with them.

This may be partly because he had health problems, since he usually breathed hard. He eventually disappeared from M Group. In contrast, the other 3 young adult males apparently intended to establish themselves in the adult male coalition network, and actively approached and groomed the older males.

V. SUMMARY

1. There was a linear rank order among the M Group males. The dominance relation tended to be expressed by the subordinates rather than by the dominants. In particular, pant-grunts/barks were focused on the alpha male.

2. Adult male chimpanzees assembled as a "core" of the unit-group. The proximity tended to be maintained by the subordinates. The dominants tended to take the leadership in ranging.

3. Male relationships were characterized by high tolerance and intense competition. Prime males frequently approached, followed, and groomed one another. Mutual grooming and meat-sharing also occurred among them. These symmetrical interactions might allow adult males to coexist easily within a unit-group. On the other hand, they compete, for rank or priority in mating.

4. Young adult males assumed ambivalent attitudes toward the prime males; they feared older males, but were also attracted by them. Young males established their positions in the "male cluster" and reached high ranks in their twenties.

Acknowledgements

This study was financially supported by the Monbusho International Scientific Program (#56041019 and 57043014 to T. Nishida).

8

Alpha Males' Interactions and Social Skills

Kenji KAWANAKA

Laboratory of Anthropology, Faculty of Science, Okayama University of Science

I. INTRODUCTION

The social unit of chimpanzees — called a unit-group (Nishida, 1968) or a community (Goodall, 1973b) — is most probably a patrilineal group (Itani, 1980a). Although many males of various age classes have been observed to emigrate from their natal unit-groups, no adult male has ever been seen to immigrate into a unit-group (Uehara, 1981; Goodall, 1983, 1986; Hiraiwa-Hasegawa, Hasegawa, and Nishida, 1984; Kawanaka, 1984, 1988; Nishida *et al.*, 1985; Takahata, 1986). All the males belonging to a unit-group were born and grew up in that group, acquiring their own social position therein.

It is known that the alpha position confers a great reproductive advantage (Nishida, 1979b, 1983a; Tutin, 1980; de Waal, 1982; Hasegawa and Hiraiwa-Hasegawa, 1983; Takasaki, 1985), and that an alpha male may be the core individual who integrates the males in a unit-group (Kawanaka, 1989). It has also been reported that coalitions between particular males, or the intervention and manipulation of a third party, play an important part in turning over the alpha male (Riss and Goodall, 1977; de Waal, 1982; Nishida, 1983a). However, social relationships between the alpha male and other males have not been fully elucidated.

This paper describes social interactions between the alpha and other males on the basis of the data obtained in three observation periods. By examining the differences and similarities among the results for the three periods, this study discusses the characteristics of interactions among males and the alpha male's role in establishing these characteristics.

II. METHODS

Observations were made in three periods: July 1978 – June 1979, August

—December 1983, and August—December 1986. The study unit-group was the M Group, consisting of about 100 animals (see Chap. 3). This study focuses on social interactions between the alpha and other males from early or late adolescence to old age (Table 8.1). JG occupied the alpha status in 1978–79 in the group we observed. Just after this observation period, however, NT took over the status, and still maintained it in 1989 (Chap. 7; Nishida, 1981; Takahata, 1986; Tsukahara, unpublished).

The age estimates of Hiraiwa-Hasegawa, Hasegawa, and Nishida (1984) are followed, and Goodall's (1983) terminology of male age classes is used: infancy (0–5 years), juvenile (5–7 years), early adolescence (8–12 years), late adolescence (13–15 years), young adult (16–20 years), prime

Table 8.1 The M Group males from early adolescence to old age, and their birth year, age classes, and rank categories.*

Name (abbr.)		Birth[1] year	1978–79		1983		1986	
			Age[2] class	Rank categ.	Age[2] class	Rank categ.	Age[2] class	Rank categ.
Kalindimya	(KL)	1930s?	—		—		—	
Mimikire	(MK)	1930s?	OA	Middle	**		**	
Kagimimi	(KI)	1940s?	OA	High	OA	High	OA	Middle
Rashidi	(RA)	1940s?	OA	Middle	**		**	
Kajugi	(JG)	1950s?	MA	Alpha	**		**	
Bakali	(BA)	1954?	PR	High	MA	High	MA	High
Ntologi	(NT)	1955?	PR	High	MA	Alpha	MA	Alpha
Lubulungu	(LU)	1957?	PR	Low	PR	Middle	MA	Low
Musa	(MU)	1957?	PR	Low	PR	Middle	MA	Low
Kalunde	(DE)	1963?	YA	Low	YA	Middle	PR	High
Lukaja	(LJ)	1965?	LA		YA	Low	PR	High
Kasulamemba	(SM)	1966?	LA		YA	Low	**	
Kasangazi	(KZ)	1968?	—		LA		YA	Low
Shike	(SU)	1970?	—		LA		YA	Middle
Kimaro	(KR)	1971?	—		EA		**	
Bembe	(BE)	1972?	× ×		EA		LA	
Hitto	(HT)	1972?	× ×		—		LA	
Nsaba	(NS)	1973	× ×		EA		LA	
Aji	(AJ)	1973?	× ×		EA		LA	
Mtwale	(TW)	1975?	× ×		EA		EA	
Hiro	(HR)	1975?	× ×		—		—	
Jilba	(JI)	1975?	× ×		—		—	
Toshibo	(TB)	1976	× ×		× ×		EA	
Masudi	(MA)	1977	× ×		× ×		EA	

* –: Data unavailable; **: disappeared; × × : did not reach early adolescence. [1]?: Presumed. [2]OA: old age; MA: middle-aged; PR: prime; YA: young adult; LA: late adolescence; EA: early adolescence.

(21–26 years), middle-aged (27-*circa* 32 years), and old age (*circa* 32 years to death). Young adult or senior males are called adults.

Adult males are divided into four rank categories following Bygott (1979): alpha, high-ranking, middle-ranking, and low-ranking males. The dominance rank of males was determined by interactions accompanied by pant-grunt vocalizations and agonistic behaviors (for details, see Hayaki, Huffman, and Nishida, 1989).

Since one old-aged male (KL) was not well habituated to the observers through the three observation periods, the data for him are omitted in the following analysis.

1. Data Collection and Analysis in 1978–79

Ad lib. sampling method (Altmann, 1974) was used for collecting the data, since the M Group chimpanzees were not well habituated in this period. After the chimpanzees were found, they were given pieces of sugar cane 10–20 cm long to lure the unhabituated ones out of the bush so as to be able to check as many individuals as possible. Data collection started after the bait was entirely consumed.

For analysis, each hour of the day was divided into halves as observation units, and the data were lumped for every unit. All individuals seen in a unit were regarded as being in proximity with one another. The proximity index of each dyad was calculated by the formula $[C/(A+B+C)] \cdot 100$, where A is the number of units in which only individual a was seen, B is the number of units in which only individual b was seen, and C was the number of units in which a and b were seen together. Grooming relations in each dyad were also scored for each observation unit. Dyadic grooming indices were calculated by the formula [(number of units in which grooming of each dyad was observed)/(proximity index of the dyad)] · 100.

A dendrogram for association among males drawn from the proximity indices and dyadic grooming relations were used for examining the interactions among males in this period.

2. Data Collection and Analysis in 1983

Focal animal sampling method (Altmann, 1974) was used for data collection in this period. The alpha male (NT) was chosen as the target for observation throughout this period; he was followed for about 293 hours. His behavior and interactions with others were recorded as they occurred, along with the starting and ending times. The individuals seen within 10 m of the target were regarded as being in his proximate vicinity, and movements in and out of this vicinity were recorded together with their times.

The interactions between the alpha and other males during this period were examined based on the following three data sets: (1) the alpha's dyadic proximity relations with each of the other males, (2) the dendrogram for association among males other than the alpha male in his proximate vicinity (within 10 m), and (3) the alpha's dyadic grooming relations with other males. The proximity indices of dyads seen in the vicinity of the alpha male were calculated by the formula $[F/(D+E+F)]\cdot 100$, where D is the total time when only individual d was seen in the proximity of the alpha male, E is the total time when only individual e was seen in the proximity of the alpha male, and F is the total time when d and e were seen together in the proximity of the alpha male. The dendrogram drawn from the above indices does not indicate the relationships among all the M Group males, but it is useful for analyzing the relationships between the alpha and other males.

3. Data Collection and Analysis in 1986

Data were collected using the focal animal sampling method. This study focused on the alpha male (NT), who was followed for about 60 hours (see Kawanaka, 1989). The proximity relations between the alpha and other males were examined by recording, for each minute, all individuals seen within 10 m of the target. The proximity indices between males other than the alpha male were calculated by the same formula as that used in 1983, while D, E, and F represented the numbers of the above-mentioned records instead of the total time. Based on these proximity indices, the dendrogram for association among males other than the alpha male was drawn as in 1983. Grooming among males was recorded as it occurred.

III. RESULTS

1. Social Interactions between the Alpha and Other Males in 1978–79

1) Dendrogram for Association among Males

The alpha male (JG) formed a cluster together with KI (old age), LU (prime), and BA (prime), while NT (prime; second-ranking) and MU (prime) formed another cluster. Two other old-aged males (RA, MK) and three young adult or adolescent males (LJ, SM, and DE) were found separately outside these clusters (Fig. 8.1).

Although the males of prime or senior age classes did not associate equally with one another, most of them (except for the two old-aged males) attached themselves to one of the two clusters. However, the young adult or adolescent males did not join either cluster. This may have been due to the way the bait was used for checking the chimpanzees encoun-

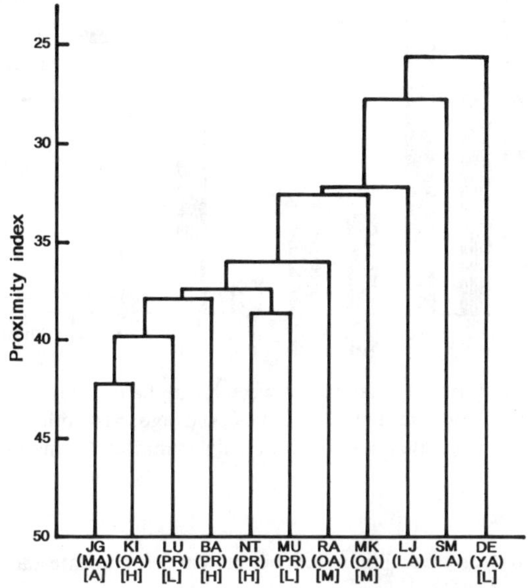

Fig. 8.1 Dendrogram for association among late adolescent or senior males in 1978–79. OA: old age; MA: middle-aged; PR: prime; YA: young adult; LA: late adolescence. A: alpha male; H: high-ranking; M: middle-ranking; L: low-ranking.

tered, but it also may suggest that prime or senior males discriminate against junior ones.

Males of a particular ranking category did not always cluster together. The alpha (JG) and second-ranking (NT) males formed separate clusters, and appeared to alienate each other. The males with whom JG clustered were subordinate to NT. The alpha male's cluster did not consist of males all of a particular age class, but includes members of age classes both senior and junior to him.

2) Dyadic Grooming Relations between Males

The alpha male (JG) had grooming interactions with high frequency with MK, KI (old age), and LU (prime), and with medium frequency with RA (old age), BA and MU (prime) (Fig. 8.2; Table 8.2). Low grooming frequencies were scored for NT (prime), DE (young adult), and SM (late adolescent). JG had grooming interactions with all late adolescent or senior males except for LJ (late adolescent), while all adult males except for NT had grooming interactions most frequently with JG (Table 8.2).

JG frequently chose his cluster members (KI, LU, and BA) as grooming

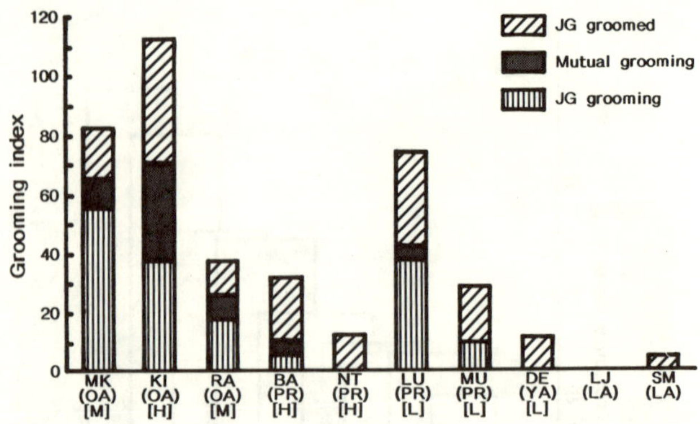

Fig. 8.2 Dyadic grooming relations between the alpha male (JG) and other males from late adolescence to old age in 1978–79. OA: old age; MA: middle-aged; PR: prime; YA: young adult; LA: late adolescence. H: high-ranking; M: middle-ranking; L: low-ranking.

Table 8.2 Matrix of grooming indices between males from late adolescence to old age in 1978–79.*

	Names, age classes, and rank categories of males[1)]										
	MK (OA) [M]	KI (OA) [H]	RA (OA) [M]	JG (MA) [A]	BA (PR) [H]	NT (PR) [H]	LU (PR) [L]	MU (PR) [L]	DE (YA) [L]	LJ (LA)	SM (LA)
MK	—	13.2	0.0	83.4	7.2	24.6	15.0	8.6	0.0	0.0	0.0
KI		—	11.0	113.7	10.1	28.5	30.5	8.0	0.0	7.8	0.0
RA			—	37.7	10.4	11.1	13.2	0.0	5.3	4.2	4.6
JG				—	32.8	12.6	75.3	28.2	11.8	0.0	5.3
BA					—	13.0	18.5	0.0	9.5	0.0	0.0
NT						—	26.5	7.8	0.0	23.5	0.0
LU							—	6.0	0.0	21.7	10.8
MU								—	0.0	3.6	3.6
DE									—	0.0	6.7
LJ										—	0.0
SM											—

* For definition of grooming indices, see text. [1)]OA: old age; MA: middle-aged; PR: prime; YA: young adult; LA: late adolescence. A: alpha male; H: high-ranking; M: middle-ranking; L: low-ranking.

partners, but rarely NT and MU, who formed a different cluster (Fig. 8.2; Table 8.2). However, JG also had grooming interactions at a high rate of frequency with the two old-aged males (MK and RA) who did not join either cluster. He devoted a longer time to grooming MK than MK did to grooming him, and about equal time to grooming and to being

Fig. 8.3 While traveling together, JG (the alpha male; right) and LU (prime and low-ranking male) make a brief stop, pant-hooting in concert. (Photo by K. Kawanaka).

groomed by KI, RA (old age), and LU (prime). In contrast, JG was groomed by BA and MU (prime) for a much longer time than he groomed them, and one-sidedly by NT (prime), DE (young adult), and SM (late adolescent).

Grooming interactions of JG with other males appeared to be connected with cluster membership and, more closely, with the partners' age classes. JG frequently chose his seniors as grooming partners, devoting a longer or equal time to grooming such partners. In contrast, it was only LU among his juniors who had frequent grooming interactions with JG and devoted about equal time to grooming and to being groomed by JG.

JG was especially intimate with LU (Fig. 8.3), while he was especially estranged from NT. JG's and NT's clusters were in closer proximity with each other than with the two independent old-aged males and the young adult or junior males. However, JG and NT, the two highest-ranking males, did not appear to intimately associate with each other. The low score of proximity between these two males and the completely one-sided grooming of JG by NT probably indicate that they were antagonists.

2. Social Interactions between the Alpha and Other Males in 1983

1) Dyadic Proximity Relations

Proximity relations between the alpha male (NT) and each of the other males varied greatly (Fig. 8.4). NT was observed together with two prime males (LU and MU) for more than 20% of the total time he was observed. Medium proportions (10–20%) were scored for KI (old age) and LJ

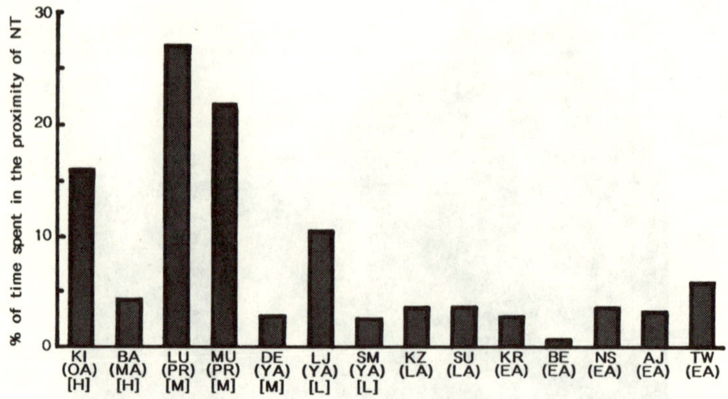

Fig. 8.4 Dyadic proximity relations between the alpha (NT) and other males from early adolescence to old age in 1983. OA: old age; MA: middle-aged; PR: prime; YA: young adult; LA: late adolescence; EA: early adolescence. H: high-ranking; M: middle-ranking; L: low-ranking.

(young adult). The scores for the remaining males, including BA (middle-aged; second-ranking), DE, and SM (young adult), were generally small, less than 5%.

NT was often in close proximity with most prime or senior males, but infrequently close to most young adult or adolescent males. Among prime or senior males, however, he was particularly estranged from the second-ranking male (BA), while of the young adult or junior males he associated particularly often with LJ.

2) Association among Males in the Proximity of the Alpha Male

LU, MU, and KI, who were frequently seen near the alpha male (NT), formed a cluster, which an early adolescent male (TW) joined (Fig. 8.5). The other two males — LJ, for whom the fourth highest proportion of proximity with NT was scored, and SU (late adolescent) — also attached themselves to this cluster. Another cluster was formed by BA (middle-aged), SM (young adult), and NS (early adolescent). The remaining males, including DE (young adult), were outside the clusters.

Although only prime or senior males had joined either of the two clusters in 1978–79, in 1983 all the adult males except DE (young adult) had joined one of the clusters seen in the vicinity of the alpha male (NT), and a few late or early adolescent males separately joined one or the other of them. This difference may be due to the use in 1978–79 of bait, which was not used in 1983.

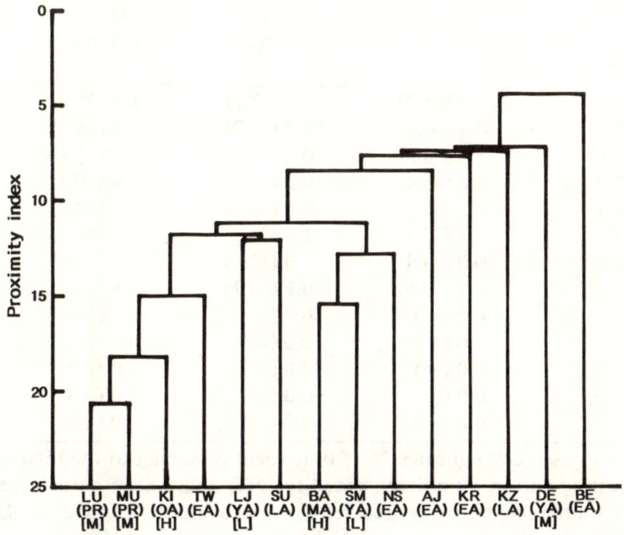

Fig. 8.5 Dendrogram for association among early adolescence or senior males in the proximity (within 10 m) of the alpha male (NT) in 1983. OA: old age; MA: middle-aged; PR: prime; YA: young adult; LA: late adolescence; EA: early adolescence. H: high-ranking; M: middle-ranking; L: low-ranking.

Of the four males who were frequently seen in the proximity of the alpha male (NT) during this period, only MU had clustered with NT in 1978–79. KI and LU had been part of a cluster with the former alpha male (JG) in 1978–79. These two males changed their associations between the two periods. In contrast, BA, who had been a member of JG's cluster along with KI and LU in 1978–79, was not frequently seen in the vicinity of NT in 1983.

3) Dyadic Grooming Relations between the Alpha and Other Males

NT had grooming interactions with all seven adult males, but with no adolescent males except for KZ (late adolescent) (Table 8.3). NT scored large proportions of grooming time for four of the eight males (LU, MU, KI, and LJ; Fig. 8.6), but small proportions for the remaining four males (BA, DE, SM, and KZ). The percentage of time spent grooming for each dyad appeared to be parallel with that of each dyad's percentage of time spent in proximity to NT.

However, the ratio between the time NT spent grooming his partner

Table 8.3 Dyadic grooming relations between the alpha (NT) and other males in 1983.*

Name (age class: rank categ.)[1)]	NT grooming	Mutual grooming	NT groomed	Total
KI (OA: H)	2.8 (63.7)	0.3 (7.1)	1.3 (29.2)	4.4
BA (MA: H)	0.4 (34.3)	0.2 (19.3)	0.5 (46.5)	1.1
LU (PR: M)	3.4 (50.3)	0.6 (8.5)	2.8 (41.1)	6.7
MU (PR: M)	2.2 (45.4)	0.2 (4.3)	2.4 (50.3)	4.9
DE (YA: M)	0.2 (46.5)	0.0 (8.6)	0.2 (44.9)	0.5
LJ (YA: L)	0.7 (20.1)	1.1 (31.7)	1.7 (48.2)	3.6
SM (YA: L)	0.0 (9.4)	0.0 (11.6)	0.1 (78.9)	0.2
KZ (LA)	0.1 (9.6)	0.1 (9.7)	0.4 (80.7)	0.5
SU (LA)	0.0 (—)	0.0 (—)	0.0 (—)	0.0
KR (EA)	0.0 (—)	0.0 (—)	0.0 (—)	0.0
BE (EA)	0.0 (—)	0.0 (—)	0.0 (—)	0.0
AJ (EA)	0.0 (—)	0.0 (—)	0.0 (—)	0.0
TW (EA)	0.0 (—)	0.0 (—)	0.0 (—)	0.0

* Number in each cell indicates % of time spent grooming to the total resting time (excluding eating and moving time) scored for NT, and that in parentheses indicates % to the total time spent grooming scored for each dyad. [1)]OA: old age; MA: middle-aged; PR: prime; YA: young adult; LA: late adolescence; EA: early adolescence. H: high-ranking; M: middle-ranking; L: low-ranking.

Fig. 8.6 While resting, NT (the second from the left) is grooming LU (left). KI (right) and LJ are also resting in NT's vicinity. (Photo by K. Kawanaka)

and the time he spent being groomed showed a different tendency (Table 8.3). NT groomed KI (old age) for a longer time than he was groomed by KI, and devoted about equal time to grooming and being groomed by BA (middle-aged), LU, MU (prime), and DE (the oldest of the young adult males). He was groomed for a longer time by LJ (young adult), and for a much longer time by SM (young adult) and KZ (late adolescent). No dyad had entirely one-sided grooming interactions such as those seen between JG and NT in 1978–79.

The grooming/groomed ratios appeared to be closely related to the partners' age classes, but not to dyadic proximity relations, cluster membership, total time spent grooming for each dyad, or male rank categories. The alpha (NT) and second-ranking (BA) males appeared to keep away from each other, as regards proximity relations and total time spent grooming. However, they groomed each other almost equally or reciprocally when they met.

3. Social Interactions between the Alpha and Other Males in 1986

1) Dyadic Proximity Relations

As in 1983, in 1986 it was the four adult males who were seen to spend frequent time in the proximity of the alpha male (NT) (Fig. 8.7), although some of them showed different association patterns from those seen in 1983. Three males (KI, LU, and LJ) scored large proportions of proximate time, as in 1983. MU markedly decreased his score for proximity to NT, and spent nearly as little time as did most of the young adult or adolescent males. In contrast, BA markedly increased his score, and spent nearly as much proximate time as LU. This was the first time BA showed a close proximity to NT.

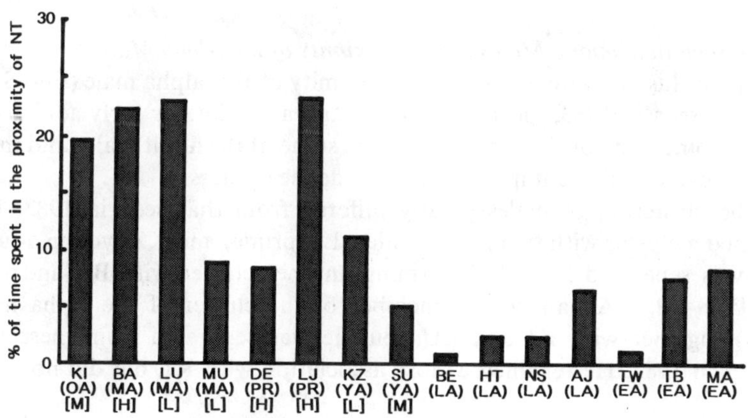

Fig. 8.7 Dyadic proximity relations between the alpha (NT) and other males from early adolescence to old age in 1986. OA: old age; MA: middle-aged; PR: prime; YA: young adult; LA: late adolescence; EA: early adolescence. H: high-ranking; M: middle-ranking; L: low-ranking.

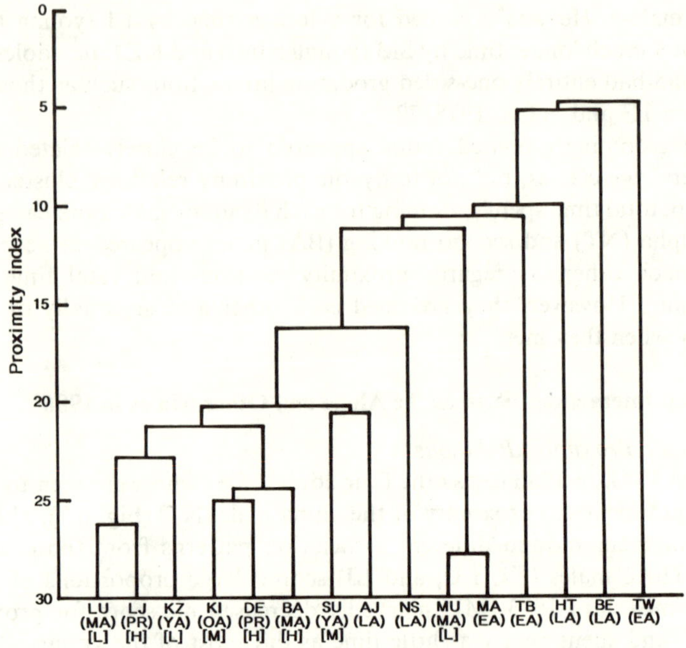

Fig. 8.8 Dendrogram for association among early adolescent or senior males in the proximity (within 10 m) of the alpha male (NT) in 1986. OA: old age; MA: middle-aged; PR: prime; YA: young adult; LA: late adolescence; EA: early adolescence. H: high-ranking; M: middle-ranking; L: low-ranking.

2) Association among Males in the Proximity of the Alpha Male

Three clusters were found in the proximity of the alpha male (Fig. 8.8). As was seen in 1983, most adult males and a few late or early adolescent males joined one of these clusters, while some of the adult males and most late or early adolescent males were outside the clusters.

The clustering of males greatly differed from that seen in 1983. LU formed a cluster with two junior males, LJ (prime) and KZ (young adult). KI was separated from LU, forming another cluster with BA and DE. In 1978–79, BA had been a member of the cluster of the alpha male (JG) together with KI and LU, but he was separated from these two males in 1983. He regained a close association with KI, but did not with LU.

3) Dyadic Grooming Relations between the Alpha and Other Males

As in 1983, the alpha male (NT) had grooming interactions involving a

Table 8.4 Dyadic grooming relations between the alpha (NT) and other males in 1986.*

Name (age class: rank categ.)[1)]	NT grooming	Mutual grooming	NT groomed	Total
KI (OA: M)	1.5 (43.6)	0.3 (9.3)	1.7 (47.1)	3.5
BA (MA: H)	1.8 (29.5)	0.4 (6.1)	4.0 (64.4)	6.1
LU (MA: L)	4.6 (64.8)	0.9 (12.9)	1.6 (22.3)	7.1
MU (MA: L)	1.2 (72.9)	0.1 (8.4)	0.3 (18.7)	1.7
DE (PR: H)	0.7 (20.4)	0.9 (24.5)	1.9 (55.1)	3.5
LJ (PR: H)	1.8 (32.4)	1.0 (18.6)	2.7 (49.6)	5.5
KZ (YA: L)	0.4 (22.9)	0.1 (3.2)	1.5 (73.9)	2.0
SU (YA: M)	0.4 (48.6)	0.0 (—)	0.4 (51.4)	0.8
BE (LA)	0.0 (—)	0.0 (—)	0.0 (—)	0.0
HT (LA)	0.0 (—)	0.0 (—)	0.0 (—)	0.0
NS (LA)	0.0 (—)	0.0 (—)	0.0 (—)	0.0
AJ (LA)	1.2 (31.0)	0.3 (7.5)	2.5 (61.5)	4.0
TW (EA)	0.0 (—)	0.0 (—)	0.0 (—)	0.0
TB (EA)	0.1 (21.3)	0.0 (—)	0.3 (78.7)	0.4
MA (EA)	0.2 (55.5)	0.0 (—)	0.2 (44.5)	0.4

* Number in each cell indicates % of time spent grooming to the total resting time (excluding eating and moving time) scored for NT, and that in parentheses indicates % to the total time spent grooming scored for each dyad. [1)]OA: old age; MA: middle-aged; PR: prime; YA: young adult; LA: late adolescence; EA: early adolescence. H: high-ranking; M: middle-ranking; L: low-ranking.

considerable amount of time with most of those for whom he scored high frequency for proximity (Table 8.4). He devoted a large proportion of time to grooming with BA, who markedly increased his proportion of proximity, and a small proportion of time to grooming with MU, who began to stay away from him. However, NT's proportion of time spent grooming with other males was not always in parallel with that of time spent in proximity. He did not devote much time to grooming with KI, who stayed in close proximity, devoting as much time to grooming with DE, who was not so frequently seen nearby.

The ratio between the time NT spent grooming his partner and the time he spent being groomed did not show a clear relation with the partners' age classes, as was seen in 1983 (Table 8.4). NT devoted a longer time to grooming LU and MU (middle-aged), about equal time to grooming and being groomed by KI (old age), LJ (prime), and SU (young adult), and a longer time being groomed by BA (middle-aged), DE (prime), KZ (young adult), and AJ (late adolescent). The ratio may not be always related only to partners' age classes. Entirely one-sided grooming interactions between the alpha and another male were not recorded for any dyad.

IV. DISCUSSION

Although differences were seen in the three observation periods in every respect—dyadic proximity relations, dyadic grooming relations, and polyadic association patterns—some common characteristics can also be mentioned.

First, it was mostly adult males that the successive alpha males intimately associated with. Such age differences in association patterns reconfirm the conclusions by Hayaki, Huffman, and Nishida (1989) and Kawanaka (1989). Second, the alpha males did not have equal relations with every such male, but chose partners for intimate association, as Takahata (1986, see Chap. 7) and Kawanaka (1989) pointed out. The alpha males' partners were usually restricted to a few males, fewer than half of the adult males. The partners chosen by the alpha males were restricted neither to a particular age class nor to a particular rank category.

Adult males formed several clusters, some of which appeared not only to be separate from but also to be at variance with the alpha male's cluster. However, even the males belonging to such clusters usually had reciprocal or symmetrical grooming interactions with the alpha male, although they might be infrequent. Itani (1984b) states that male chimpanzees within the same unit-group can interact with one another in ways that disregard their ranks. It may be further said that they can interact with one another in ways that disregard their rivalry with one another.

However, not every male could always interact in such ways. In 1978–79, the alpha (JG) and second-ranking (NT) males appeared to avoid each other, and NT one-sidedly groomed JG when they encountered. They may not have been able to disregard their rivalry. If a dominant male needs to withhold himself in order to interact reciprocally with his subordinate, the above example may indicate that JG could not do so when he encountered NT.

JG left the association network of the M Group males just after my observation period in 1978–79, and NT became the group's alpha male (Nishida, 1981, see Chap. 7). Although JG was observed to stay within the M Group's range for about three years, he ceased to associate with the males, and eventually disappeared (Takahata, 1986). We do not know what caused JG's expulsion, but it may have been related to the assumption of alpha status by NT. However, not every male who yielded his alpha status to another became an outcast from his unit-group. Some old-aged males of the M Group are thought to have been alpha males in the past and to have remained there after losing their status (e.g., Hayaki, Huffman, and Nishida, 1989). Some examples of such males have also been recorded at Gombe (e.g., Goodall, 1986). No data were obtained for elucidating the factors differentiating the relationships between the ex-alpha and other males. JG's outcast status might have been partly a result of his intolerance of NT, the successor to his alpha status.

The males who clustered with JG when he was the alpha male were separated into two groups after NT became the alpha male: one consisting of KI and LU, who came to associate intimately with NT, and the other comprising BA, who continued to stay away from NT. However, BA had changed again in 1986, coming into close proximity with NT. Thus, fluidity in males' associations is conspicuous. The intimate relations among males are not formed solely by factors such as sibships, which often stabilize social relationships (see also Nishida, 1983a).

Although male relationships are often characterized by opportunism (e.g., de Waal, 1982), LU maintained the most intimate relations with the two successive alpha males (see also Chap. 7). He is small of build (Uehara and Nishida, 1987), and has been outranked by junior males such as DE, LJ, and SU (Hayaki, Huffman, and Nishida, 1989; Kawanaka, 1989). The highest position he reached was the fifth-ranking in 1981–83 (Takahata, 1986; personal observation), and he has recently declined in rank.

In chimpanzee society, where a patrilineal group is the social unit and grown-up males cannot transfer to a non-natal unit-group, it should be crucial for males to gain and to maintain their own social position within their natal unit-group, even if they cannot attain a high rank or alpha status. LU's life history demonstrates that not all male chimpanzees attain alpha status. Although it is likely that a male chimpanzee's rank has an inverse U-shaped relationship with age, as Bygott (1979) stated, the peak may not necessarily be alpha status.

LU's relations with the alpha males were not such that he had to groom them one-sidedly. He had symmetrical grooming interactions with the alpha males; at times he was even groomed for a longer time than he spent grooming them. From these interactions, it cannot be thought that LU one-sidedly followed or submitted himself to the alpha males. In addition, the alpha males devoted a longer time to grooming some old-aged males, even though they were not so frequently in each other's proximate vicinity (see also Chap. 7). Similar results were also reported from Gombe: although the higher the males were ranked, the longer they were groomed, only the alpha male groomed his partners for a longer time than he was groomed (Simpson, 1973). These data indicate that it is the alpha male who actively desires such associates. If this is the case, the alpha male's intention should be explored for an understanding of the relationships among male chimpanzees in a unit-group.

Hayaki, Huffman, and Nishida (1989) state that male chimpanzees usually interact in polyadic settings, and describe several examples in which a male attacked a dominant group-mate in the presence of the alpha male. They also state that the dominance relationships among males are usually unclear because of such polyadic encounters, where most interactions are affected by the males other than those directly involved (see also Nishida and Hiraiwa-Hasegawa, 1987).

In the polyadic setting that includes the alpha male, a male ranking just below the alpha male is sometimes attacked by his subordinate but cannot cope with such an attack in spite of his superiority to the attacker. Such a situation may be stressful or uneasy for the male being attacked. By accompanying old-aged or other males declining in rank, the alpha male may produce such a stressful setting for his high-ranking competitors. This may be the alpha male's political way of manipulating the other males and thereby nipping future rivalry in the bud. If so, it may be only to the observers that the dominance relationships among males are unclear; each male present in the polyadic setting may play his own part, well aware of the dominance relationships among all the participants.

The alpha male's manipulation appears usually to be accepted by the others. This may result in symmetrical interactions among males who act as if disregarding their dominance relationships and rivalries. However, this is not always the case. The one-sided interactions between the alpha and second-ranking males (JG and NT) seen in 1978–79 may be an example of the exceptions. JG could not get along with NT, probably because NT had gained enough power to be at variance with JG or not to give in to his manipulation. The cases described by de Waal (1982) and by Nishida (1983a) are additional exceptions, in which the lowest-ranking males did not play the part of an attacker to the male ranking next to the alpha male. They were old-aged or declining in rank, and manipulated their dominants, including the alpha males. Both cases broke out in small-sized unit-groups comprising only three grown-up males. In order for the alpha male to effectively manipulate other males, a unit-group may need to contain a set of males of various age classes including both those declining and those rising in rank. JG's outcast position and disappearance after he had once attained the alpha status may be an unusual case: if all past prime males disappeared, there would be a lack of old-aged and rank-declining males.

Male chimpanzees repeat polyadic encounters within their natal unit-group, where they continuously maneuver to gain and to retain social position. This may result in their high level of sociability, which has been emphasized as one of the most conspicuous characteristics of their society since the earlier stages of the study (Nishida, 1968, 1979b).

V. SUMMARY

1. The social interactions of two successive alpha males of a chimpanzee unit-group were described on the basis of data obtained at the Mahale Mountains National Park, Tanzania, during three periods: July 1978 – June 1979, August–December 1983, and August–December 1986.

2. The alpha males intimately associated only with a few—less than half—of the adult males. Such alpha males' partners were restricted neither to a particular age class nor to a particular rank category.

3. Adult males formed several clusters, some of which not only stayed separate from but also were at variance with the alpha male's cluster. However, even the males belonging to such clusters mostly had symmetrical grooming relations with the alpha male. The males belonging to the same unitgroup were able to interact with one another, as if disregarding their rivalry.

4. The membership of clusters repeatedly changed, and fluidity in males' associations was conspicuous. This indicates that intimate relations among males are not formed solely by such factors as sibships.

5. Although the alpha male's intimate partners changed from period to period, he usually associated with males who were declining in the dominance rank. These males did not seem to follow him one-sidedly; the alpha male chose such males intentionally. He groomed them as long as or even longer than he was groomed by each of them.

6. It was often observed that low-ranking males attacked males who dominated them in rank order in the presence of the alpha male. The choice of males declining in rank as his associates may be the alpha male's maneuver for keeping the males ranking just below him in a stressful or uneasy situation, thereby nipping future rivalry in the bud.

Acknowledgements

I am grateful to: T. Nishida for his guidance in the field since the very beginning of my study at Mahale and for his critical comments on the earlier draft of this manuscript; S. Uehara, Y. Takahata, H. Hayaki, H. Takasaki, K. Hunt, and R.R. Olomi for their cooperation in data collection; H. Takasaki for his valuable comments on and interest in this work; and my wife Hatsuko and our two growing children for all that they have done for me. This study was supported by the Japan International Cooperation Agency and by grants under the Monbusho International Scientific Research Program (#58041025, 59043022 to T. Nishida, and #61041071, 62043047 to T. Kano).

9

Social Context of Pant-grunting in Young Chimpanzees

Hitoshige HAYAKI

Department of Zoology, Faculty of Science, Kyoto University

I. INTRODUCTION

This study reports on episodes in which young chimpanzees directed or received pants, pant-grunts, pant-barks, or pant-screams. The pant-grunt is usually uttered in the context of reunion, and is always directed by a subordinate toward a more dominant individual (Bygott, 1979; Noë, de Waal, and van Hooff, 1980; de Waal, 1982; Goodall, 1986). Pant-grunts are usually soft sounds, although they sometimes grade into pant-barks or pant-screams. Panting also occurs in a greeting or during a grooming session (Goodall, 1986).

This study describes the instances in which these vocalizations occurred, and analyzes the context and outcome of such events. One goal of the study is to demonstate the social meaning of pant-grunting in the unique fission-fusion society of chimpanzees.

II. METHODS

The study took place in two periods; August 1983–January 1984, and August–December 1985. The subjects of the study were 10 young chimpanzees (6 males and 4 females) of M Group. Data were collected using the focal animal sampling method (Altmann, 1974). Table 9.1 shows the length of time each individual was tracked. Total observation time was about 502 hrs. Of the 10 focal animals, TW was tracked only in 1985, and *SH* (italics indicate female) only for 6.5 hours in 1983.

The following terms are defined for the analysis (Fig. 9.1).

Encounter: a pair *encounters* when they approach within 10 m of each other, after having been separated by 10 m or more for longer than one minute.

Table 9.1 Focal animals and duration of tracking.

Name (abbr.)[1]		Sex	Birth yr.	Duration of tracking (min)		
				1983	1985	Total
Kasangazi	(KZ)	male	1968?	1655.8	1631.1	3286.9
Shike	(SU)	male	1970?	1564.9	1718.1	3283.0
Nsaba	(NS)	male	1973	1770.0	1468.9	3238.9
Aji	(AJ)	male	1973?	1557.2	1738.1	3295.3
Mtwale	(TW)	male	1975?	–	1748.4	1748.4
Toshibo	(TB)	male	1976	1508.5	1607.8	3116.3
Pulin	(*PU*)	female	1971?	1959.3	1666.1	3625.4
Wantumpa	(*PA*)	female	1975	1648.6	1714.4	3363.0
Lusya	(*LS*)	female	1975	1518.7	1617.1	3135.8
Shinako	(*SH*)	female	1978	389.2	1632.3	2021.5
Total				13572.2	16542.3	30114.5

[1] Italics indicate female.

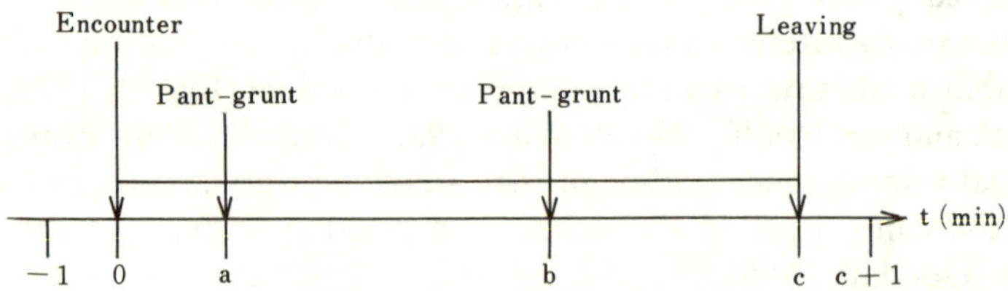

Fig. 9.1 Schematic representation of a pant-grunt episode.

Leaving: a pair *leaves* one another when they are no longer within 10 m for more than one minute.

Episode: an *episode* is made up of those social interactions which occur between encounter and leaving.

Pant-grunt Event: an event in which the focal animal emits or receives pant, pant-grunt, pant-bark, or pant-scream vocalizations. These vocalizations are denoted by pant-grunt in this study.

Pant-grunt Episodes: episodes in which pant-grunt events occurred. Eight pant-grunt episodes were unusual in that the participants did not come within 10 m of each other.

In this study, 410 pant-grunt episodes were observed (Table 9.2). Of these episodes, 283 were observed completely from encounter to leaving. In the other 127 episodes, the focal animal and its partner approached within 10 m of each other one minute or less after the start of observation, or they were no longer within 10 m one minute, or less, before the end of observation.

Table 9.2 Completeness of observation of pant-grunt episodes.*

Complete/Incomplete	1983	1985	Total
Complete	96 (4)	187 (3)	283 (7)
Incomplete 1	24	44	68
Incomplete 2	19 (1)	27	46 (1)
Incomplete 3	4	9	13
Total	143 (5)	267 (3)	410 (8)

* Numerals indicate number of episodes. Numerals in the parentheses indicate number of episodes in which the participants did not stay within 10 m of each other.
Incomplete 1: episodes in which leaving was not observed.
Incomplete 2: episodes in which encounter was not observed.
Incomplete 3: episodes in which neither leaving nor encounter was observed.

This study follows Goodall (1983) with regard to definition of age-sex classes. However, since age classes of some individuals change between the two study periods, the following age classification was used for each sex in some analyses: (1) individuals older than the oldest focal animal (KZ or *PU*), (2) individuals whose age was between the oldest and the youngest focal animal (TB or *SH*), and (3) individuals younger than the youngest focal animal. The first group corresponds basically to the adult age class; the second, to the adolescent and juvenile classes; and the third, to the juvenile and infant classes.

III. RESULTS

1. Frequencies of Pant-grunts

The 410 pant-grunt episodes contained 458 pant-grunt events (Table 9.3). Focal animals emitted pant-grunts in 364 events, and received pant-grunts in 91 events. In 3 events, both the focal animal and his/her partner uttered pant-grunts simultaneously. Focal males and females both emitted pant-grunts more frequently than they received them.

Most pant-grunts were directed toward adult males (Fig. 9.2a), especially toward NT, who was the alpha male. In 321 of 364 events (88.2%), pant-grunts were directed toward adult males, while NT was the recipient in 144 events (44.9%). The number of pant-grunts directed toward NT by focal females was higher than the total directed toward all other adult males. For focal males, the reverse was true (Table 9.4). This may be due to the fact that the young males were in the process of establishing their positions in adult male-male coalitions.

Age changes of pant-grunting for each sex were not so clear. Young females rarely pant-grunt before the first estrus. *SH* did not pant-grunt in 1983, and did so infrequently in 1985. *LS* and *PA*'s frequencies were

Table 9.3 Number of pant-grunt episodes and pant-grunt events.*
(a) Focal males

Focal animals	Direct	Receive	Both[1]	Total
⟨1983⟩				
KZ	14 (19)	10 (10)		24 (29)
SU	18 (20)	16 (17)		34 (37)
NS	16 (18)	0 (0)	1 (1)	17 (19)
AJ	22 (29)	1 (1)		23 (30)
TB	10 (10)	0 (0)		10 (10)
⟨1985⟩				
KZ	16 (24)	10 (10)	1 (1)	27 (35)
SU	18 (18)	35 (39)		53 (57)
NS	8 (8)	3 (3)		11 (11)
AJ	18 (21)	5 (5)		23 (26)
TW	32 (39)	2 (2)		34 (41)
TB	32 (35)	3 (3)		35 (38)
Total	204 (241)	85 (90)	2 (2)	291 (333)

(b) Focal females

	Direct	Receive	Both[1]	Total
⟨1983⟩				
PU	20 (20)	0 (0)	1 (1)	21 (21)
PA	9 (10)	0 (0)		9 (10)
LS	5 (5)	0 (0)		5 (5)
SH	0 (0)	0 (0)		0 (0)
⟨1985⟩				
PU	18 (19)	0 (0)		18 (19)
PA	20 (20)	0 (0)		20 (20)
LS	37 (41)	0 (0)		37 (41)
SH	8 (8)	1 (1)		9 (9)
Total	117 (123)	1 (1)	1 (1)	119 (125)

* Numerals indicate number of episodes in which the focal animal directed or received pant, pant-grunt, pant-bark or pant-scream. Numerals in parentheses indicate number of events.

[1] Both participants pant-grunted simultaneously.

Table 9.4 Pant-grunt events which focal animals directed at adult males.

Recipient	Focal male	Focal female	Total
Alpha male	93	51	144
Non-alpha adult males[1]	136	41	177
Total	229	92	321

$\chi^2=5.830$, df$=1$, $p<0.05$

[1] KZ, who was an adult male in 1985, was excluded. See Methods.

also low in 1983. Both individuals experienced their first estrus in 1985, and their frequencies of pant-grunting increased (Fig. 9.2a). *LS* pant-

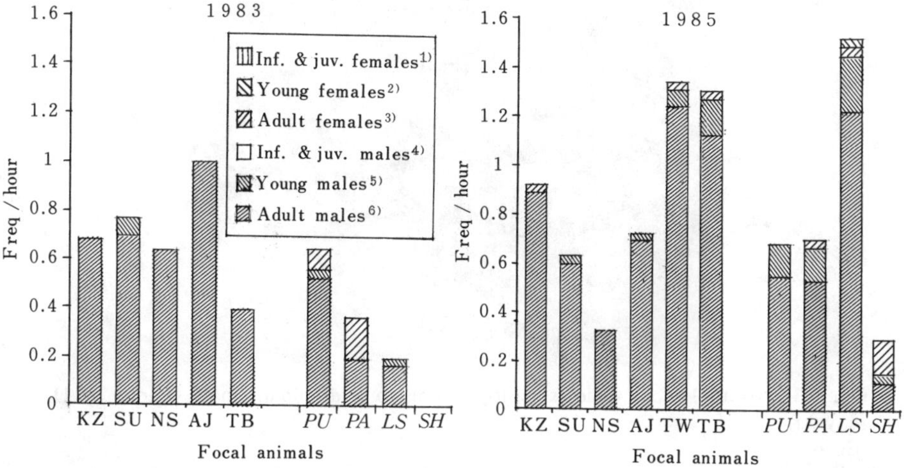

Fig. 9.2a Frequencies of pant-grunt events by the focal animals in each study period. The following age/sex classes were used.

1) Infant and juvenile females younger than *SH*.
2) Young females aged between *SH* and *PU*.
3) Adult females older than *PU*.
4) Infant and juvenile males younger than TB.
5) Young males aged between TB and KZ.
6) Adult males older than KZ.

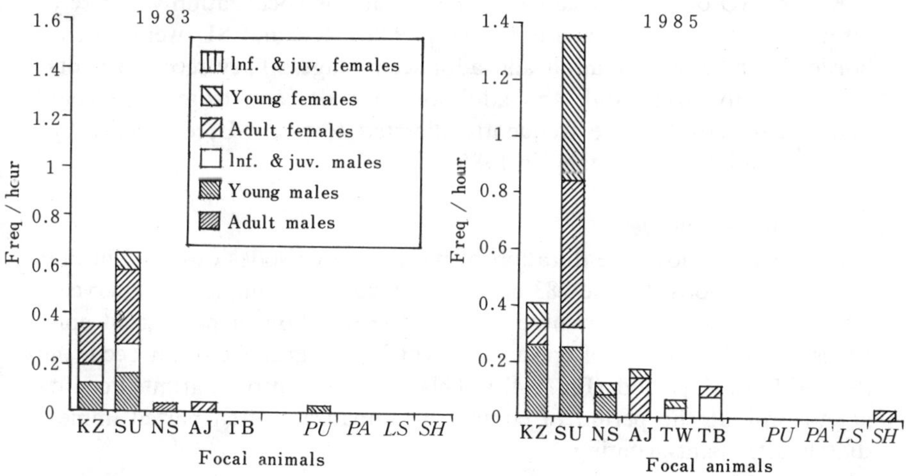

Fig. 9.2b Frequencies of pant-grunt events directed toward focal animals in each study period. Age/sex classes follow Fig. 9.2a.

Fig. 9.3 An adolescent male is pant-grunting to a dominant male. (Photo by H. Hayaki)

grunted very frequently in 1985, probably because she was orphaned. AJ and TW, who were male orphans, also pant-grunted frequently. It is noteworthy that TB pant-grunted frequently in 1985. At that time, he tended to stay with his mother and younger sister on the periphery of M Group and rarely moved with the adult males (his co-attendance index with adult males was very low; Hayaki, 1988). He frequently pant-grunted at adult males when he joined them after a long separation.

KZ and SU often received pant-grunts, but this vocalization was rarely directed at the other focal animals (Fig. 9.2b). KZ and SU were on the borderline between the adult and adolescent stages. They received pant-grunts mainly from adult and adolescent females and from adolescent males. Pant-grunts were frequently directed toward SU, especially by adult and adolescent females in 1985.

2. Pant-grunt Episodes

Figure 9.4a shows the duration of pant-grunt episodes observed in the two study periods. Of the 283 episodes which were completely observed, 100 (35.3%) lasted one minute or less. The duration of pant-grunt episodes is similar to that of episodes in which pant-grunts did not occur in the 1985 study period (Fig. 9.4b). This suggests that pant-grunts do not act to prolong the period of contact. Chimpanzees often parted immediately after pant-grunting.

3. Pant-grunt Events on Encounter

In 351 of the 410 episodes, the encounter between two chimpanzees

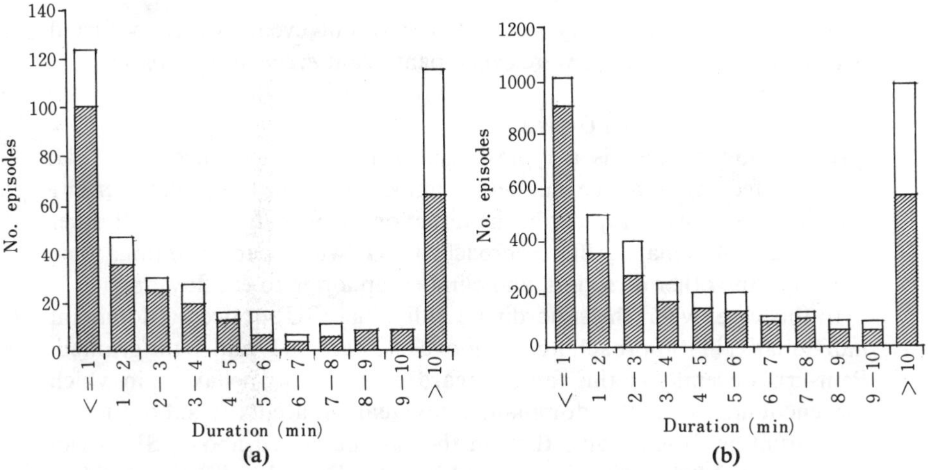

Fig. 9.4 Distribution of durations of episodes. Shaded blocks indicate episodes observed completely from encounter to leaving. Unshaded blocks indicate episodes observed incompletely.
(a) Pant-grunt episodes observed in the two study periods. (b) Episodes in which pant-grunting did not occur in the 1985 study period.

was observed. Figure 9.5 shows the time between the encounter and the first pant-grunt event (time *a* in Fig. 9.1). Pant-grunt events occurred one minute or less after the encounter in 275 events (78.3%). Most occurred immediately after the encounter, including events in which the subordinate individual began to pant-grunt at the dominant before approaching

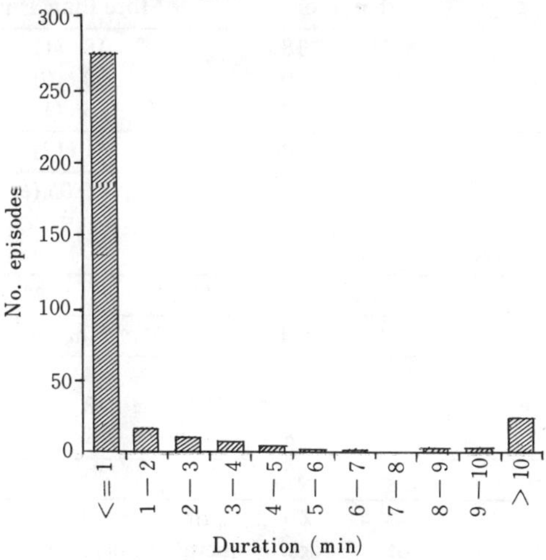

Fig. 9.5 Distribution of the length of time between the encounter and the first pant-grunt event.

within 10 m. In this study, the 258 pant-grunt events which occurred within 30 sec of encounter were called pant-grunt *events on encounter*.

1) Pant-grunt Events as Greetings

The following case is a typical pant-grunt event on encounter. SU, who was feeding in a tree, suddenly climbed down to the ground, crying "hyaoo." He moved rapidly in the direction of NT (alpha male), 2 other males, and 4 females. SU approached NT while pant-grunting, then moved in the other direction, and climbed up a tree to eat flowers.

In this case, when the subordinate individual (SU) noticed a dominant individual (NT) approaching, he moved toward NT and pant-grunted. Pant-grunt events like this can be regarded as greeting behavior, in which the encounter with the dominant individual induced the subordinate's pant-grunting. It is notable that, in the case described above, SU began to approach NT before being noticed by him. The subordinate individual often actively sought out the dominant in order to greet him. Of 258 pant-grunt events on encounter, 226 (87.6%) were classified as greetings. One-hundred eighty-nine of these 226 episodes were observed completely from encounter to leaving; 84 (44.4%) lasted only one minute or less.

Focal animals pant-grunted in 158 of the 189 complete episodes. Pant-

Table 9.5 Partners and duration of episodes in pant-grunt greetings.*
(a) Focal males directed.

Recipient	Duration		Total
	1 min or less	More than 1 min	
Alpha male	18	18 (1)	36 (1)
Non-alpha adult males[1]	10	30 (10)	40 (10)
Others	2	3 (1)	5 (1)
Total	30	51 (12)	81 (12)

$\chi^2=5.09$, df=1, $p<0.05$ (excluding "Others")

(b) Focal females directed.

Recipient	Duration		Total
	1 min or less	More than 1 min	
Alpha male	20	14	34
Non-alpha adult males[1]	9	18	27
Others	8	8	16
Total	37	40	77

$\chi^2=3.92$, df=1, $p<0.05$ (excluding "Others")

* Numerals indicate number of episodes. Numerals in parentheses indicate number of episodes in which grooming occurred after pant-grunt events.

[1] KZ was excluded. See Table 9.4.

grunt episodes usually ended within one minute when pant-grunts were directed toward the alpha male. In contrast, episodes usually lasted for more than one minute when directed toward non-alpha adult males who were older than KZ (Table 9.5). Grooming occurred in 10 of the episodes in which focal males pant-grunted toward non-alpha adult males, but rarely occurred when pant-grunts were directed toward the alpha male. Thus, pant-grunts directed toward the alpha male tended to occur in the absence of practical interactions; these vocalizations seemed to be more formal than those directed toward the other adult males.

2) Pant-grunt Events Following the Display of a Dominant

In 26 of the 258 pant-grunt events which occurred on encounter, a subordinate individual pant-grunted at a dominant who had begun a charging display. Subordinates also sometimes pant-grunted toward nearby dominant individuals who performed a charging display. These were not regarded as greeting behavior.

3) Pant-grunt Events in other Contexts

The contexts of the remaining 6 pant-grunt events which occurred on encounter were somewhat complicated. Two events involved an individual who was attacked by a dominant individual. In one event, he pant-grunted at the other dominant adult male. In the other event, he was panted at by a subordinate female. Infants were involved in 2 more events. In one event, an adult female approached and panted at a dominant male who was hugging her infant. In the other event, a subordinate infant panted at a dominant one who had mounted him and was thrusting in the ventro-ventral posture. In the other 2 events, the dominant individual panted at the subordinate, or both individuals panted simultaneously.

4. Pant-grunt Events not on Encounter

Of the 458 pant-grunt events which were observed, 179 (39.1%) could not be regarded as having occurred on encounter. These pant-grunt events took place in the following contexts.

1) Pant-grunt Events Following the Display of a Dominant

Twenty-eight pant-grunt events occurred when a dominant male in the vicinity of a subordinate began to perform pant-hoots, a charging display, or showed aggression. For example, TW was licking a stone when KI (an adult male) pant-hooted and began throwing a stone in a charging display. TW ran pant-grunting at KI, then immediately fled and sat some distance away.

This type of pant-grunt event seems to be a kind of appeasement behavior in order to restrain aggression by a dominant. However, as in the above example, a subordinate individual often pant-grunted at a dominant individual who had not charged at him. This was the case in 42 of the 54 events of this type (26 on encounter, plus 28 not on encounter) which were observed.

2) Pant-grunt Events Occurring with the Participants Moving Together

Forty-one pant-grunt events took place while the participants were moving together. Usually, these events occurred when a subordinate individual noticed a higher-ranking individual approaching. For example, a juvenile male, TB was walking behind SM and ahead of BA, who were adult males. When TB stopped and looked back, he noticed BA approaching. TB bobbed his head and pant-grunted at BA, then began to move once more between the two adults. In these episodes, the participants usually continued to move together after pant-grunting.

Seven pant-grunt events in motion were special cases. In one case, a subordinate female panted when her genitals were inspected by adult males while she was moving. In the other 6 events, a subordinate individual began to feed after pant-grunting at a dominant with whom he had been moving. In such cases, the subordinate seemed to have excused himself for beginning to eat before the dominant.

3) Pant-grunt Events Occurring before Moving

Forty-one pant-grunt events occurred just before one or both participants began to move. In 28 events, a subordinate individual pant-grunted at a dominant, who had begun to move. In 25 of these cases, the participants moved after the pant-grunt events. However, in 3 cases, the dominant individual stopped moving while receiving the subordinate's pant-grunt.

In the remaining 13 events, a subordinate individual pant-grunted at a higher-ranking individual who was in the vicinity, before the latter began to move. In some cases, the subordinate pant-grunted when the pair resumed moving after a short rest. In such cases, the subordinate pant-grunted at the dominant as if he had apologized for keeping the dominant waiting. In 2 of the 13 events, only the subordinate who had pant-grunted moved away.

4) Pant-grunt Events Occurring during a Grooming Session

Eleven pant-grunt events occurred during a grooming session. In this situation, panting rather than pant-grunting occurred, often when the dominant individual shifted his body. Adolescent males were often tense when they engaged in grooming with adult males. During focal sampling, only the subordinate was observed to pant at the dominant. In *ad lib.* observation, however, both participants were sometimes seen panting at the same time.

5) Pant-grunt Events Occurring during Meat Eating

Chimpanzees became tense during meat-eating situations. When approaching and begging, they often pant-grunted at a dominant individual who had meat. However, the difficulty in tracking focal animals in such situations meant that only 3 such pant-grunt events were observed in focal sampling. If adult males had been tracked, much more such events could have been observed.

6) Pant-grunt Events Occurring when Approaching an Individual during Feeding or Resting

Thirty-eight pant-grunt events occurred when one individual approached another in the context of feeding (16 events) or resting (22 events). In 24 events, the subordinate individual approached the dominant; while in 14 events, the dominant approached the subordinate. Participants often engaged in grooming after a pant-grunt event which occurred in the context of resting (13 of 22 events). When feeding, subordinate individuals sometimes became tense, responding to every little movements of the dominant. For example, TW and an adult male, MU, were feeding in the same tree. When MU moved slightly toward him, TW pant-grunted at the older male. After pant-grunting, TW remained tense, scratching himself, and did not resume feeding until MU moved away.

7) Pant-grunt Events Occurring Suddenly during Feeding or Resting

Thirteen pant grunt events could not be classified into the contexts discussed above. In 2 events, a subordinate individual pant-grunted at a higher-ranking individual before beginning to feed. In 4 events, a subordinate suddenly interrupted his feeding and pant-grunted at a dominant who was also feeding nearby. In 7 events, a resting subordinate suddenly pant-grunted, then immediately resumed resting. These events resemble those which occurred when one individual approached another, although in these cases neither of the participants approached. The subordinates seemed to have been hypersensitive in these instances, and apparently pant-grunted to appease the more dominant individual. The subordinate may have pant-grunted at the dominant when he/she felt that aggression from the higher-ranking individual was imminent.

8) Pant-grunt Events in which the Vocalization was Emitted by the Dominant Individual

Three events were observed in which both participants pant-grunted simultaneously at one another. One of these events occurred on encounter. In these cases, the dominant individual seemed to pant-grunt for reassurance. Further, three additional events were observed in which the dominant individual pant-grunted at the subordinate. In one event, which occurred on encounter, a dominant male panted after threatening a subordinate male. In another event, a dominant female panted when she approached a subordinate female who was eating a rare food. In a similar event, a dominant female panted when she approached the subordinate female who was fishing for ants. At the approach of the dominant individual, however, the lower-ranking individual screamed and fled. The dominant individual's infant then screamed as well, and the dominant held him in her arms and pant-grunted again at the subordinate.

5. Summary of Contexts in Which Pant-grunt Events Occurred

Table 9.6 summarizes the contexts in which pant-grunt events occurred. Although pant-grunt events usually occurred on encounter, a subordinate individual often pant-grunted at a dominant who was in the vicinity. Focal males pant-grunted as frequently in non-greeting contexts as in greeting contexts (110 events in greeting, and 122 in non-greeting). They often pant-grunted at non-alpha adult males in non-greeting contexts (Fig. 9.6a). On the other hand, focal females pant-grunted mainly in greeting contexts (80 events in greeting, and 35 in non-greeting, Fig. 9.6b). These

Table 9.6 The contexts in which pant-grunting occurred.

Contexts	No. of pant-grunt events	
(a) On encounter		258
1. Greeting	226	
2. The dominant displaying	26	
3. Others	6	
(b) Not on encounter		179
1. The dominant displaying	28	
2. Moving together	41	
3. Before moving	41	
4. During a grooming session	11	
5. Meat eating	3	
6. Approaching a nearby individual	38	
7. Others	17	
Unknown		21
Total		458

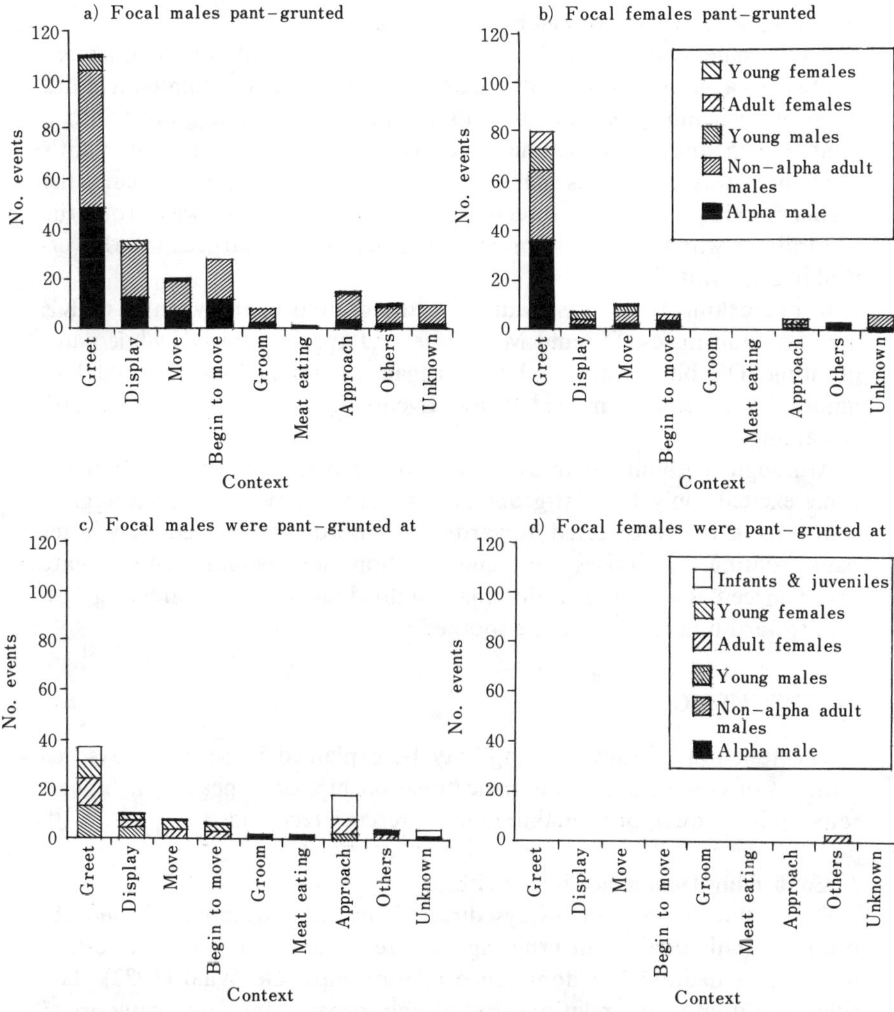

Fig. 9.6 Frequencies of pant-grunt events for each age/sex class of partners in each situation. Age/sex classes follow Fig. 9.2. Contexts correspond to those listed for Table 9.6 as follows. Greet: (a) –1, display: (a) –2 and (b) –1, move: (b) –2, before moving: (b) –3, groom: (b) – 4, meat eating: (b) – 5, approach: (b) – 6, others: (a) 3 and (b) –7.

results suggests that focal males paid more attention to the non-alpha adult males than did focal females.

Focal males received pant-grunts more frequently in non-greeting than in greeting contexts (36 events in greeting, and 49 in non-greeting, Fig. 9.6c). Focal females rarely received pant-grunts in any context (Fig. 9.6d)

6. Pant-grunt Events Followed by Aggression

Adult males who were the recipients of pant-grunts often appeared excited, swaggering about with their hair erect. As the dominant adult male became more excited, the subordinate's pant-grunt graded into a pant-scream and he sometimes fled while screaming. The dominant's excitement was sometimes manifested as a charging display. Usually the higher-ranking individual directed his display, moving away from the subordinate who had pant-grunted, but sometimes he charged at the lower-ranking individual.

In one example, AJ was feeding on the ground in the vicinity of KZ and the adult males, DE and MU. When AJ approached DE while pant-grunting, DE bit his arm. AJ then began screaming loudly, while DE, ignoring his scream, moved away. Eventually, AJ followed DE, still screaming.

Although a dominant individual who received pant-grunts often became excited, only 11 pant-grunt events were observed in which aggression was actually directed toward the subordinate immediately after pant-grunting. This raises a difficult question: why would a subordinate chimpanzee pant-grunt at a dominant individual who may direct aggression toward him rather than be soothed?

IV. DISCUSSION

The function of pant-grunting may be explained in several ways: as a means of confirming a dominance relationship, of appeasing a higher-ranking individual, or of making one's presence recognized.

1. Confirming Dominance Relationships

Pant-grunts are almost always directed toward a dominant by a subordinate individual. Pant-grunting is, thus, deeply involved in establishing and maintaining dominance relationships. De Waal (1982) classified the dominance relationships of chimpanzees into two categories: formal and real dominance. Real dominance is based on aggressive encounter, the outcome of which is often variable, depending on the context. For example, the presence or absence of coalition partners may alter the outcome of an aggressive dominance-related encounter. Formal dominance, on the other hand, is rigid; that is, the outcome of an encounter is completely predictable. Formal dominance is manifested by the pant-grunt; in an encounter between a dominant and a subordinate individual, the pant-grunting of the subordinate is "context free" (de Waal, 1987). Therefore, it may be said that chimpanzees confirm dominance relationships in an almost ritualized manner through the use of pant-grunts.

On the other hand, it is likely that pant-grunts perform other functions rather than confirming dominance. In this study, most pant-grunts were directed toward adult males, especially the alpha male, by females or young males. Thus, pant-grunts occurred between pairs whose dominance relationships were already apparent (Hayaki, Huffman and Nishida, 1989). Since their dominance relationships are already obvious, it is unlikely that they must be confirmed by pant-grunting This suggests that pant-grunting, although it has the effect of confirming dominance, may serve some other functions.

It is important to note that adult males usually do not pant-grunt at other adult males with the exception of the alpha male. Because pant-grunting confirms dominance, restricting this vocalization to the clearly dominant alpha male may allow the non-alpha males to keep their dominance relationships ambiguous (Hayaki, Huffman, and Nishida, 1989).

2. Appeasement

Pant-grunts often occurred in a moment of tension or excitement, sometimes as a response to an aggressive display. In such cases, pant-grunting seems to act as an appeasement behavior; that is, the subordinate individual pant-grunts in order to restrain or deflect the aggression of the dominant (Goodall, 1968; Nishida, 1970; Kawanaka, 1989). Since chimpanzees may be able to anticipate aggression by a more dominant individual, pant-grunting may allow a subordinate to appease a higher-ranking individual and deter aggression before it actually takes place.

Many pant-grunts occurred during encounters. Bygott (1979) reports that attacks and displays frequently take place in the context of a reunion, that is, between two individuals who meet after having been out of each other's sight and proximity for at least half an hour. The subordinate individual may, therefore, pant-grunt on encounter because he/she anticipates aggression, regardless of whether the dominant has given aggressive signals. Pant-grunting thus acts as a type of "anticipatory appeasement."

This hypothesis also explains why subordinates suddenly pant-grunt at a dominant individual who is in the vicinity. Lower-ranking individuals might pant-grunt at a dominant in anticipation of aggression. Subordinate individuals, especially the adolescent males, sometimes look hypersensitive when in the vicinity of dominant adult males. In such situations, the subordinate may occasionally misinterpret even a slight movement of the dominant individual as a sign of imminent aggression (Hayaki, 1985a; Goodall, 1986).

The appeasement hypothesis explains why most pant-grunts are directed at the alpha male, as he is the most aggressive individual. Because adult males are more aggressive than adult females, more appeasement pant-grunts would be directed toward them. These suggestions agree with earlier conclusions of Nishida (1970) and Bygott (1974).

On the other hand, in many "ritualized" pant-grunt events, the dominant individual looked excited, often making himself bigger by standing up with hair erect, and sometimes performing an aggressive display (Bygott, 1974; Noë, de Waal, and van Hooff, 1980; de Waal, 1982). Furthermore, in this study, a dominant individual sometimes attacked a subordinate who was pant-grunting. This indicates that pant-grunting does not always succeed as an appeasement or in deterring aggression. In fact, the subordinate individual would not have been attacked if he/she had not pant-grunted.

3. Individuals Making Their Presence Recognized

Another problem with the appeasement hypothesis is related to the "psychology" of pant-grunting. In a situation where the dominance relationships are clear, the subordinate individual usually behaves quietly in a self-effacing manner. The dominant individual, on the other hand, displays his strength, throwing out his chest, and squaring his shoulders. The co-existence of two individuals of different rank is maintained mainly by the self-restraint of the subordinate (Itani, 1988). If so, pant-grunting is a puzzling behavior because a subordinate individual makes a point of approaching and uttering loud sounds directed toward a higher-ranking individual. The subordinate who is pant-grunting seems to be too conspicuous. Van Hooff (1973), in his analysis of the sequential structure of chimpanzee behavior, labeled pant-grunts and other behaviors contingent on pant-grunts as "excitement". It seems strange that the one who appeases gets excited, especially since this may also excite the dominant individual.

Moreover, "anticipatory appeasement" does not explain why a subordinate who anticipates aggression does not flee. In many encounters, a subordinate began approaching a more dominant individual before the latter had even noticed his presence. Since they could remain at a distance from the dominant individual, the question arises why they felt it necessary to approach and pant-grunt.

De Waal (1982) reported on changes in the pattern of pant-grunting in 6 years during which three adult males (Yeroen, Luit, and Nikkie) exchanged alpha status. When Yeroen and Luit were alpha males, they received pant-grunts more frequently than other males in the group. In contrast, Nikkie did not receive additional pant-grunts when he was the alpha. It seems that the frequency with which an animal receives pant-grunts does not entirely correspond with its dominance rank, but is related more to the individual's influence or social power. This suggests that pant-grunts are directed toward an individual primarily because he is influential, not because he is higher-ranking than the one who is pant-grunting. Usually, a chimpanzee will pant-grunt only at the most dominant individual present, even if there are other individuals who are higher-ranking than he/she is. Why do subordinate individuals not pant-grunt at the other dominant individuals? It may be because, in most situations, only the most dominant individual is crucially influential.

Young chimpanzees are clearly subordinate to all adult males. The presence of adult males often influences and constrains the behavior of younger individuals, especially of young males. For example, young males rarely copulate with females when adult males are in the vicinity (Hayaki, 1985a). Young chimpanzees occasionally have to confirm themselves to be tolerated by adult males. In particular, when the situation is changing, they must confirm that they are still tolerated by the most influential individual in the changed circumstances.

By pant-grunting young chimpanzees check whether the influential figure will allow them to coexist in the situation. They attempt to make their presence recognized by the individual with the most social power at the time.

The chimpanzee unit-group is a fission-fusion society, splitting into temporary subgroups or parties (Nishida, 1968; Sugiyama, 1968; Goodall, 1968). The membership of a party constantly changes within a unit-group. This means that the most influential individual in a party often changes. Pant-grunting as a greeting behavior may have developed as a result of this unique aspect of chimpanzee society. Pant-grunting may thus be similar in origin to human greetings, since human hunter-gatherer societies also have fission-fusion characteristics (Tanaka, 1980).

V. SUMMARY

1. The episodes in which young chimpanzees directed or received pants, pant-grunts, pant-barks, or pant-screams were analyzed.

2. Most pant-grunts were directed toward adult males, especially toward the alpha male. Late adolescent and young adult males sometimes received pant-grunts from adult and adolescent females.

3. Pant-grunting did not act to prolong the social interaction. Both participants typically left immediately after pant-grunting.

4. Although pant-grunting typically occurred as greetings, 39% of pant-grunt events did not occur in encounters. Young males pant-grunted in non-greeting situations more frequently than young females.

5. Although pant-grunting has an effect of confirming dominance, this may not be the main function of the behavior, since young individuals usually pant-grunt at individuals who are obviously dominant to them. The "anticipatory appeasement" hypothesis that pant-grunts function to appease a dominant individual in anticipation of aggression is also insufficient in explaining some important characteristics of pant-grunting. It is hypothesized, in this study, that young chimpanzees pant-grunt in order to make their presence recognized by the crucially influential individual in the situation and check how tolerant this individual is at the time.

Acknowledgements

This study was financially supported by grants under the Monbusho International Scientific Research Program (#58041025, 59043022, 60041020, 61043017 to T. Nishida).

10

Age Differences in Ant-eating by Adult and Adolescent Males

Kenji KAWANAKA

Laboratory of Anthropology, Faculty of Science, Okayama University of Science

I. INTRODUCTION

Characterized as omnivores, chimpanzees have a diversified food repertoire, and do not specialize in any feeding habit (e.g., Itani, 1974; Wrangham, 1975; Teleki, 1981; McGrew, 1983; Nishida and Uehara, 1983). While living largely on fruits and other plant foods, they also feed to a considerable extent on animal matter such as vertebrate meat and insects.

In the Mahale Mountains National Park, Tanzania, chimpanzees have been recorded to prey on about 20 species of vertebrates and at least 25 species of insects (Nishida and Uehara, 1983). Predation on vertebrates and tool-using for obtaining insects have been studied from various standpoints (e.g., Nishida, Uehara, and Nyundo, 1979; Kawanaka, 1982a; Nishida and Hiraiwa, 1982; Uehara, 1982, 1986; Takahata, Hasegawa, and Nishida, 1984; see also McGrew, 1983). However, consumption of insects unaccompanied by tool-using, especially of social and colonial ants, has been rarely studied, even though the bulk of insect prey is collected without tool-using and is considered nutritionally significant (Nishida, 1980c; Nishida and Hiraiwa, 1982; Uehara, 1986).

This paper describes ant consumption unaccompanied by tool-using by male chimpanzees of various age classes. Age variation in this activity is discussed and is compared with differences in social aspects. Lastly, by comparing ant-eating with consumption of other animal foods, this study assesses the importance of ant-eating to the growing male chimpanzees of Mahale.

II. METHODS

1. Procedure

The study was conducted from August 8 to December 5, 1986. The study unit-group was the M Group consisting of about 100 animals (see Chap. 3). Twelve of the 19 males, aged early adolescence to middle-aged, were chosen as the targets of observation, and each was followed for about 25 hours or more using the focal animal sampling method (Altmann, 1974). Thisstudy follows Hiraiwa-Hasegawa, Hasegawa, and Nishida (1984) for age estimation, and uses the terminology for male age classes from Goodall (1983); i.e., infancy (0–5 years), juvenile (5–7 years), early adolescence (8–12 years), late adolescence (13–15 years), young adult (16–20 years), prime (21–26 years), middle-aged (27-*circa* 32 years), and old age (*circa* 32 years to death).

The target's behavior and its interactions with others were recorded, as well as the starting and ending times of each behavioral sequence whenever possible. Proximity relations among males were examined by recording each minute all individuals seen within 10 m of the target. Each of these records is called an observation unit. Age differences in proximity relations and interactions with others have been described elsewhere (Kawanaka, 1989).

This study focuses on 10 out of the 12 targets whose time spent feeding exceeded 20% of the total observation time (Table 10.1). Continuous collection of ants without an interruption longer than one minute was counted as a bout. When a bout contained interruptions shorter than one minute, each unbroken session was called a sub-bout. Bout length was then measured by the total time of sub-bouts, excluding the duration of interruptions.

2. Ant Prey and Techniques Used to Obtain Ants

Mahale's chimpanzees prey on insects representing at least 25 species of 15 genera (Nishida and Hiraiwa, 1982; Uehara, 1982, 1986; Nishida and Uehara, 1983). Of these, two types of ants are consumed in the greatest proportion. One is "smaller ants" consisting mainly of five species of *Crematogaster* which build their nests in dead branches of trees or in dry, hollow grass stems. The other is "larger ants" consisting mainly of four species of *Camponotus* that live in tree-trunks, in branches, or in grass stems. Both types are abundantly and evenly distributed throughout the Mahale chimpanzees' range, regardless of vegetation type.

Chimpanzees have two techniques for obtaining these ants. The first is "ant-eating" (Fig.10.1), in which they break off a dead branch or dry grass stem, crack it using their anterior teeth, and lick eggs, larvae, and pupae, as well as imagos. This technique is used mainly to collect smaller ants. Larger ants are taken by this technique also, but not as intensively as smaller ants (Uehara, 1986). The second technique is "ant-fishing", which involves tool-using and is used to obtain larger ants (in particular soldiers

Table 10.1 Basic data about ant-eating and ant-fishing by male chimpanzees of different age classes obtained at Mahale in August–December, 1986.

Name of target (Birth yr)[1] Age class[2]	A[3] Obser-vation time	B[3] Time spent in eating	C[3] Time spent in ant-eating	D Bout no.	B/A % of time spent in eating	C/B % of time spent in ant-eating	D/B Avg. bout no. per hour	C/D[3] Avg. bout length	E[3] (E/B) Time spent in ant-fishing (%)
MA ('77) E	1461.9	499.9	47.8	24	34.2	9.6	2.9	2.0	20.1 (4.0)
TW ('75?) E	1528.7	575.3	50.5	19	37.3	8.8	2.0	2.7	3.2 (0.6)
AJ ('73?) L	1521.7	511.0	44.4	13	33.6	8.7	1.5	3.4	0.0 (—)
NS ('73) L	1575.3	655.7	38.0	13	41.6	5.8	1.2	2.9	0.0 (—)
BE ('72?) L	1504.5	419.2	13.3	4	27.9	3.2	0.6	3.3	11.9 (2.8)
SU ('70?) Y	1522.9	440.2	12.3	6	28.9	2.8	0.8	2.1	0.0 (—)
KZ ('68?) Y	1812.4	666.8	3.1	3	36.8	0.5	0.3	1.0	0.0 (—)
LJ ('65?) P	1536.2	501.1	5.4	5	32.6	1.1	0.6	1.1	0.0 (—)
DE ('63?) P	1523.8	383.8	0.0	0	25.2	0.0	—	—	0.0 (—)
NT ('55?) M	3556.6	923.0	5.5	6	26.0	0.6	0.4	0.9	0.0 (—)

[1] ?: presumed.

[2] E: early adolescence; L: late adolescence; Y: young adult; P: prime; M: middle-aged.

[3] Minutes.

Fig. 10.1 A middle-aged male cracks a dead branch, and is about to lick out the ants inside. (Photo by K. Kawanaka)

and workers) living in tree-trunks; for details of this technique, see Nishida and Hiraiwa (1982).

During focal sampling, nine of the ten targets were seen to perform ant-eating. However, only three of them, all early or late adolescent males (MA, TW, BE), were observed to perform ant-fishing (Table 10.1). They devoted less time to ant-fishing than to ant-eating. It appears that male chimpanzees perform ant-fishing much less frequently than ant-eating, and that 25 hours of observation time are too little to obtain sufficient data on ant-fishing by males. This study, therefore, deals with ant consumption by ant-eating. The types of ants taken by this technique could not be identified in many cases; hence the following analysis does not discuss separate ant species.

In many cases of ant-eating, chimpanzees first brush the crawling imagos away, and then lick the remaining components of the nest, or stop licking even if many imagos remain. Ant-eating thus appears to be a nutrition-oriented behavior, aimed primarily at taking eggs, larvae, and pupae, rather than imagos (Nishida and Hiraiwa, 1982). Ant nutrients have scarcely been analyzed thus far, but it is likely that chimpanzees obtain amino acids and some trace minerals from the insect tissues, as well as substantial amounts of vitamin A and some members of the B complex from some types of larvae (Bourne, 1971).

III. RESULTS

1. Number of Bouts

In the age classes examined, males perform ant-eating less frequently

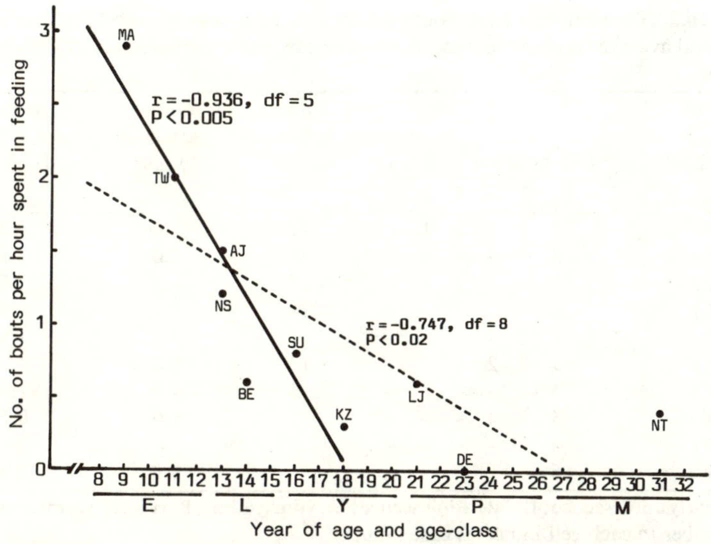

Fig. 10.2 The relationship of number of ant-eating bouts per hour with male's age. Solid line indicates regression line for seven targets between early adolescence and young adult, and dashed line for all ten targets. E = early adolescence, L = late adole scence, Y = young adult, P = prime, and M = middle-aged.

as they grow older (Table 10.1, Fig. 10.2). Thus, bout numbers per hour for all ten targets are inversely correlated with age ($r = -0.747$, df = 8, $P < 0.02$). This inverse correlation is much more highly significant ($r = -0.936$, df = 5, $P < 0.005$) for the seven targets aged early adolescence to young adult. These results indicate that the number of ant-eating bouts per hour by males decreases from early adolescence, reaching its lowest point at young adulthood. This lower level is characteristic of prime and senior age classes as well.

The number of sub-bouts in a bout averaged from one to two with a range of one to six (Table 10.2). Bouts consisting of one or two sub-bouts were most frequently observed for most of the targets. Although the bouts of junior targets appear to contain more sub-bouts, the sub-bout number per bout does not significantly correlate with the target's age (Kendall $\tau = -0.460$, $S = -20$, $P > 0.05$).

2. Percentage of Time Spent in Ant-eating

For all the targets, the proportion of time spent in ant-eating relative to total feeding time correlates inversely with age ($r = -0.810$, df = 8, $P < 0.005$; Table 10.1, Fig. 10.3). This indicates that time spent ant-eating decreases with age. For the seven targets aged early adolescence to young

Table 10.2 Distribution of sub-bout number in a bout, average sub-bout number per bout, and average sub-bout length, in ant-eating by male chimpanzees of different age classes.

Name of target (Age class)[1]	Sub-bout no. in a bout[2]						Total sub-bout no.	Average sub-bout no. per bout	Average sub-bout length[3]
	1	2	3	4	5	6			
MA (E)	15	2	2	3	1	1	48	2.0	1.0
TW (E)	8	6	4	–	–	–	38	2.0	1.3
AJ (L)	11	1	–	1	–	–	17	1.3	2.6
NS (L)	10	–	2	1	–	–	20	1.5	1.9
BE (L)	2	1	1	–	–	–	7	1.8	1.9
SU (Y)	3	–	2	1	–	–	13	2.2	1.0
KZ (Y)	3	–	–	–	–	–	3	1.0	1.0
LJ (P)	3	1	1	–	–	–	8	1.6	0.7
DE (P)	–	–	–	–	–	–	0	—	—
NT (M)	6	–	–	–	–	–	6	1.0	0.9

[1] E: early adolescence; L: late adolescence; Y: young adult; P: prime; M: middle-aged.
[2] Number in each cell indicates case number.
[3] Minutes.

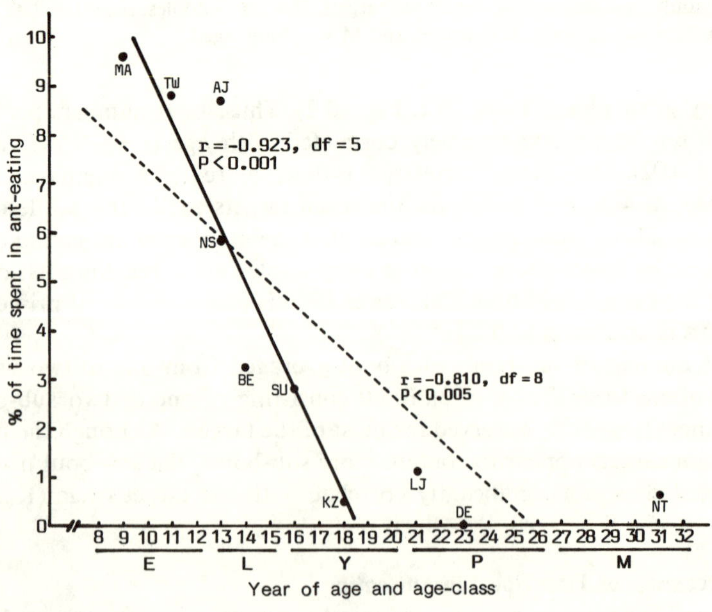

Fig. 10.3 The relationship of percentage of time spent in ant-eating with male's age. Solid line indicates regression line for seven targets between early adolescence and young adult, and dashed line for all ten targets. E = early adolescence, L = late adolescence, Y = young adult, P = prime, and M = middle-aged.

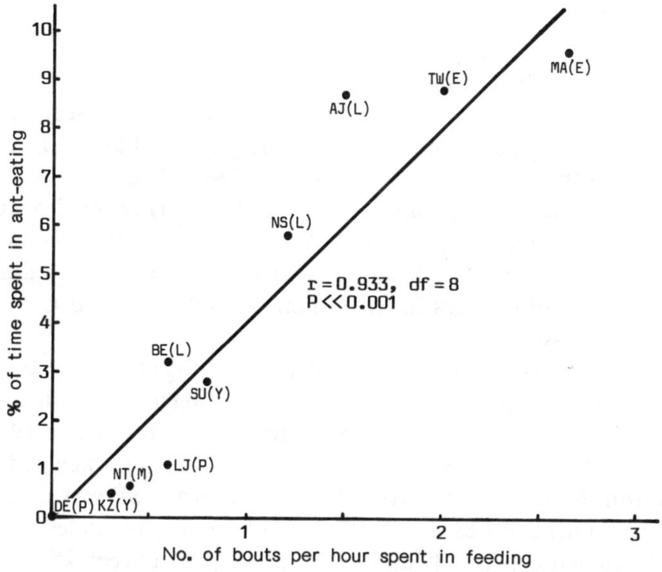

Fig. 10.4 Linear correlation between percentage of time spent ant-eating and number of ant-eating bouts per hour. E = early adolescence, L = late adolescence, Y = young adult, P = prime, and M = middle-aged.

adult, this inverse correlation is more highly significant ($r = -0.923$, $df = 5$, $P < 0.001$). Thus, like the number of bouts per hour, the time spent in ant-eating by males decreases from early adolescence reaching the lowest level at young adulthood. This level is then retained into senior age.

The percentage of time spent in ant-eating is highly significantly correlated with the average number of bouts per hour for all targets ($r = 0.933$, $df = 8$, $P \ll 0.001$; Fig. 10.4). Thus the relative level of ant-eating by males of early adolescent or senior age classes can be assessed by either of these two variables. In this study, levels of ant-eating are evaluated by the percentage of time spent in this activity.

The pattern of age differences in average bout length differs from those of the above two variables (Table 10.1). The maximum average bout length was achieved by late adolescent targets. Bout length in this age class averaged about three minutes, while early adolescent and some young adult targets averaged two to three minutes per bout. The remaining targets spent about one minute or less per bout. The maximum average sub-bout length was also recorded for late adolescent targets (Table 10. 2). These targets spent about two minutes per sub-bout, while the others spent only about one minute.

IV. DISCUSSION

1. Age Differences in Ant Intake

Age differences were noted in the time devoted to ant-eating. Thus, three early to late adolescents (MA, TW, AJ) devoted about 15 times as much time to this activity as prime or middle-aged targets. Three additional late adolescents or young adults (NS, BE, SU) devoted 5–10 times as much time as seniors (Table 10.1). Although age differences in intake efficiency are unknown, these differences in time spent ant-eating may roughly indicate differences in the volume of ants obtained by males of different age classes.

Useful data are presented by Uehara and Nishida (1987) for appraising the volume of ants consumed per 1 kg of body weight. They describe the body weights of two male siblings, SB (presumed to be born in 1958) and MS (born in 1966), of the K Group of Mahale. They were weighed in four seasons from November 1977 to October 1978, when SB was 19–20 years old (young adult) and MS was 9–10 years of age (early adolescent). SB's weight fluctuated between 39 and 44 kg, and MS's between 25 and 31 kg. SB was thus 1.4 or 1.5 times as heavy as MS in every season. If these weight differences are typical of male chimpanzees of Mahale, age differences in volume of ants per kg body weight can be estimated. Thus, early and some late adolescent males may consume about 20 times as many ants as prime or senior males per kg of body weight, while the other late adolescent and some young adults consume about 7 to 15 times.

Pusey (1978) states that male chimpanzees of the Gombe National Park, Tanzania, show a spurt in weight gain at 110–160 months of age, i.e., from early adolescence to the first half of late adolescence. It is in this life stage that males devote more time to ant-eating than their seniors, and are surmised to consume a much greater volume of ants. Pusey also states that males attain adult weight at about 180 months of age. This occurs at the end of late adolescence, when ant intake by males decreases to almost its lowest level.

2. Relation of Ant-eating with Some Social Aspects

Sex differences in animal food consumption by adult chimpanzees are reported from Gombe: males concentrate on vertebrates and females on social insects (Wrangham, 1975; McGrew, 1979). Such differences are attributed to sex differences in grouping and ranging patterns. Similar sex differences are reported from Mahale (Uehara, 1986), while male/female differences in sociability have been investigated with regard to sex differences in tool use by chimpanzees of the Tai National Park, Ivory Coast (Boesh and Boesh, 1981, 1984). The following discussion examines the role of several of these social factors in explaining age differences in ant-eating by male chimpanzees of Mahale.

1) Ranging Patterns

At Gombe, males are more gregarious and travel over a wider range than females (especially non-cycling females), who tend to remain within limited core areas and to live in solitude (Wrangham, 1975, 1979b). Since termite-fishing and ant-dipping (Gombe chimpanzees' major techniques used to obtain social insects) are performed at prey's nests, which are unevenly distributed, sex differences in predation on social insects may be attributed to these differences in grouping and ranging patterns. Females, for example, can repeatedly and much more frequently visit the known nests than males (McGrew, 1974, 1979; Wrangham, 1975).

At Mahale, it is unknown whether the size of ranges covered by males varies according to their age. However, all the targets, regardless of age, appeared to join parties almost equally in their daily travel (Kawanaka, 1989). This implies that the ranges covered by males of various age classes do not much differ from one another. Given this, and the even distribution of ant prey in Mahale, it is unlikely that the age differences in ant-eating among males are related to their ranging patterns.

2) Proximity Relations with Other Males

Prime or middle-aged targets were seen in the proximity (within 10 m) of at least one male of the same or senior age class for more than 60% of their observation units, while young adult and some late adolescent targets spent less than 40% in such proximity (Kawanaka, 1989). The remaining late or early adolescent targets, however, spent more than 40% of their time in proximity to senior males; some of these individuals spent nearly as much time in proximity as did prime or middle-aged males. Thus, males do not simply change their proximity relations with prime or senior males according to age.

For all targets, the percentage of time spent in ant-eating is not correlated with time spent in the proximity of at least one prime or senior male (Fig. 10.5a). However, when the seven targets aged early adolescence to young adult are analyzed separately, these two variables are significantly correlated ($r = 0.760$, $df = 5$, $P < 0.05$). The percentage of time spent ant-eating for the same seven targets is also significantly (but inversely) correlated with time spent alone; i.e., with observation units when no one else was seen within 10 m of the targets ($r = -0.783$, $df = 5$, $P < 0.05$; Fig. 10.5b). Once again, the correlation disappears when all ten targets are analyzed.

Both ant-eating and their proximity to adult males decline from early

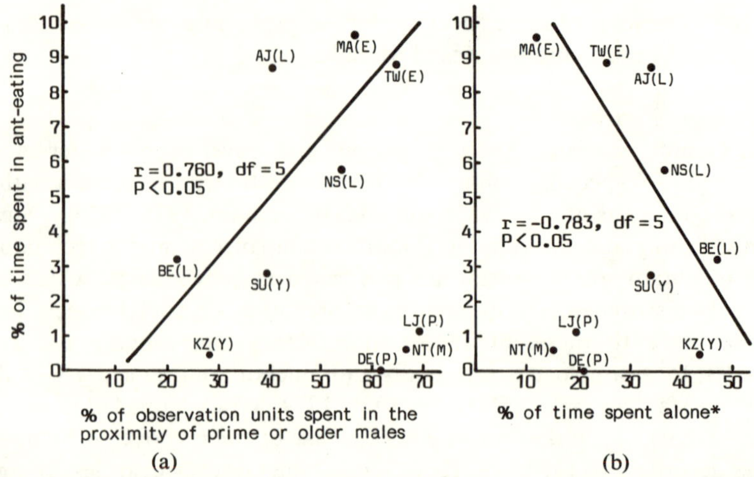

Fig. 10.5 The relationships of percentage of time spent in ant-eating with (a) percentage of observation units spent in the proximity (within 10 m) of prime or senior males, and (b) percentage of time spent alone. Solid lines indicate regression lines for seven targets between early adolescence and young adult. E = early adolescence, L = late adolescence, Y = young adult, P = prime, and M = middle-aged. *: see text.

adolescence, reaching the lowest level at young adult. Prime or senior males increase their intimate association with one another without increasing their level of ant-eating activity.

3) Frequency of Pant-grunt Vocalization

Young adult and senior targets pant-grunt almost exclusively to the alpha male, but rarely to others. By contrast, late or early adolescent targets pant-grunt to their seniors, as well as to the alpha male, but never among themselves (Kawanaka, 1989). Males of young adult or senior age classes thus clearly differ from their juniors in this respect (see also Hayaki, Huffman, and Nishida, 1989).

Frequency of pant-grunting does not appear to relate to percentage of time spent in ant-eating. For example, five targets from early adolescence to young adult (MA, AJ, NS, SU, and KZ) exhibited similar frequencies of pant-grunting [0.57 (NS) - 0.87 (AJ)], but the percentage of time spent ant-eating varied widely from 9.6% (MA) to 0.5% (KZ) (Fig. 10.6). These two variables are not significantly correlated for all ten targets ($r = 0.550$, $df = 8$, $P > 0.05$), or for seven targets from early adolescence to young adult ($r = 0.178$, $df = 5$, $P > 0.1$).

4) Grooming Relations with Other Males

Targets who had grooming relations with other males chose prime or

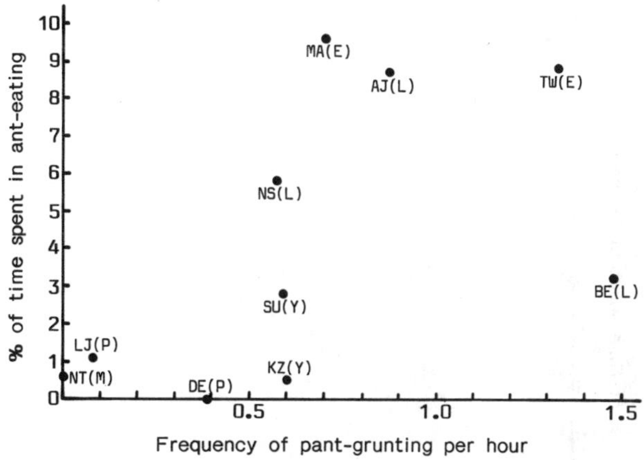

Fig. 10.6 The relationship of percentage of time spent in ant-eating with frequency of pant-grunting per hour. E = early adolescence, L = late adolescence, Y = young adult, P = prime, and M = middle-aged.

senior males as partners almost exclusively (Kawanaka, 1989). Prime or middle-aged targets spent about 15% or more of their observation time in grooming with such partners, while junior targets spent about 6% at the most (Fig. 10.7).

Although the percentage of time spent grooming by young adult or junior targets varies from 0.4% (BE; late adolescence) to 6.3% (SU; young adult), the dots for all the targets in Fig. 10.7 can be regarded as being distributed in the shape of an "L". Males between early adolescence and young adulthood decrease time spent ant-eating, while scarcely changing that spent grooming with their seniors. In contrast, prime or senior males increase time spent grooming, while maintaining the lowest level of ant-eating activity.

In sum, four social variables were examined in relation to ant-eating activity by male chimpanzees; none of the results demonstrated that the more intimately males interact with one another, the less time they devote to ant-eating. For young adult or junior targets, the percentage of time spent ant-eating correlates positively with time spent in the proximity of their seniors, and inversely with time spent alone. However, such males reduce their levels of both sociability and ant-eating activity with age. It seems unlikely that one of the two variables causes the other, but probably both variables decrease in parallel with one another. Thus, the low level of ant-eating activity by prime or senior males cannot be attributed to a high level of sociability.

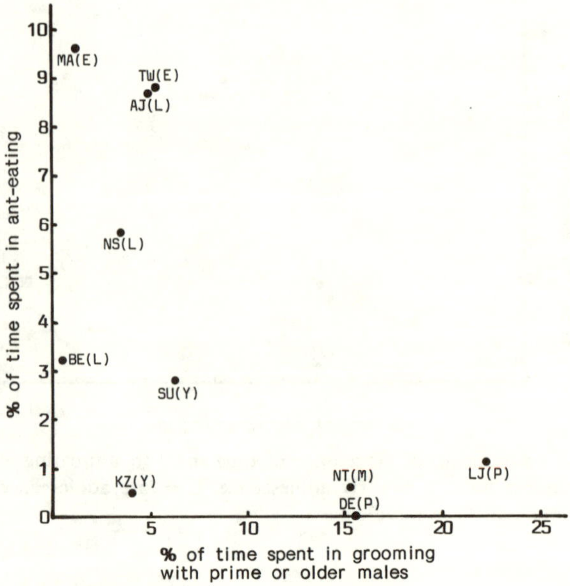

Fig. 10.7 The relationship of percentage of time spent in ant-eating with percentage of time spent in grooming with prime or senior males. E = early adolescence, L = late adolescence, Y = young adult, P = prime, and M = middle-aged.

3. Comparison with Other Animal Foods Feeding Activities

1) Larger Ants

Examining fecal samples from adult chimpanzees of the M and K Groups of Mahale, Uehara (1986) finds sex differences in the intake of the larger *Camponotus* ants (which are obtained mainly by ant-fishing) for the M Group, but not for the K Group. These sex and group differences are due to the fact that only samples from the M Group females contain these ant remains at much larger percentages than those from the remainders, although even samples from the M Group females contain larger ant remains less frequently than smaller ones. Probably both adult males and females at Mahale can fill their requirement for larger ants with a low level of intake. The high level of intake by the M Group females may be better explained by some aspect other than basic nutritional need.

Uehara (1986) suggests that sex differences in sociability may be one cause of differences in larger ant intake between males and females of the M Group. Thus, although both ant-eating (of smaller ants) and ant-fishing (of larger ants) are performed with the aim of preying on ants, probably these two activities differ in their motivations, as well as in nutritional meaning, as is suggested by Nishida and Hiraiwa (1982).

2) Vertebrate Meat

It is well documented that adult male chimpanzees consume more vertebrate meat than do females (e.g., Teleki, 1973; McGrew, 1979; Takahata, Hasegawa, and Nishida, 1984; Uehara, 1986). Two studies examined the differences among males in meat intake. Wrangham (1975) indicates that meat intake significantly correlates with age but is less closely related to rank, while Takahata, Hasegawa, and Nishida (1984) indicate that it significantly correlates with rank. The latter study also shows that young adult or late adolescent males obtain less meat than adult females [Fig. 4 in Takahata, Hasegawa, and Nishida (1984; p. 226)].

Such differences in meat intake by age or rank were seen in a predatory episode observed at Mahale on November 18, 1986 (by K. Hunt, R. Olomi, Moshi, B., Hamisi, B., and K.K.). In this instance, three red colobus monkeys of one troop were captured at one time by the M Group chimpanzees. All the males, from prime to old age, were observed to eat meat while possessing the carceasses, or to obtain some meat from the possessor. Several adult females were also seen to obtain a portion from the possessor. However, no young adult or adolescent males were observed to obtain any meat except for small pieces which they recovered from the ground.

Probably, meat intake by males is kept to a very low level up to young adult, but increases thereafter as individuals advance in years or improve their rank. Thus, the pattern of age differences in meat intake differs from that in ant-eating. The intake of animal matter through ant-eating does not appear to be replaced gradually by meat-eating. It decreases from early adolescence to young adult regardless of vertebrate meat intake. By the time males reach prime age and increase their meat intake, several years have passed since ant-eating activity had decreased and leveled off. The total intake of animal matter by ant- and meat-eating combined does not remain constant from early adolescence onward, and is at a minimum for the young adult age class. Probably, the main animal food items taken differ before and after this minimum; it is thus unlikely that ant- and meat-eating complement each other.

Teleki (1973, 1975, 1981) repeatedly emphasizes that meat-eating by chimpanzees involves not only ecological but also social aspects. Prime males that show an increase in meat intake also begin to associate intimately and symmetrically with other males of the same or senior age classes, and hence can be considered to have attained social maturity (Kawanaka, 1989). This also indicates the importance of the social aspects of meat-eating.

However, this does not necessarily mean that prime or senior males receive an equal share of meat in every predatory episode (e.g., Teleki,

1973; Kawanaka, 1982a; Takahata, Hasegawa, and Nishida, 1984). All such males present at the scene when prey is captured do not always take part in the interactions over the prey. In addition, not a few males fail to obtain a portion even if they do join such interactions. Nevertheless, they get along well nutritionally without meat. With regard to nutrition, vertebrate meat may not be an irreplaceable, but rather an extra, food item even for prime or senior males. At least it may not be an item which every individual needs periodically and at similar intervals, although this does not decrease the significance of meat-eating in other respects.

In contrast, ant-eating involves the collection of prey which is available everywhere in the Mahale chimpanzees' range. It is performed almost every day during the entire year (Nishida, 1980c; Nishida and Hiraiwa, 1982; Uehara, 1982, 1986). Furthermore, as an individual activity which is unaccompanied by social interactions such as simultaneous or collective chasing and sharing of the prey, ant-eating should be less intensely affected by social factors than meat-eating.

4. Importance of Ants to Growing Male Chimpanzees

Teleki (1981) estimates that Gombe's chimpanzees may obtain vertebrate meat at a rate of about 110–180 g per chimpanzee per hour in a predatory episode. Comparing this with an estimated rate of about 70 g of safari ants in ant-dipping (McGrew, 1974), he states that "predation on mammals seems to be a higher-yield activity than collection of safari ants" (p. 328). However, the technique used in ant-dipping is much less efficient than that used in ant-eating. Furthermore, by ant-eating they can consume eggs, larvae, and pupae, which they cannot take by ant-dipping. Nishida and Hiraiwa (1982) estimate that the efficiency of intake by ant-fishing may be still lower than that by ant-dipping. In terms of efficiency, therefore, ant-eating seems superior to both ant-dipping and ant-fishing, and not at all inferior to predation on vertebrates.

If nutrients from animal matter are essential to the physical development of growing male chimpanzees, those supplied through ant-eating must be greatly important to Mahale's individuals. Early or late adolescent males who are rapidly gaining weight obtain animal matter almost exclusively through ant-eating. Growing females of Mahale appear to engage in ant-eating more frequently than adult females, and also more than males of about the same age (Hiraiwa-Hasegawa, 1989; Uehara, unpublished). Thus, smaller ants may also be important to the physical development of the females of Mahale.

Uehara (1986) reports that, for both unit-groups of Mahale, fecal samples from anestrous females tend to contain smaller ant remains at higher percentages than those from males, but that neither sex nor group differences reach statistical significance. Although fecal sample analysis is limited in its ability to quantify the amount of foods consumed, these results suggest that adults of Mahale do not differ as strikingly by either sex or group in smaller ant intake as they do in larger ant and vertebrate meat intake. Although adult chimpanzees need not consume as many smaller ants as growing ones, they may be a significant ingredient in the adult's diet as well.

The possible importance of smaller ants to growing males of Mahale suggested in this study may not be readily applicable to chimpanzees in general. It remains to be clarified whether animal matter comparable to the smaller ants consumed by growing males of Mahale is taken by similar individuals in other populations. Although chimpanzees in general specialize in consuming social and colonial insects, the actual prey and the techniques of predation vary among populations (Nishida, 1980c, 1987b; McGrew, 1983).

At Gombe, for example, although both types of ants on which Mahale's chimpanzees mainly feed also occur, *Crematogaster* ants are taken only rarely, and *Camponotus* ants have never been seen to be taken by individuals of any age-sex class (Goodall, 1968; Wrangham, 1975; McGrew, 1983). The bulk of insects collected by Gombe's chimpanzees consists of weaver and safari ants and termites; safari ants and termites being obtained by tool-using (Goodall, 1968; McGrew, 1974, 1979). Thus, if insect intake by tool-using amounts to a significant proportion of animal matter consumed by growing chimpanzees of Gombe, and if such insects are important for their physical development, then tool use is a critical survival skill. In contrast, Mahale's chimpanzees do not appear to rely so intensively on tools for obtaining insects. Such local variations in the meaning and intensity of tool-using are interesting from an evolutionary perspective.

The nutritional importance of ants to chimpanzees must be re-examined after ant nutrients and their physiological significance are analyzed in the future (see also Nishida and Hiraiwa, 1982; Uehara, 1986).

V. SUMMARY

1. Age differences in the level of ant-eating by male chimpanzees, unaccompanied by tool-using, were studied at the Mahale Mountains National Park, Tanzania, during a 4-month period in 1986.

2. Males markedly reduce ant-eating activity, in parallel with the level of sociability, from early adolescence onward, reaching the lowest level at young adult. While retaining the low level of ant-eating, prime or senior males increase their association with other males of the same or senior

age classes. These males' low ant-eating activity cannot be attributed to their high level of sociability.

3. Ant-eating is compared with other types of animal food consumption. The published data indicate that vertebrate meat intake by males is kept to a very low level up to young adult, but increases after they reach prime age. Therefore, young adult or junior males probably decrease ant-eating regardless of their level of meat-eating. Thus, these two activities do not appear to complement each other.

4. It is suggested that ant nutrients are important for growing males of Mahale. These males obtain animal matter almost exclusively through ant-eating, and consume ants at levels about 20 times greater (per kg of body weight) as do prime or senior males.

Acknowledgements

I would like to make grateful acknowledgements to: K. Hunt of University of Michigan and R. R. Olomi of the University of Dar es Salaam for allowing me to use the data obtained under joint observations; S. Uehara of Sapporo University for the generous offer of his unpublished data, interest in this work, and precious comments; T. Nishida, H. Takasaki, and M.A. Huffman of Kyoto University for helpful comments on earlier drafts of this manuscript; and my wife Hatsuko and our two growing children for all that they have done for me. This study was supprted by grants under the Monbusho International Scientific Research Program (#61041071, 62043047 to T. Kano of Kyoto University).

11

A Preliminary Report on Age Differences in Plant-feeding Behavior among Adult and Adolescent Males

Shigeo UEHARA

Faculty of General Education, Sapporo University

I. INTRODUCTION

Various ecological data pertaining to wild chimpanzees (*Pan troglodytes*) have been accumulated by many researchers at widely separated sites in Africa (e.g., food list: Izawa and Itani, 1966; de Bournoville, 1967; Suzuki, 1969; Wrangham, 1975; Sabater Pi, 1979; Nishida and Uehara, 1983; Sugiyama and Koman, 1987; McGrew, Baldwin, and Tutin, 1988). Wrangham (1975, 1977) studied the relationship of adult male feeding behavior to the environment and to ranging and grouping patterns in Gombe National Park, Tanzania. Also at Gombe, Wrangham and Smuts (1980) analyzed sex differences in patterns of association and dispersion. They compared the amount of time spent feeding by adult males and females in relation to party formation. Few systematic attempts, however, have been made to compare plant-feeding behavior in general among chimpanzees of different age-sex classes.

Different energetic or nutritional requirements may exist between immature and mature chimpanzees (see Chap. 10). Different energetic requirements may also exist among individuals of different body size in the same age-sex category. On the other hand, similar energetic or nutritional requirements may be satisfied by consuming different food items (McGrew, 1979; Boesch and Boesch, 1984; Uehara, 1986). To meet their diverse requirements, chimpanzees may adopt various feeding tactics. This preliminary report examines two tactics, about which little has been studied, in the plant food consumption of male chimpanzees of different ages. The tactics dealt with are rate of feeding and consumption of fallen fruits.

II. METHODS

Observations were made on M Group chimpanzees between Septem-

Table 11.1 Six target male chimpanzees observed between September and December 1987.

Name	Abbreviation of name	Birth year*	Age class	Observation time (sec)
Ntologi**	NT	1955?	Middle-aged	113,868
Musa	MU	1957?	Middle-aged	113,421
Kasangazi	KZ	1968?	Young adult	114,344
Shike	SU	1970?	Young adult	113,567
Aji	AJ	1973?	Late adolescent	113,873
Tos hibo	TB	1976	Early adolescent	113,038

*: Presumed year is indicated by ?
**: The alpha male.

ber 15 and December 27, 1987. At this time, M Group consisted of about 100 chimpanzees (Chap. 3; see also Hiraiwa-Hasegawa, Hasegawa, and Nishida, 1984). For the age estimation of chimpanzees, see Hiraiwa-Hasegawa, Hasegawa, and Nishida (1984), and for the terminology of age classes, see Goodall (1983).

Six males were chosen as targets (Table 11.1), and each target was followed for about 31.5 hours during the study period using the focal animal sampling method (Altman, 1974). An attempt was made to select the least or the second least observed individual as the target for a particular day, and to follow him as long as possible. This could not always be realized, however, as described later. The target's behavior and direct interactions with other chimpanzees were recorded, along with their starting and ending times whenever possible. A "focal record" began at the start of the first activity after the target was located, and ended at the finish of the last activity before he was missed. Additional data on grass or herb stem consumption were collected in December by the *ad lib.* sampling method (Altman, 1974).

The present report deals only with feeding behavior. Any object observed to be ingested by the chimpanzees was defined as food, and a food type was defined as a particular part (e.g., fruit, leaf, pith of stem) of a particular food species (Wrangham, 1977). Continuous collection and consumption (including manipulation or processing) without interruption of the same food type in a settled position (e.g., sitting, lying down, hanging) was counted as a feeding bout. A feeding bout was regarded as ending when the target finished eating, switched food type, or began to walk or brachiate, even if the food was retained in the hand or mouth.

The amount of bait given to the subject chimpanzees was reduced from 1975 onwards, and artificial feeding has been completely abandoned since 1987 (see Chap. 1).

III. RESULTS

1. Time Spent Feeding

At Mahale, continuous observation of a target throughout the daytime was difficult, primarily because of the dense undergrowth in the study area. In addition, focal data could not be collected evenly across the four months (Table 11.2) of the study, even though the chimpanzees formed large subgroups continuously throughout this period. For example, no focal observations could be made between the 11th and 30th of October when most M Group members exclusively utilized the mountainous area in the southern part of their range. Thus, there was not much point in a direct comparison of the activity budgets of the six targets. Neither was there much point in directly comparing the time spent feeding on each food type. In the end, certain feeding behaviors which did not exhibit conspicuous seasonal changes were analyzed in this study.

2. Rate of Feeding on Grass or Herb Stems

Feeding rate is difficult to measure in the forest, mainly because of problems with visibility, even if the target chimpanzee is habituated to the observer. When the animals are feeding on the ground, it is relatively easy to measure the time spent in the consumption of a particular food type, as well as the volume consumed (e.g., Nakagawa, 1989).

The soft pith of the stem of some monocotyledonous herbs or grasses is a common food of the chimpanzees of Mahale (Nishida and Uehara, 1983). During the present study, all the targets were observed to feed on the pith of *Aframomum* spp. (Zingiberaceae) and *Pennisetum purpureum* Schumach (Gramineae) (Table 11.3). These food types are utilized quite regularly throughout the year (Nishida and Uehara, 1983). In addition, a fairly stable amount seems to be consumed from each stem of each species regardless of the season.

In general, the pith is eaten on the spot where it is harvested, but on some very rare occasions, especially by juveniles and infants, it is carried some distance, or even into a tree, before consumption. Although there are at least three species of *Aframomum* eaten by the chimpanzees of Mahale (Nishida and Uehara, 1983), it is convenient in the field to divide them into two categories according to habitat; forest species and woodland species. However, this distinction does not seem necessary as far as feeding rate is concerned (see Fig. 11.2). *P. purpureum* occurs in large quantities in the secondary grassland. Direct feeding competition is negligible for these plants, since both are very abundant.

Usually the pith of the lower parts of mature *Aframomum* or immature

Table 11.2 Observation time and feeding time of the six target chimpanzees between September and December 1987.

Target name		September		October		November		December		Total	
NT	O.T.	32,649		4,790		40,894		35,535		113,868	
	T.S.F.	4,468	(13.7)	300	(6.3)	8,006	(19.6)	5,940	(16.7)	18,714	(16.4)
MU	O.T.	22,941		31,444		33,510		25,526		113,421	
	T.S.F.	4,362	(19.0)	5,402	(17.2)	9,903	(29.6)	4,537	(17.8)	24,204	(21.3)
KZ	O.T.	11,080		49,637		21,832		31,795		114,344	
	T.S.F.	2,219	(20.0)	6,310	(12.7)	6,168	(28.3)	6,360	(20.0)	21,057	(18.4)
SU	O.T.	25,391		10,364		24,599		53,213		113,567	
	T.S.F.	5,640	(22.2)	1,898	(18.3)	3,737	(15.2)	10,814	(20.3)	22,089	(19.5)
AJ	O.T.	28,917		21,791		30,697		32,468		113,873	
	T.S.F.	1,677	(5.8)	5,158	(23.7)	7,440	(24.2)	6,281	(19.3)	20,556	(18.1)
TB	O.T.	17,465		13,944		36,271		45,308		112,988	
	T.S.F.	8,472	(48.5)	3,731	(26.8)	7,331	(20.2)	17,335	(38.3)	36,869	(32.6)
	T.S.F.*	3,935	(22.5)	3,731	(26.8)	7,331	(20.2)	13,171	(29.1)	28,168	(24.9)

O.T.: Observation time (sec).
T.S.F.: Time spent feeding (sec). Figures in parentheses are percentages.
T.S.F.*: Time spent feeding except for *Camponotus* ant-fishing (sec).

Table 11.3 Time spent feeding on the pith of *Aframomum* spp. and *Pennisetum purpureum* by the six target chimpanzees between September and December 1987.

Target name		September		October		November		December		Total	
NT	I	4,468		300		8,006		5,940		18,714	
	II	–		–		1,887	(23.6)	112	(1.9)	1,999	(10.7)
	III	–		–		723	(9.0)	–		723	(3.9)
MU	I	4,362		5,402		9,903		4,537		24,204	
	II	85	(1.9)	–		3,810	(38.5)	1,952	(43.0)	5,847	(24.2)
	III	–		–		2,247	(22.7)	–		2,247	(9.3)
KZ	I	2,219		6,310		6,168		6,360		21,057	
	II	–		–		615	(10.0)	2,802	(44.1)	3,417	(16.2)
	III	–		–		–		901	(14.2)	901	(4.3)
SU	I	5,640		1,898		3,737		10,814		22,089	
	II	–		–		–		43	(0.4)	43	(0.2)
	III	244	(4.3)	–		–		148	(1.4)	392	(1.8)
AJ	I	1,677		5,158		7,440		6,281		20,556	
	II	75	(4.5)	–		245	(3.3)	916	(14.6)	1,236	(6.0)
	III	60	(3.6)	1,570	(30.4)	980	(13.2)	–		2,610	(12.7)
TB	I	8,472		3,731		7,331		17,335		36,869	
	II	–		80	(2.1)	621	(8.5)	2,281	(13.2)	2,982	(8.1)
	III	71	(0.8)	–		341	(4.7)	293	(1.7)	705	(1.9)

I: Total feeding time (sec); II: Time spent feeding on the pith of *Aframomum* spp. (sec); III: Time spent feeding on the pith of *Pennisetum purpureum* (sec). Figures in parentheses are percentages of the total feeding time.

Fig. 11.1 A late adolescent male (AJ) feeding on the pith of *Pennisetum purpureum*. (Photo by S. Uehara)

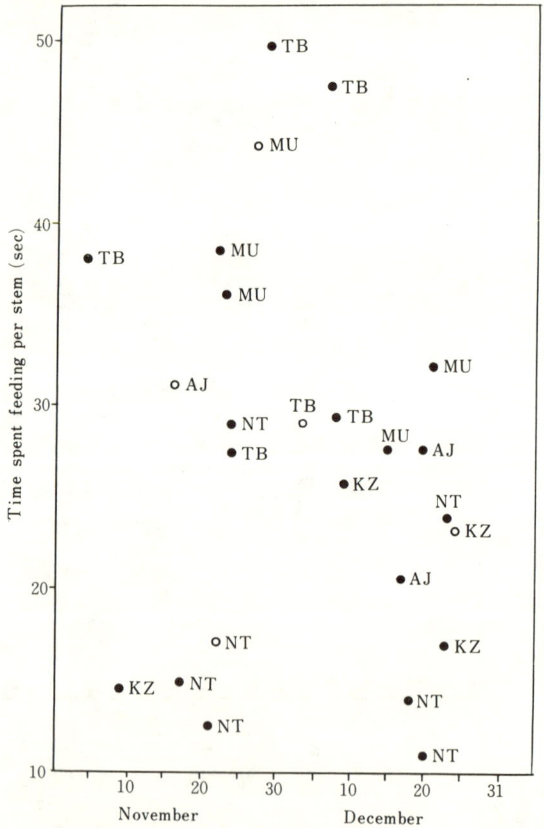

Fig. 11.2 Mean daily feeding rates on *Aframomum* spp. stem. Solid circle: forest species; Open circle: woodland species.

Pennisetum stems was eaten, but occasionally the pith of the distal part of mature *Pennisetum* was collected by holding several stems together and trampling them down to the ground. In either case, the chimpanzee always pulled up one stem, held it horizontally with both hands, and then bit into it with an opposing set of canines. He then pulled his head back or sideways, at the same time jerking his hands away from his body (Fig. 11.1). This combined motion exposed the pith, which was eaten directly from the stem or, on rare occasions, pulled out with one hand to be eaten.

On average, 2.5 *Aframomum* stems were eaten consecutively per bout (N=176 bouts, range: 1–15 stems) and 5.0 *Pennisetum* stems per bout (N=66 bouts, range: 1–42 stems). Since stems were consumed one by one, the feeding rate could be measured, as long as the bout continued, by counting the number of seconds between the pulling up of new stems. When mature *Pennisetum* was eaten, the time spent trampling the stems down to the ground was excluded.

The data on *Aframomum* and *Pennisetum* consumption which were collected between November and December were analyzed in order to minimize the possible effects of plant phenology. This was necessary because the wet season of Mahale in 1987 started in September and settled in completely in October (Appendix 1).

1) Feeding Rate on Aframomum *Stems*

The mean feeding rates on *Aframomum* stems, calculated on a daily

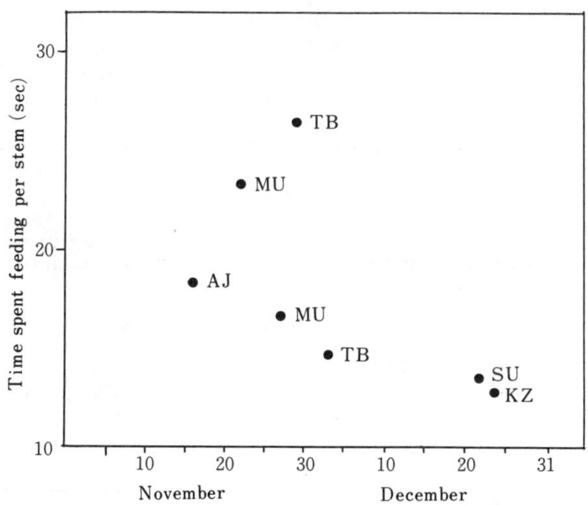

Fig. 11.3 Mean daily feeding rates on *Pennisetum purpureum* stem.

basis for each target, are shown in Fig. 11.2. SU's feeding rate could be measured for only two stems on the same day and was excluded from the analysis. Although the mean daily feeding rates of each target fluctuated from one day to another, no apparent seasonal change could be recognized (Fig. 11.2). Moreover, the mean daily feeding rates of MU and TB always tended to be relatively low, while those of NT and KZ were always relatively high, and those of AJ were always intermediate (Fig. 11.2). Therefore, the data for each target were pooled for the two months and the mean feeding rates calculated from the pooled data were compared statistically (Table 11.4).

Variances of the feeding rates between each target were first compared using an *F*-test. If the two variances were equal, the mean feeding rates of each target were then compared using a *t*-test. If the two variances were unequal they were compared using an Aspin-Welch test. The results are summarized in Tables 11.5a and 11.5b.

As the tables show, the mean feeding rates on *Aframomum* stems are, except for MU, ordered by age; although the differences between NT and KZ and between MU and TB are not statistically significant.

Table 11.4 Feeding rate on the pith of *Aframomum* spp. by the five target chimpanzees between November and December 1987.

Target name	N	Mean*	Range*	Unbiased variance	Standard error
NT	22	18.7	10 – 30	41.94	1.38
MU	139	33.8	9 – 89	210.33	1.23
KZ	150	20.4	7 – 59	68.67	0.68
AJ	46	24.0	10 – 60	99.31	1.47
TB	83	35.3	13 –100	246.15	1.72

*: Time spent feeding per stem (sec).

Table 11.5a Comparison of variances of feeding rates between each of the five target chimpanzees: *Aframomum* spp. stem.

	NT	MU	KZ	AJ	TB
NT	—	unequal**	equal	equal	unequal**
MU		—	unequal**	unequal**	equal
KZ			—	equal	unequal**
AJ				—	unequal**
TB					—

*: $p<0.05$; **: $p<0.01$.

Table 11.5b Comparison of the mean feeding rates between each of the five target chimpanzees: *Aframomum* spp. stem.

	NT	MU	KZ	AJ	TB
NT	—	A-W**	*t*	*t**	A-W**
MU		—	A-W**	A-W**	*t*
KZ			—	*t**	A-W**
AJ				—	A-W**
TB					—

t: *t*-test; A-W: Aspin-Welch test; *: p<0.05; **: p<0.01.

2) *Feeding Rate on* Pennisetum *Stems*

The mean feeding rates on *Pennisetum* stems, calculated on a daily basis for each target, are shown in Fig. 11.3 NT's feeding rate could be measured for only two stems on the same day and was excluded from the analysis. Although the feeding rates of each target on *Pennisetum* were sampled on only one or two days (Fig. 11.3), the same statistical procedures as those used for *Aframomum* stem-eating were employed (Tables 11.6, 11.7a, and 11.7b). The mean feeding rate of AJ was calculated by dividing the total bout length by the total number of stems eaten, and accordingly was unsuitable for the statistical analysis (Table 11.6).

Table 11.6 Feeding rate on the pith of *Pennisetum purpureum* by the five target chimpanzees between November and December 1987.

Target name	N	Mean*	Range*	Unbiased variance	Standard error
MU	66	16.8	9–35	25.43	0.62
KZ	158	12.8	7–31	14.23	0.30
SU	11	13.5	7–22	22.87	1.44
AJ	52	18.3	–	–	–
TB	33	19.2	7–42	68.30	1.44

*: Time spent feeding per stem (sec).

Table 11.7a Comparison on variances of the feeding rates between each of the four target chimpanzees: *Pennisetum purpureum* stem.

	MU	KZ	SU	TB
MU	—	unequal**	equal	unequal**
KZ		—	equal	unequal**
SU			—	equal
TB				—

*: p<0.05; **: p<0.01.

Table 11.7b Comparison of the mean feeding rates between each of the four target chimpanzees: *Pennisetum purpureum* stem.

	MU	KZ	SU	TB
MU	—	A-W**	*t**	A-W
KZ		—	*t*	A-W**
SU			—	*t**
TB				—

t: *t*-test; A-W: Aspin-Welch test; *: $p<0.05$; **: $p<0.01$.

As the tables show, the mean feeding rates on *Pennisetum* stems are also ordered by age, again with the exception of MU. Differences between KZ and SU and between MU and TB are not statistically significant.

Given the data available, the inescapable conclusion is that, except for a middle-aged individual MU, the feeding rate on both *Aframomum* and *Pennisetum* stems is higher in adult than in adolescent males.

3. Consumption of Fallen Fruits

Seven species of fruits which were usually harvested in the trees were eaten after falling to the ground by at least one of the six targets. Of these, five species [*Uapaca nitida* Muell. Arg. (Euphorbiaceae), *Garcinia huillensis* Oliv. (Guttiferae), *Myrianthus holstii* Engl. (Moraceae), *Parkia filicoidea* Oliv. (Leguminosae), and *Landolphia owariensis* P. Beauv. (Apocynaceae)] had relatively short fruiting seasons of only one or two months. Only one or two targets were observed to feed for any considerable length of time on the fruit of each of these species. It was impossible, therefore, to collect sufficient data to compare fallen fruit-eating among the target animals for these species.

In contrast, fruits of the remaining two species were gathered, both in the tree and on the ground, for three or four months consecutively. *Pycnanthus angolensis* (Welw.) Warb. (Myristicaceae: tree) was available between September and November, and *Saba florida* (Benth.) Bullock (Apocynaceae: woody vine) between September and December (Table 11.8). In the former species, the fleshy aril along with the enclosed nut-like seed was swallowed. In the latter, the juicy flesh (sarcocarp) was preferred if the fruit was ripe, but the immature seeds were extracted and eaten if the fruit was unripe.

Table 11.8 shows the time spent in feeding by the target chimpanzees on intact or fallen fruits of *P. angolensis* and *S. florida* in each month. AJ's feeding time on *P. angolensis* fruit amounted only to 425 sec in total, and was excluded from the analysis.

Although this is a very simple comparison, it indicates that among

Table 11.8 Monthly distribution of fallen fruit-eating (*Pycnanthus angolensis* and *Saba florida*) by the six target chimpanzees between September and December 1987.

Target name		September	October	November	December	Total
P. angolensis						
NT	A	1,359	–	3,129	–	4,488
	B	0	–	0	–	0
	B/A (%)	0.0	–	0.0	–	0.0
MU	A	808	2,796	458	–	4,062
	B	32	182	101	–	315
	B/A (%)	4.0	6.5	22.1	–	7.8
KZ	A	–	1,976	1,403	–	3,379
	B	–	0	0	–	0
	B/A (%)	–	0.0	0.0	–	0.0
SU	A	95	902	108	–	1,105
	B	0	0	0	–	0
	B/A (%)	0.0	0.0	0.0	–	0.0
TB	A	293	889	1,943	–	3,125
	B	0	0	36	–	36
	B/A (%)	0.0	0.0	1.9	–	1.2
S. florida						
NT	A	1,593	183	1,120	3,523	6,419
	B	77	183	0	28	288
	B/A (%)	4.8	100.0	0.0	0.8	4.5
MU	A	–	208	1,133	1,323	2,664
	B	–	208	0	74	282
	B/A (%)	–	100.0	0.0	5.6	10.6
KZ	A	1,052	251	965	1,527	3,795
	B	0	0	98	0	98
	B/A (%)	0.0	0.0	10.2	0.0	2.6
SU	A	1,341	990	–	828	3,159
	B	0	0	–	0	0
	B/A (%)	0.0	0.0	–	0.0	0.0
AJ	A	565	–	658	818	2,041
	B	0	–	0	0	0
	B/A (%)	0.0	–	0.0	0.0	0.0
TB	A	1,250	1,022	1,604	1,778	5,654
	B	0	0	0	206	206
	B/A (%)	0.0	0.0	0.0	11.6	3.6

A: Time spent feeding on the fruit in total (sec).
B: Time spent feeding on fallen fruits (sec).

the four adult males MU was the individual who was the most heavily dependent on the fallen fruit of these two species.

IV. DISCUSSION

Both NT and MU were estimated to be middle-aged. However, MU's teeth were considerably worn while NT's were not. The difference was already discernible in 1982 (Hiraiwa-Hasegawa and Hasegawa, unpublished). MU did not show any apparent slowing down of his activity (except for the feeding rate described above) but in social interactions he was already dominated in rank by the two younger males, KZ and SU (Uehara, unpublished). Moreover, he was groomed far more frequently and for a longer time than he groomed others (Uehara, unpublished), which is a trait characteristic of old adult chimpanzees (see Chap. 12). [MU did not show such a tendency in 1981 (see Chap. 7).] It is likely that MU (who is somewhat exceptional in his low feeding rate on *Aframomum* and *Pennisetum* pith and in his strong dependency on fallen fruits) is regarded as an aged chimpanzee, although there still remains a possibility that individual indiosyncrasy is responsible for his anomalous behavior.

During a field study of 14 adult male chimpanzees at Gombe, Wrangham (1975) measured rates of feeding on three species of fruit. He found no evidence of individual differences in feeding rate between two males who were eating the same food at an abundant source. The quality of the fruit (e.g., size, ripeness, etc.), which is not always identical, clearly affects feeding rate, making comparison among individual chimpanzees difficult in some cases. In contrast, the pith of *Aframomum* and *Pennisetum* seems to be relatively homogeneous in quality. These foods, therefore, may be more useful for a comparison of feeding rate.

The series of movements used to expose the pith (see above) were performed quite continuously and were never repeated while a stem was being consumed. In contrast, the time spent chewing before pulling up another stem seemed to be variable. It thus appeared that, as far as the six target chimpanzees were concerned, the difference in rate of feeding on pith was caused mainly by variation in the length of chewing time and not by differences in feeding skill. This conclusion is supported by the difference in variances between *Aframomum*- and *Pennisetum*-eating (Tables 11.4 and 11.6). In the latter species, the chimpanzees always selected the softer parts of the stem as the plant grew taller. This might result in the shortening of chewing time in general.

The effect of social rank is often thought to be the cause for differences in feeding rate (e.g., the toque monkey: Dittus, 1977; the vervet monkey: Whitten, 1983). This does not seem to be the case, however, since feeding competition is negligible for the food types analyzed in the present study.

The mean feeding rates on *Aframomum* and *Pennisetum* stems were higher in adult males than in adolescent males. This is puzzling since there seems to be no reason why an adolescent male that is still growing in weight (Goodall, 1983) must chew this food for a longer time. Differences in bite force which may exist between adult and adolescent males do not seem to be important for consumption of the soft pith. Unfortunately, this problem can not be solved with the current data, but the following speculations may be worth testing in future studies.

Individual adult male chimpanzees at Gombe spent less time in feeding as party (subgroup) size increased. It appears that frequent social interactions, especially among adult males, tend to reduce feeding efficiency (Bygott, 1974; Wrangham, 1975). The present study did not collect data comparing the feeding rate when the target chimpanzees were alone to the rate when they were in a group. Such data might determine if an increase in feeding rate may function to improve feeding efficiency. Further studies are necessary in order to determine the extent to which adult male chimpanzees can really regulate their total food intake by controlling their feeding rate. If this is the case, the difference in feeding rate between adult and adolescent males may also be due to differences in frequency of social interactions, since the latter tend to lead relatively more solitary lives, even if they are travelling with their seniors (Kawanaka, 1989; Uehara, unpublished).

Further studies may indicate changes in the feeding rate as chimpanzees age. The case of MU suggests that the rate decreases in older chimpanzees, and that physical deterioration is the major reason for this decline.

Fallen fruits are not always inferior to intact ones in quality. Chimpanzees prefer fallen fruits of some species whose fruits seem to become more palatable, at least to human observers, after falling from the trees (Nishida and Uehara, 1983). However, for many other species (such as *P. angolensis* and *S. florida*) whose fruits are usually harvested in the trees, the use of fallen fruits may be regarded as a tactic, especially among aged chimpanzees, to save energy, to compensate for physical deterioration, or to avoid the risk of falling. Huffman (Chap. 12) described two old chimpanzees who seemed to have some difficulty in climbing up trees or vines, and who would sometimes wait on the ground while others fed high up in the canopy. It is apparent that consumption of fallen fruits, if they are available, is important to such individuals.

V. SUMMARY

1. Two possible feeding tactics in plant food consumption were studied among six male chimpanzees of different ages. The study took place in the Mahale Mountains National Park, Tanzania, during a four-month period in 1987.

2. Feeding rates on the stems (pith) of the monocotyledonous herb *Aframomum* spp. and the grass *Pennisetum purpureum*, were measured and were higher in adult males than in adolescent males. The difference in feeding rate was caused mainly by variation in the length of chewing time, not by variation in feeding skill.

3. Since there seems to be no biological reason why adolescent males must chew for a longer time, social factors such as the frequency of interactions (which is higher among adult males) may be responsible for the difference. It is likely that increasing the feeding rate functions to improve feeding efficiency, and that adult males can regulate their total food intake to some extent by controlling their feeding rate.

4. The time spent feeding on intact or fallen fruits of *Pycnanthus angolensis* (tree) and *Saba florida* (woody vine) was compared among the target chimpanzees. Only one middle-aged individual, who showed various signs of advanced aging, was constantly dependent on fallen fruits.

5. Although chimpanzees prefer fallen rather than intact fruits of some species, fallen fruit-eating of many other species (including *Pycnanthus* and *Saba*) whose fruits are usually harvested in the trees, may be regarded as a tactic, especially in aged chimpanzees, to save energy, to compensate for physical deterioration, or to avoid the risk of falling.

Acknowledgements

The present study was financed by grants under the Monbusho International Scientific Research Program to T. Nishida (#62041021 and 63043017). T. Nishida, H. Takasaki, K. Takasaki, S. Kobayashi, M. Huffman, and T. Tsukahara cooperated in the field. Rashidi K., Ramadhani N., Hamisi B., and Mohamedi S. assisted in data collection. T. Hasegawa took the trouble to undertake the statistical analysis. M. Hiraiwa-Hasegawa and T. Hasegawa permitted me to use their unpublished data. H. Takasaki, T. Hasegawa, T. Nishida, M. Yuri, and an anonymous reviewer commented on an earlier version of the manuscript. K. Kawanaka provided word processing techniques. The preparation of this paper was financed in part by a 1989 Research Grant from Sapporo University. To these people and intstitutions, I make grateful acknowledgement.

12

Some Socio-behavioral Manifestations of Old Age

Michael A. HUFFMAN

Department of Zoology, Faculty of Science, Kyoto University

I. INTRODUCTION

The study of aging in nonhuman primates has gained momentum over the past several years in an attempt to find animal models which may provide insight into the aging process of humans (Bowden and Jones, 1979). A number of these studies have focused on physiological aspects of aging such as reproductive senescence (e.g., Graham, 1979, 1986 for *Pan troglodytes*; Wolfe and Noyes, 1981 for *Macaca fuscata*), and physical degeneration and its behavioral manifestations (e.g., Turnquist, 1986; DeRousseau, Bito, and Kaufman, 1986 for *M. mulatta*). Other studies have looked at the aging process from the viewpoint of the individual's interactions with other group members (e.g., Hrdy, 1981 for *Presbytis entellus* and *Macaca* spp.; Itoigawa, 1982; Nakamichi, 1984 for *M. fuscata*; Hauser and Tyrrell, 1984 for *M. fuscata* and *M. arctoides*) and even at the effects of aging on group social structure (e.g., Itani, 1980b for primates in general; Yamagiwa, 1987 for *Gorilla gorilla*).

With the exception of research on the reproductive physiology of captive chimpanzees, most of these studies have been done on macaque species. The scarcity of chimpanzee studies is understandable given the fact that only a few wild populations have been studied long enough to reasonably estimate the ages of their most senior members (Gombe: Goodall, 1983, 1986; Mahale: Hiraiwa-Hasegawa, Hasegawa, and Nishida, 1984).

In captivity, chimpanzees have been reported to survive into their 50s with persistent menstrual cycling in females until death (Graham, 1979). At Gombe, the estimated maximum age at death is 33 years for males and 37 years for females (Teleki, Hunt, and Pfifferling, 1976). Cycling ceased in the two oldest females 2–3 years before death (Goodall, 1986). With old age defined as starting at around 33 years at Gombe and 41 years at Mahale, few individuals appear to survive into old age in the wild*(Goodall, 1986; Hiraiwa-Hasegawa, Hasegawa, and Nishida, 1984). Consequently, little information is available on the social and sexual behavior of old chimpanzees in the wild. This paper examines several aspects of their social behavior.

II. METHODS AND MATERIALS

This study was conducted on the M Group chimpanzees (*Pan troglodytes schweinfurthii*). At the time of this study, the group consisted of approximately 100 individuals. The study was divided into two 5 month periods (August-December) in 1985 and 1987. Observations were carried out by focal animal sampling (Martin and Bateson, 1986), following the target individual for as long as possible. During focal observations, the proximity of all visible individuals within 10 m of the focal animal was noted and all interactions directly involving that individual were recorded. The system of age estimation and classification used here follows Hiraiwa-Hasegawa, Hasegawa, and Nishida, 1984; i.e., juvenile (5–7 years),

Table 12.1 Study subjects of M Group in 1985 and 1987.

Name and age class	Parturition status	Estimated year of birth	Focal obs. 1985	Time(min) 1987
Old Adult				
Kalindimya (KL)	N.A.	late 1930s	0	–
Wanaguma (*NG*)	parous	late 1930s	0	923.4
Ganmo (*GA*)	parous	late 1930s	0	0
Wankuma (*UM*)	parous	early 1940s	1820.1	–
Kagimimi (KI)	N.A.	early 1940s	1346.2	–
Lunkuma (*LN*)	parous	mid 1940s	0	0
Mpoko (*MP*)	parous	1945?*	0	0
Eagnes (*EA*)	parous/dependant†	1945?*	0	0
Happy (*HA*)	parous/dependant†	1945?*	0	0
Prime Adult				
Wantendele (*WD*)	parous/dependant†	1961	0	1868.8
Gweklo (*GW*)	nulliparous	1962	1800.6	1788.4
Fatuma (*FT*)	parous/dependant	1963	1799.6	1852.1
Lukaja (LJ)	N.A.	1965	1820.4	1185.1
Young Adult				
Plum (*PL*)	parous/dependant†	1971	1836.4	–
Pulin (*PU*)	nulliparous	1970	1851.6	1808.5
Total			12274.9	9426.3

Female names are italicized, *Classified old adult as of 1987.
Parous with dependent offspring: † lactating.–disappeared from group, N.A., not applicable. Relative age order based on estimates of Hiraiwa-Hasegawa, Hasegawa, and Nishida (1984).

adolescent (males: 8–15 years; females: 8–14 years), young adult (males: 16–20 years; females: 15–20 years), prime adult (21–40 years), and old adult (over 41 years).

As shown in Table 12.1, three old adult (1 male and 2 females) and six young and prime adult individuals (1 male and 5 females) were selected as focal animals. The young and prime adult females were selected for comparison according to their various reproductive states, i.e., nulliparous young adult (PU) and prime adult (GW), multiparous prime adult with dependent young (WD, FT), and primiparous young adult with infant (PL). In each study period, approximately 30 hours of observations were recorded per individual, with the exception of KI and NG who disappeared for several weeks at the end of their respective study periods. Also included in this table are the 6 other old adults observed in M Group during this study. All individuals are listed in order of estimated age.

During the two study periods, a total of nine old adult individuals were observed; six in 1985 and six in 1987, or 6–7% of M Group. Of these, the two males (KL, KI) and one female (UM) had disappeared from the group by the beginning of the 1987 study period.

The following methods and definitions were used in analyzing the data collected.

1. Activity Patterns

The 361.69 observation hours of focal animals were used in the analysis of activity patterns. Due to difficulty in locating the group or target individuals, focal observations were not equally distributed throughout the day. The percentage of total observation time was distributed among six periods as follows: 07:00–10:00, 12.4%; 10:00–12:00, 25.7%; 12:00–14:00, 28.4%; 14:00–16:00, 22.5%; 16:00–18:00, 9.0%; 18:00–20:00, 2.0%.

Focal data for each individual were divided into the following four activity categories: *Forage*—Feeding in a fixed location, or moving between feeding spots in a tree or on the ground, where movement without feeding is less than 60 sec. *Rest*—Sitting or laying down quietly and/or engaging in brief social interactions (e.g., play, appeasement, aggression etc.). *Groom*—Grooming activity usually occurs during resting periods, but is here calculated separately (see below for definitions). *Travel*—Walking, taking few short (≤ 60 seconds) rests, including movement between locations of other activities such as foraging, resting, or grooming.

2. Attendance Records

For each day of observation, those chimpanzees which were present were recorded on an "attendance list" (see Hayaki, 1988). For females, it was also noted whether or not they were in estrus or were carrying an infant.

Attendance lists compiled by researchers over six 6-month periods (August–January) between 1981 and 1987 were used.

From these records, the Relative Attendance Index (RAI) was calculated (Hayaki, 1988). This index is a measure of individual attendance in the core sub-group of adult males and is calculated as follows:

$$\text{RAI} = \frac{\text{No. of attendance days of each individual}}{\text{Average No. of attendance days of adult males}}$$

The average number of attendance days for the adult males in M Group (for August–January) was 123.25 in 1981–82, 108.88 in 1983–84, 47.25 in 1984–85, 92.71 in 1985–86, 48.71 in 1986–87, and 92.43 in 1987–88. The averages for 1981–82 and 1983–84 correspond with Hayaki (1988) for adult males KI, BA, NT, LU, MU, DE, LJ, and SM. From 1984–85, KI's frequency was omitted from the calculation due to his declining attendance, and in 1987–88 KZ was added.

3. Proximity and Association

The amount of time other individuals spent within 10 m of the focal individual was recorded. From these records, the Proximity Index (PRI) was calculated (Hayaki, 1988). The PRI shows the extent to which A remains in proximity to B, when B was recorded the same day on the attendance list:

$$\text{PRI (A, B)} = \frac{\text{PR(A, B)}}{\text{T(A, B)}} \times 100$$

PRI (A,B) is the proximity between A and B, PR (A,B) is the total duration of proximity between A and B when A was the focal animal, and T (A,B) is the total duration of focal observation time for A on days when both A and B were listed on the attendance list. Pairs for which T (A,B) was less than 600 min were excluded from the analysis (after Hayaki, 1988).

From PRI, the Association Index (ASI) was calculated. This index was developed for use when focal observation time is too short to use percentages (Hayaki, 1988), and is a measure of the proportion of days two individuals were together irrespective of the time spent together.

$$\text{CAI(A,B)} = \frac{\text{No. of days when both A and B were on attendance list}}{\text{No. of days when at least A or B were on attendance list}}$$

CAI (A,B) is the Co-attendance Index.

$$\text{ASI (A,B)} = \text{CAI (A,B)} \times \text{PRI (A,B)}$$

4. Grooming

All grooming involving focal individuals was recorded with details of duration and direction, i.e., groomer, groomee, mutual grooming. A *grooming sequence* is defined as the period of grooming between two individuals during which there were rests of no more than 60 seconds. A *grooming bout* is the period in which direction of grooming does not change. For example, when A stops grooming B and B begins to groom A, or when they both start simultaneously to groom each other (*mutual grooming*), one bout has ended and another has started. In order to evaluate the overall tendency for reciprocity in an individual's grooming behavior, the sum of all grooming (time, bout number) for each focal individual was categorized by direction of grooming and calculated using Nishida's (1988) *reciprocity index* (RI for total time and number of bouts):

$$\text{RI} = \frac{\text{A} - \text{B}}{\text{A} + \text{B} + 2\text{C}}$$

where A is the total time or number of bouts in which the focal individual was the groomer, B the total duration or number of bouts in which it was the groomee, and C the total duration or number of events in which mutual grooming was observed. According to this formula, a score of +1 means one-sided grooming by the focal individual, −1 means one-sided grooming by the grooming partners of the focal individual, and 0 means perfectly reciprocal grooming. Nishida (1988) defines grooming as virtually reciprocal when the score falls between −0.2 and +0.2.

5. Notes on Statistical Analysis

Because the exact year of birth was not known for old adults, when calculating correlations (r_s) involving age (Table 12.3), a relative age rank order was assigned and individuals of similar age were each assigned an average of their ranks. Non-parametric statistics were used in the analysis of activity and social interaction patterns (Wilcoxon 2-sample test: normal approximation with continuity correction of 5; Table 12.2, Fig. 12.2).

III. RESULTS

1. Patterns of Activity and Association

1) Activity

In general, a sub-group tended to alternate between traveling, social-rest and foraging activities throughout the day. However, a peak in social-resting was reached between 12:00 and 14:00. There were two peaks of foraging activity, a lower one in late morning and a higher one in early evening up to the time the animals moved into their sleeping nests. This general activity pattern is similar to that previously reported at Mahale for K Group, and at Gombe and Kibale (Nishida, 1974; Ghiglieri, 1984; Goodall, 1986). No statistically significant differences were found between old adult and young or prime adult focal individuals.

Table 12.2 shows the distribution of the time as a percentage of total observation time spent in the above mentioned activities by focal individuals during the two study periods. Among the four individuals observed during both years (LJ, FT, PU, and GW), some scores did not greatly change (±1–9%) while others did (±15–18%). In particular, the total time spent in each activity by PU and LJ changed over the two years. The drastic drop in time spent grooming by LJ in 1987 can largely be attributed to the fact that he was recovering from an illness contracted in 1986 which was accompanied by a dramatic drop in rank (H. Takasaki and K. Hunt, personal communications; Huffman, personal observations). It is interesting to note that LJ's grooming score in 1987 is comparable to that

Table 12.2 Distribution of time (%) spent in various activities by young, prime, and old adult males and females.

Age class	Name	Forage	Travel	Groom	Rest	Total time (min)
Males						
1985						
oa	KI	.22	.16	.21	.40	1346.2
pa	LJ	.31	.15	.24	.30	1820.4
1987						
pa	LJ	.39	.24	.06	.31	1185.1
Females						
1985						
oa	*UM*	.35	.19	.03	.42	1820.1
pa	*FT* I.	.28	.26	.12	.34	1799.6
ya	*PL* I.+	.27	.14	.15	.44	1836.4
pa	*GW* II.	.23	.20	.12	.45	1800.6
ya	*PU* II.	.27	.20	.11	.42	1851.6
1987						
oa	*NG*	.21	.21	.07	.51	923.4
pa	*WD* I.+	.36	.20	.12	.32	1868.8
pa	*FT* I.−	.23	.22	.15	.40	1852.1
pa	*GW* II.	.30	.18	.08	.44	1788.4
ya	*PU* II.	.44	.19	.10	.27	1808.5

Female names are italicized, oa: old adult, pa: prime adult, ya: young adult I: with independent young, +: lactating, −: pregnant, II: nulliparous.

of the old females NG and UM. The variation observed in PU over the two years with regard to foraging and resting may have been influenced by the fact that she frequently foraged alone and thus had fewer opportunities for social behavior in 1987.

There were no statistically significant differences in the proportion of time spent in each activity in relation to either age or sex. This was true for each year individually and for both years combined. However, in 1987 among all focal individuals, NG spent the most time resting. In 1985, prior to LJ's illness, males spent a greater percentage of their total activity time in grooming than did females. These differences, however, did not reach statistical significance. For both years combined, young and prime adult females spent a significantly greater proportion of their time grooming than did old adult females (Wilcoxon, $p < 0.01$, two-tailed test).

While few significant differences were recognized between old adult and adult males and females, it was obvious that old adults, KL, NG, UM, and KI, moved more slowly than others. KL, NG, and UM especially seemed to have a harder time climbing up trees or vines, and moved slowly and cautiously in the trees. They would sometimes wait on the ground to rest while others fed high up in the canopy. However, like other focal individuals, their feeding activity was related to the location of the food source; there was no greater tendency for older animals, for example, to forage for fallen canopy fruits on the ground.

Interestingly, when NG and UM were feeding in the trees, they tended to move away from others in the same tree or into a nearby tree to feed alone. Male brachiation-displaying through the trees is dangerous, and NG and UM may have moved in order to avoid falling or to avoid competition over sometimes limited feeding space. However, old adults were not always at a disadvantage in competing for food resources. During meat eating episodes, for example, KL, KI, NG, and UM were usually allowed access to parts of the carcass, while the attempts of many others were briskly rebuffed.

NG, on two occasions, and UM, on one, were observed moving ahead of the sub-group while foraging alone. When the sub-group stopped to rest or feed, the old females continued ahead for 100 m or more before stopping to forage. They continually paused to listen to pant-hoots (which are an indicator of group movement) and adjusted their movement accordingly. This may have been done to insure that they did not stray from the group while allowing for adequate foraging time.

2) Attendance

Table 12.3 gives the Relative Attendance Index (RAI) scores for 14

Table 12.3 Relative Attendance Indices (RAI).

Age class	Name	1981–82	1983–84	1984–85	1985–86	1986–87**	1987–88
oa	*KL*	.43	.17	.48	.42	.21	–
oa	*NG*	.63	.91	.85	.94	.57	.44
oa	*GA*	.26 b*)	.23 c)	.25	.15	0	.05
oa	*UM*	.58	.97	.66	.83	.23	–
oa	*KI*	.91	.72	.91	.47	.48	–
oa	*LN*	.39	.19	.06 a)	.46 a)	.18 a)	.24
oa	*MP*	.17	.16	.19	.14	0	.16
pa	*EA*	.01	.01 c)	.06 b*)	.15 b)	.18 b)	.10 b)
pa	*HA*	0	.06 b)	.11 b)	.21 b)	.21 b)	.68
pa	*WD*	0	.45	.82 b)	.33 c)	.96	.91 b*)
pa	*GW*	.70	.69	.78	.81	1.02	.68
pa	*FT*	.74 b*)	.53 b)	.82	.89	.98 b*)	.53 c)
pa	*LJ*	1.05	1.05	1.15	1.11	1.08	.77
ya	*PL*	.88	1.05	1.02 b*)	1.01 b)	.96 b)	–
ya	*PU*	.66	.57	.89	1.19	1.33 a)	.83 a)
						#	#

oa: old adult, pa: prime adult, ya: young adult (PL, young adult 1983–, PU young adult 1984–), a) isolation: pregnancy, miscarriage, non-cyclic period, b) birth*, lactation, c) death of infant, **data from only four months, –: disappears from group, #: significant correlation (r_s) at $p < 0.01$.

individuals over a 7 year period. They are listed in descending order of estimated age. For females only, a significant negative correlation was found between relative age and RAI scores during 1986–87 and 1987–88 ($r_s = -.76$, $n = 12$, $p < 0.01$; $r_s = -.67$, $n = 10$, $p < 0.01$ respectively). However, no such correlations were found for the other periods sampled. Male sample size was too small for independent analysis, but a noticeable difference was observed between the older KL and KI up until 1985–86 at about the time that they both disappeared. This suggests a gradual decline with advancing age. For both males and females, there are other factors which may cause variation in core group attendance, including reproductive status, social relations with certain members of the core group (see below), health, and shyness towards observers.

Between 1986 and 1987, a marked decrease in time spent with the core sub-group for the old adults KL and NG took place despite their strong links to the core-subgroup (see below). The fact that NG exhibited complete cessation of her sexual cycle from 1986 on (see Chap. 2), suggests that this may also be a factor in determining the degree of association with the core sub-group. During the 1987–88 study period in particular, NG was not seen frequently in the group when her son NT was in consort. A combination of KL's advanced age (born *circa* late 1930s) and his shyness towards humans was probably partially responsible for his lower, yet stable, RAI scores.

Among the old adults, fluctuations in high and low scores from year to year may in part be related to short term illnesses (Goodall, 1986) and/or food-abundance related ranging patterns which may make it more difficult for these individuals to keep up with the group during some periods.

A few chance sightings of KL, KI, NG, and UM were made when they were not traveling with the core sub-group. Pairs of individuals (KI–KL, NT–NG, GK–NG, and UM–LA) were observed moving independently of the group, sometimes for two or more days. It appears that these old adults remained in the company of other particular individuals (kin and non-kin) when not moving with the core sub-group, rather than becoming completely solitary.

3) Proximity

Over the two study periods, focal animals were observed to have been in proximity (within 10 m or less) to an average of 45.8 different individuals (infants excluded) (S.D. = 3.6, n = 13). For 77% (459/596) of these pairs, the duration of focal observations on co-attendance days was over 600 minutes. These pairs were used in the following analysis. Forty-two percent (195/459) had PRI scores lower than 5.0 (mean = 15.00 partners per focal individual, S.D. = 5.4, range 6–25) and only 3.7% (17/459) had scores higher than 30.0 (mean = 1.9 partners per focal individual, S.D. = .8, range 0–4). KI had the least number of partners with scores under 5.0, followed by NG and PL. KI also had the greatest number of partners with scores over 30.0; namely NT, LJ, TW, and DE. UM had no partners with scores over 30, and NG had only one, NT. In both years, the highest PRI scores occurred between mother-offspring dyads (FT–PN: 76.2, 94.7; WD–MA: 46.8; NG–NT: 40.3), the alpha male and an old adult male (KI-NT: 65.7), and two ex-K Group females (GW and CH: 45.9, 66.8) who transferred into M Group at about the same time (Nishida, Hiraiwa-Hasegawa, and Takahata, 1985).

Figure 12.1 shows the distribution of the average PRI scores with different age-sex classes for focal individuals in 1985 and 1987. Mother-offspring pairs were excluded from this analysis. In general, during both years, adult males (young, prime, and old combined) tended to be found together with other males, while young and prime adult females tended to have had higher average proximity scores with other females (all age classes combined).

The old adult females UM and NG differed from young and prime adult females by having higher average PRI scores with prime adult males

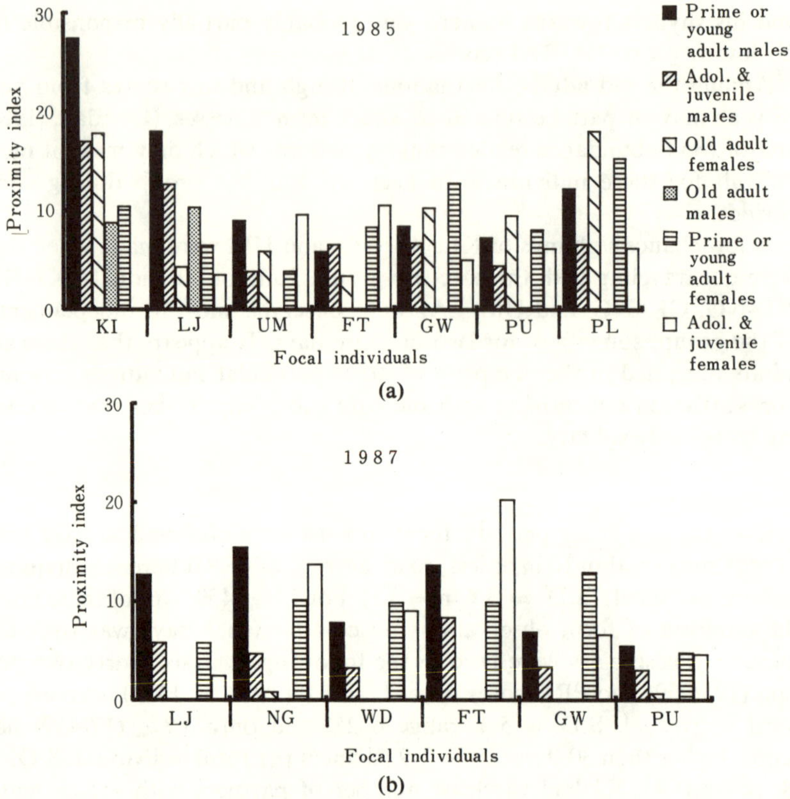

Fig. 12.1 Distribution of the average proximity index score over age-sex class for nine focal individuals in (a) 1985 and (b) 1987. Offspring are excluded.

(excluding proximity between NG and her prime adult son NT). It is interesting to note that NG and UM's average PRI scores with adolescent and juvenile females are also higher than those of young or prime adult females without dependent offspring of their own (PU, GW).

In 1985, KI was found most frequently with prime adult males, followed by old adult females (namely NG and UM), and adolescent males. For LJ, it was adolescent males, followed by old adult males. The above data reflect the tendency for certain old adult males and females to travel with the young and prime adult male core-subgroup (Fig. 12.1a). In 1987 (Fig. 12.1b), LJ's average PRI scores with young and prime adult and adolescent males declined noticeably, reflecting a decline in the overall time he spent with the male core-group.

4) Association

Figure 12.2 shows the ten individuals most frequently found to associate

(ASI) with each focal animal in 1985 a) and 1987 b). The pattern is similar to that recognized above for PRI. Old adult and prime adult males associated with more males, while old adult and adult females associated with more young and prime adult females than with individuals of the opposite sex. ($\chi^2 = 21.68$, df = 1, $p < 0.001$). No significant differences were recognized between old adults and young and prime adults of either sex. However, the highest ASI scores for the females NG and UM were

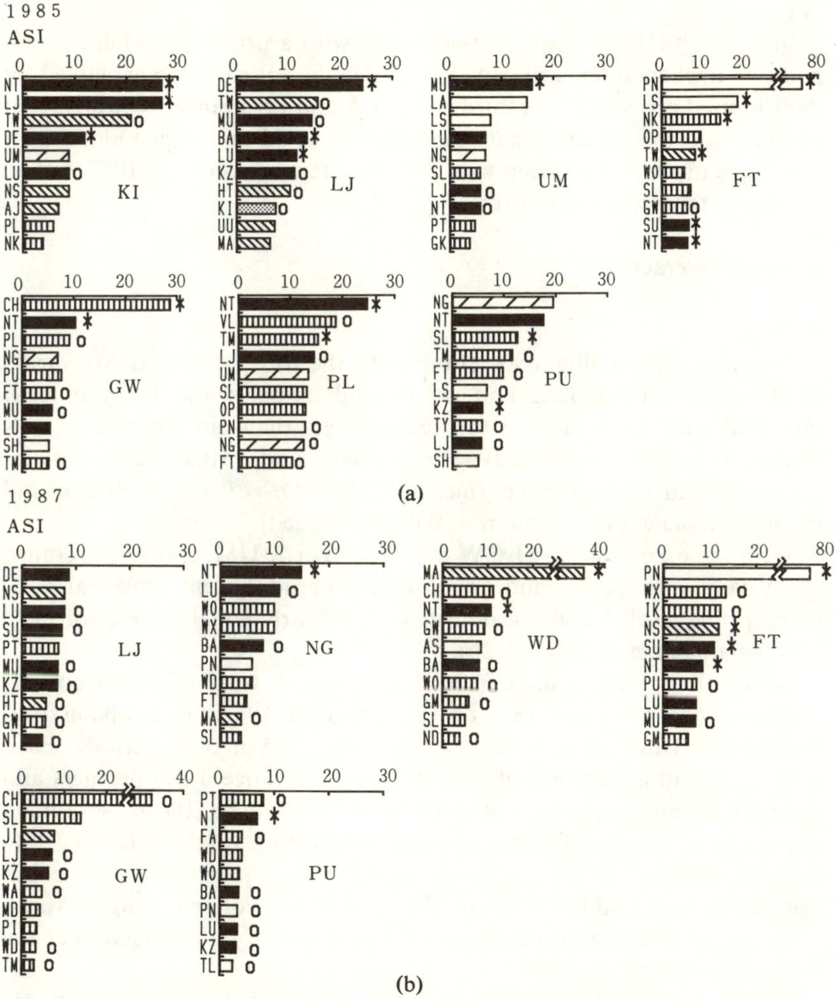

Fig. 12.2 Association index scores (ASI) of the top ten partners for nine focal individuals in (a) 1985 and (b) 1987.
o: grooming sequences observed between pair, * grooming sequences observed between pair on more than three days. Age-sex class labels are the same as in Fig. 12.1.

only half that of the male KI, suggesting a lower level of sociability in old adult females than old adult males.

NG had little social contact with young or prime adult females. Furthermore, it seems likely that the contact she did have with females of these age classes was dependent on her proximity to her prime adult son. Likewise, NG's association with the two young offspring (PN, MA) of FT and WD are probably linked to the mothers' association with NG's son NT.

In contrast, UM's frequent association with a prime adult male (MU) and two orphaned females (LA and LS) could not easily be explained by their mutual proximity to a third party. LA and LS seemed to be attracted to UM, and one or the other sometimes traveled or foraged alone with her. Both orphans were seen with NG on a few occasions in 1987, but to a lesser extent than with UM.

2. Social Interactions

1) Grooming

Patterns of grooming associations with the top partners (ASI) varied in the group. The young and prime adult females and LJ combined, groomed with more of their top ten partners than did old adults (Wilcoxon, $p < 0.01$, two-tailed). Only young and prime adult females groomed with more partners (mean = 7.0, S.D. = 0.7, n = 8) than did old adult females (Wilcoxon, $p < 0.05$, two-tailed).

With the exception of LJ-GW in 1987, KI, LJ, UM, and NG groomed exclusively with young and prime adult, or adolescent male partners, while prime adult females groomed with a nearly equal number of male and female partners (Fig. 12.2b).

Using all grooming data from both years, a total of 512 grooming sequences (3012 minutes) were observed between 196 pairs involving focal individuals (Table 12.4). Nine pairs were observed in both periods. Each individual had an average of 15.0 (S.D. = 5.5, range 6–22) partners and engaged in an average of 40.3 (S.D. = 20.8, range 18–76) grooming sequences. Fewer total hours of focal observation time for KI and NG may be responsible for the fewer number of sequences observed, but appears to have had less effect on the total number of grooming partners. UM, observed for approximately 30 hours, was found to have had the fewest number of partners.

Although the old adult grooming data is limited, several trends are apparent. Table 12.5 shows the distribution of total time (%) spent in grooming by the nine focal individuals according to the age–sex class of the partner. Infants are excluded and mother–offspring grooming is listed

Table 12.4 Reciprocity of grooming behavior.

Age class	Name	Year	No. partners	No. sequences (bouts)	Groomer % of time	Groomee (no. bouts)	Mutual	Total time (min)	R.I. Time	R.I. Bout
oa	*NG*	1987	8	12 (12)	.05 (1)	.95 (11)	0 –	39.9	−.90	−.83
oa	*UM*	1985	6	22 (27)	.48 (11)	.46 (13)	.06 (3)	68.5	−.009	−.06
oa	KI	1985	11	37 (82)	.27 (25)	.58 (42)	.15 (15)	294.5	−.26	−.18
pa	*WD*	1987	18	28 (36)	.35 (12)	.61 (20)	.04 (4)	117.4	−.25	−.20
pa	*GW*	1985	15	37 (88)	.55 (40)	.23 (24)	.22 (24)	187.1	+.27	+.14
		1987	11	20 (44)	.55 (23)	.24 (13)	.19 (8)	151.8	+.27	+.19
pa	*FT*	1985	20	54 (84)	.43 (41)	.47 (32)	.10 (11)	182.8	−.04	+.09
		1987	13	12 (67)	.52 (32)	.23 (21)	.25 (14)	236.9	+.22	+.14
pa	LJ	1985	21	76 (160)	.27 (62)	.28 (54)	.45 (44)	461.0	−.01	+.04
		1987	11	21 (33)	.17 (7)	.51 (17)	.32 (9)	69.7	−.25	−.24
ya	*PL*	1987	22	67 (119)	.39 (45)	.50 (59)	.11 (15)	266.0	−.11	−.10
ya	*PU*	1985	21	54 (73)	.21 (37)	.59 (22)	.20 (14)	188.3	−.32	+.17
		1987	18	40 (62)	.15 (10)	.58 (35)	.27 (17)	176.2	−.34	−.32
	sub-total		195	480 (887)				2440.1		
Mother-offspring pairs										
	NG-NT	1987	1	5 (6)	.09 (1)	.91 (11)	0	23.0	−.82	−.67
	WD-MA	1987	1	13 (19)	.55 (10)	.45 (9)	0	110.0	+.11	+.09
	FT-PN	1985	1	17 (67)	.81 (32)	.23 (21)	.12 (5)	59.5	+.65	+.33
		1987	1	9 (26)	.69 (14)	.23 (7)	.08 (5)	60.9	+.42	+.23
	sub-total		4	44 (118)				253.4		
Total			199	524 (1005)				2693.5		

Female names are italicized. oa: old adult, pa: prime adults, ya: young adult.

* R.I. equals (A−B)/(A+B+2C) where A, B, C refer to bouts or total time with focal individual as groomer (A), groomee (B) and participant in mutual grooming (C).

Table 12.5 Distribution of total time (%) spent grooming during focal observations by age-sex of partner.

Age class	Focal individual	Old Adult Male	Old Adult Female	Adult Male	Adult Female	Adolescent Male	Adolescent Female	Juvenile Male	Juvenile Female	Total time (min.)
1985										
oa	KI			.89	.06	.05				294.5
pa	LJ	.04		.82	.09	.05				461.0
oa	*UM*			.91	.06	.03				68.5
pa	*FT*		.01	.36	.12	.35	.11	.05	(.11)	182.8 ((242.3))
ya	*PL*		.02	.35	.46	.10			.07	266.0
pa	*GW*		.03	.40	.42	.15				187.1
ya	*PU*		.01	.35	.52	.07	.04		.01	188.3
1987										
pa	LJ			.81	.11	.08				69.7
oa	*NG*			.35 (.58)	.35	.13			.17	40.0 ((62.9))
pa	*WD*		.05	.20	.12	.59 (.68)	.03		.01	117.4 ((227.4))
pa	*FT*			.30	.45	.25			(.20)	236.9 ((297.8))
pa	*GW*			.40	.56	.04				151.8
ya	*PU*			.43	.24	.15	.13		.02	176.2

Female names are italicized, oa: old adult, pa: prime adult, ya: young adult, females with dependent young: *WD*, *FT*, *PL*., nulliparous females: *GW*, *PU*. Mother-offspring dyads are calculated separately (% time grooming with offspring)/((total grooming time)).

separately. For KI, the greatest proportion of time was spent grooming with young and prime adult males. This was also the case for the prime adult male LJ in both 1985 and 1987.

Old adult females differed from most of the young and prime adult females in that they, like KI and LJ, engaged in grooming with young and prime adult males more than with young and prime adult females. In contrast, prime adult females generally groomed with other adult (young and prime) females more, or just as much as, they did with adult (young and prime) males. FT and WD were exceptions to this pattern. When grooming with offspring is listed separately, NG is shown to have spent an equal amount of time grooming with young and prime adult males and females. However, it should be noted that the grooming time between NG and her son, NT, amounts to 37% (23/62.9 minutes) of her total time. UM, with no known offspring in the group, spent 95% of her grooming time (irrespective of her sexual state) with prime adult males. This was the case despite her tendency to associate more with other females.

Old adults rarely engaged in mutual grooming. They were observed mutually grooming their partners in significantly fewer sequences (18%, 14/77 sequences, 4 pair) than were prime or young adults (34%, 116/346 sequences, 34 pairs) ($\chi^2 = 6.96$, df $= 1$, $p < 0.01$). UM engaged in mutual grooming in one sequence with NT, and KI in only three sequences with prime adult males (NT, LJ, DE). NG was never observed to engage in mutual grooming; her grooming relationships were strongly one-sided. She had the lowest negative RI scores of all individuals analyzed, excluding any possible effects of grooming with her son (Table 12.4), and was observed as groomer in only 5% of her total grooming activity. Grooming was also highly one-sided between NG and her son NT.

KI had the second lowest RI scores. Like NG, he too was groomed far more frequently, and for a longer time, than he groomed others. This pattern may be common to most aging individuals, but further observations on a larger sample size are needed to confirm it.

In contrast, UM's grooming behavior could be classified as virtually reciprocal. The absence of adult offspring to depend upon and receive attention from may have led her to make a more concerted effort towards associating with others.

In contrast, young and prime adult males and females (excluding data for mother-offspring pairs) tended to be fairly reciprocal with their partners. In eight out of ten cases over the two study periods, the individual's grooming behavior could be classified as virtually reciprocal, i.e. between $+0.2$ and -0.2 with respect to either RI time or bout scores

As for the mother-offspring grooming (Table 12.4), reciprocity increased with age and was reached between WD-MA in late adolescence. However, with increasing age of the mother, reciprocity decreased and the mother tended to receive more grooming from her offspring. Nishida (1988) reported similar findings.

IV. DISCUSSION

The results presented here are in general agreement with previous studies of socio-behavioral manifestations of old age in macaque species. These reported a gradual decline in activity and/or an increasing tendency to withdrawal from social interactions (Itoigawa, 1982; Nakamichi, 1984; Hauser and Tyrrell, 1984).

The behavioral manifestations of old age reported here, especially those of females, took place in individuals in their late 30s and early 40s. This is similar to the report of Hiraiwa-Hasegawa, Hasegawa, and Nishida (1984). The age at which visible behavioral differences occurred varied from individual to individual. Relative age was not an absolute indicator of behavioral or social performance, rather, variations in life history such as the presence or absence of offspring, health, non-kin social relationships and reproductive state played an important role.

In this study, unlike previous studies, few statistically significant differences between adults and old adults were recognized with respect to general activity patterns. These results may have been influenced by the small sample size due to the small number of old adults present in M Group, and their limited observability. Furthermore, it is difficult to eliminate the influence general group activity rhythms may have on those of the individual. Nevertheless, observations of behavioral idiosyncrasies and a decline in grooming activity confirmed social changes concurrent with advanced age. Thus, certain old females spent more time in the core-subgroup interacting with prime adult males compared with other females. This led to a narrowing of their social network to grooming mainly with a few select adult males.

Changes in the behavior of old adult males tended to be more quantitative than qualitative. They remained associated with males in the core-subgroup. Goodall (1986) states that old adult males become increasingly more solitary during the last year of their lives. Despite their declining rank, the two oldest males at Mahale, KL (late 40s, possibly 50) and KI (early 40s), maintained low but stable contact with the core sub-group of adult males until 1986 when they both were last seen.

According to Goodall (1986), old adult females tend to become less gregarious and spend more time with their growing number of offspring and grandchildren. During this study, cases in which little or no known kin remained in the group presented different situations for some old females.

NG's only known surviving offspring was the alpha male, NT, and the majority of her social interactions were with him. In fact, it seems likely that NG's associations with many individuals were largely a result of her proximity to her son, who was a focus of activity within the core-subgroup (Fig. 12.3). Nakamichi (1988) notes similar patterns involving the alpha male in a troop of *Macaca fuscata*. Furthermore, NG had a cataract in one eye, which may have impaired her ability to recognize others quickly, indirectly limiting the number of individuals she chose to interact with. DeRousseau, Bito, and Kaufman (1986) suggest that some age related changes are based largely on such biological effects of aging.

UM had no known offspring remaining in the unit-group after her daughter MM (born *circa* 1971) disappeared sometime in late 1984 or 1985. The majority of her close social interactions were directed to the seventh ranking adult male MU and two orphaned females LA and LS.

Hayaki (1988) also mentions the close association between orphans and old adult and/or prime adult females. In a free-ranging troop of *M. fuscata*, Itoigawa (1982) reported similar associations between orphans and old females. Lacking offspring to associate with, such old adult females may be attracted to young orphans as a source of social interaction, and vice versa. A consequence is that both are reintegrated into the social network, partially filling the gap caused by the absence of kin.

A greater degree of tolerance is shown towards old adult males and

Fig. 12.3 The oldest female in M Group, NG, with WX, an adult female. (Photo by T. Hasegawa)

females by others. Aggression towards old adult females is reportedly low (Nishida, in press) and old adult males have been observed to threaten younger, physically more powerful males without retaliation (Hayaki, Huffman, and Nishida, 1989). This may be explained partly by the expected decline in the ability to actively participate in or influence coalitions. In other words, the behavior of old individuals may be less likely to destabilize such established social relationships. Such restraint on the part of younger adults toward old adults contributes to the maintenance of their peaceful coexistence, promoting what Itani (1988) refers to as conditional equality. A further indication of the unique status of old adults in the group is the one-sidedness of grooming directed towards them.

Further investigation of the possible differences in overall activity patterns is required, with special attention given to differences in traveling and foraging activities. As long-term field studies on various primate species continue, the exact age of older individuals and their life histories will become known. With this information, analysis of the life history factors which may affect longevity (such as feeding habits, health, reproductive success, number of surviving kin, and the age and social standing of offspring) may be possible, thus providing interesting insights into the role played by these individual differences in the aging process (e.g., Goodall, 1983; Small, 1984; Huffman and Seifu, 1989; also see reviews on life history variables: Harvey, Martin, and Clutton-Brock, 1987; Dunbar, 1988).

V. SUMMARY

1. The social behavior of old adult chimpanzees, *Pan troglodytes schweinfurthii* (*circa* 41 years or older) was investigated.

2. A gradual decrease in activity, notably traveling and grooming, was recognized in old adults, but few differences with the exception of grooming were statistically significant when compared with young and prime adult males (16–40 yrs.) and females (15–40 yrs.).

3. Old and prime adult males (20–40 yrs.) were found to associate with significantly more males (all age classes in clusive), while adult females (young, prime, and old) associated with significantly more females (all age classes inclusive).

4. Old adult males and females spent less time grooming than young and prime adult males and females (old adult females significantly less than young and prime adult females) and had fewer grooming partners, most of whom were prime adult males. Unlike young and prime adults, they rarely engaged in mutual grooming (simultaneously bi-directional) and

were groomed much more frequently and for longer periods than they groomed their partners.

5. The social sphere of both old adult males and females decreased with age, this change being more marked in females.

6. While old adult males maintained stable but increasingly lower level contact with the prime adult male core-group, old adult females interacted less with prime adult females and developed additional co-feeding and traveling relationships with female adolescent and juvenile orphans.

Acknowledgements

My appreciation goes to Drs. T. Nishida and D. Hill for their valuable comments on the writing of this manuscript and, in addition, to Drs. M. Hiraiwa-Hasegawa, T. Hasegawa, K. Kawanaka, S. Uehara, and K. Masui for the use of the attendance list recorded by them during their research periods at Mahale. The author also greatly appreciates Dr. T. Furuichi and M. Nakajima's generous help in the statistical analysis. I am extremely grateful to F. Naito for her unyielding encouragement and moral support at all stages of this study. This study was financed by a grant under the Monbusho International Scientific Research Program to T. Nishida (#60041020, 61043017, 62041021, 63043017). During the final stages of completing the manuscript, the author was funded by a postdoctoral fellowship from the Japan Society for the Promotion of Science.

13

Maternal Investment before Weaning

Mariko HIRAIWA-HASEGAWA

Department of Anthropology, Faculty of Science, University of Tokyo

I. INTRODUCTION

A vast amount of research has been conducted on mother-infant relationships and infant behavioral and social development of various species of primates in the wild (e.g., Chevalier-Skolnikoff and Poirier, 1977; Hinde, 1974). Interactions between mother and infant were often thought to be the prototype for relationships among adults, and were thus expected to shed light on the understanding of adult social behavior. In general, the ultimate goal of research was to obtain insights into the biological bases of human social behavior (Hinde, 1974). This goal was particularly emphasized in research on the chimpanzee, which is the closest relative of human beings.

However, Trivers's (1972) theory of parental investment provided studies on mother-infant relationships in primates with a new theoretical framework (Altmann, 1980; Hinde, 1984; Kurland, 1977; Nicolson, 1987). In Trivers's model, analysis of how and why a certain pattern of maternal care evolved requires collection of data regarding the costs of maternal care, offspring survival, and female lifetime reproductive success.

Earlier studies on mother–infant relationships in chimpanzees emphasized that the mother–infant bond was extremely strong, lasting into the offspring's adulthood. These studies have also shown that a great deal of variability exists in the relationship between individual mothers and their infants (Goodall, 1968, 1971). However, quantitative data describing maternal care in relation to female reproductive performance have not yet been published.

This paper attempts to quantify three different aspects of chimpanzee maternal care before weaning: nursing, infant transport, and grooming. The first two aspects were selected because they apparently inflict costs on the mother. The third is a typical primate social activity on which a considerable amount of time is spent during the day. Biologically, grooming functions to remove ectoparasites and dirt from the body surface (Hutchins and Barash, 1976). It also functions socially to maintain a bond between individuals. Because an individual's time for social behavior is limited, the time a mother allocates for grooming her infant is regarded as a form of maternal investment.

The interbirth interval is taken to reflect the costs associated with all forms of maternal care given to one offspring. The amount of the care given to male and female offspring were compared with the data on interbirth intervals analyzed in Chapter 3.

II. METHODS

The subject of this study is the chimpanzees of the M Group and the K Group. During my field work, the population size of the M Group was approximately 100, while that of the K Group was 10 (Hiraiwa–Hasegawa, Hasegawa, and Nishida, 1984). The research was carried out over a total of 30 months between July 1979 and June 1982. Observations on the two groups totalled about 1,200 hours.

I adopted the focal animal sampling method (Altmann, 1974) in collecting data on 16 mother-infant pairs. Thirteen pairs came from the M Group and 3 from the K Group. The ages of the target infants ranged from just after birth to 6 years. The combined total hours of observations using the focal animal sampling method amounted to 668.5 hours. Description and a breakdown of the observational hours on each target infant appears in Table 13.1. During the focal watches, I recorded the behavior of mother and infant into a microcassette recorder and later transcribed the data onto checksheets of one-minute intervals. The durations of nipple contact, feeding, and grooming were measured from the onset to the termination of the behavior.

Three different measurements were taken to describe nipple contact behavior: the mean length of each nipple contact bout; the number of nipple contact bouts per hour; and the mean number of minutes of nipple contact per hour of observation. A nipple contact bout was defined as the period from the time an infant touched the mother's nipple with its mouth until it released the nipple. However, when the infant started nipple contact within 5 minutes after it ceased the previous contact, it was regarded as the same bout and the durations of both nipple contacts were summed.

A travel bout was defined as consecutive locomotive activity, terrestrial or arboreal, of 5 m distance by a mother. In each travel bout, it was noted whether the infant was carried ventrally, dorsally, or travelled independently. The percentage of ventral transport, dorsal transport, and

Table 13.1 Description of the target infants.

Name of infant	Sex of infant	Name and sex of older sib	Period of observation	Total hours of observation	Survival at the end of observation
AB	f	KZ (m), TB (m)	[0–3m]	11.5	
BB	m	PA (f)	[0–5m]	16.5	
EL	f	LS (f)	[1–10m]	29.0	
WS's	m	–	[0–3m]	12.0	died
IK's	f	–	[0–3m]	5.7	died
TL	f	NS (m)	[0–1y 10m]	60.5	
AM	f	–	[0–1y 6m]	63.0	died
PN	f	–	[0–1y 11m]	111.8	
AS	f	–	[8m–3y 2m]	40.2	
RY	m	–	[1y 1m–3y 11m]	29.7	died
SH	f	KR (m)	[1y 7m–3y 11m]	47.7	
MA	m	–	[2y-4y 1m]	49.2	
TB	m	KZ (m)	[2y 10m–5y 4m]	39.6	
PA	f	–	[4y 3m–6y 5m]	24.7	
LS	f	–	[4y 9m–6y 3m]	9.6	
KB	m	–	[4y 9m–5y 11m]	118.3	died

independent travelling for the total number of travel bouts were calculated for each infant.

A chimpanzee can allocate its time for social behavior only after it has fulfilled its own maintenance requirements of feeding and ranging (Dunbar, 1988). Thus, a mother's potential socializing time (PST) was calculated by subtracting her time spent on feeding and ranging from total observational hours.

During my field work, only two habituated females of the M Group gave birth to male infants while six habituated females gave birth to female infants. Thus, data on male infants under 2 years old was limited.

The period from an infant's birth until the birth of its younger sibling was divided into the following age classes. The first two years of life were divided into four 6-month blocks since the most rapid change was expected to occur in this period. The third and fourth years were analyzed as individual one-year blocks, while the period from the beginning of the fifth year until the time of the next infant's birth was treated as one age block.

III. RESULTS

1. Nipple Contact

During the first 4 months of life, the target infants were never observed to eat solid food. They continued to show nipple contact until their

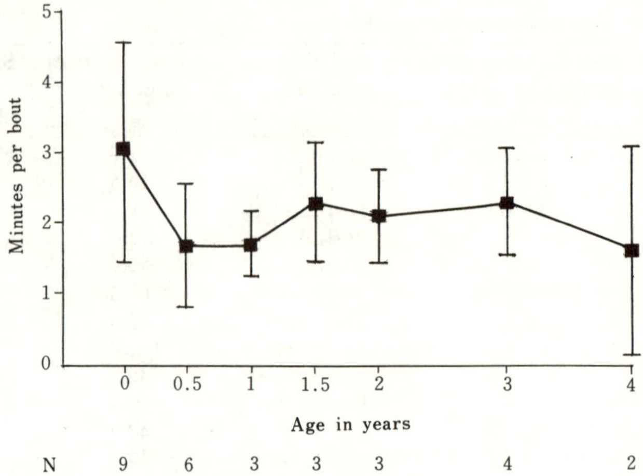

Fig. 13.1 Age changes in mean length of each nipple contact.

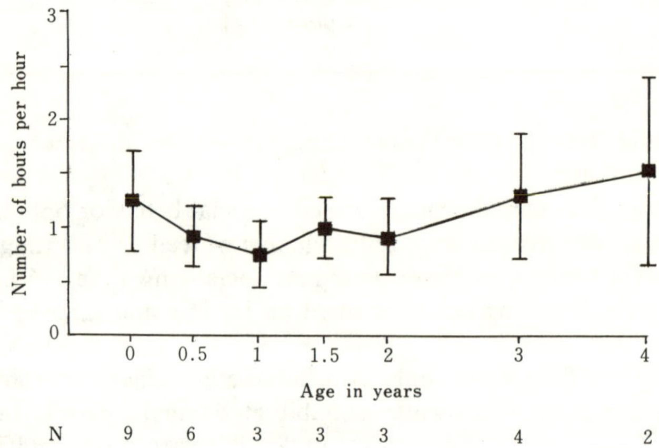

Fig. 13.2 Age changes in number of nipple contact bouts per hour.

fifth year; the oldest infant observed to have nipple contact was a 4-year 11-month old male.

The mean length of a nipple contact bout was calculated for each target infant for each age block. Mean bout length was longer during the first 6 months than in other age blocks, but did not show a regular decrease with infant age (Kruskal-Wallis test, $\chi^2 = 12.9231$, $p > 0.05$, Fig. 13.1)

The mean number of nipple contact bouts per hour of observation also did not show significant change across age blocks (Kruskal-Wallis test, $\chi^2 = 9.83161$, $p > 0.05$), though it tended to be higher in older infants (Fig. 13.2).

The mean number of minutes spent on nipple contact per hour of observation was longest during the first 6 months but there was again no significant difference among age blocks (Kruskal-Wallis test, χ^2 = 7.0239, p > 0.05, Fig. 13.3).

Suckling patterns of male and female infants of the first age block were compared. No significant difference was found in the mean length of nipple contact bouts (Mann-Whitney U-test, N (female) = 6, N (male) = 3, Ucal = 6.5, p > 0.05), and the mean number of bouts per hour of observation (Mann-Whitney U-test, N (female) = 6, N (male) = 3, Ucal = 7, p > 0.05). However, female infants spent more time on nipple contact than male infants (Mann-Whitney U-test, N (female) = 8, N (male) = 6, Ucal = 8, p < 0.05).

2. Feeding Time of Mothers

Time spent feeding is expected to reflect the mother's energetic needs (Altmann, 1980; Silk, 1986). Mothers with smaller infants might be expected to spend more time feeding than mothers with older infants. However, no significant difference was found in percentage of time spent feeding among mothers of daughters of different ages (Kruskal-Wallis test, χ^2 = 1.3450, p > 0.05). This was also the case among mothers of sons of different ages (Kruskal-Wallis test, Hcal = 2.8818, p > 0.05). Target mothers, on average, spent 28 percent of the total observational hours on feeding. When the data on infants older than 2 years were combined, mothers of daughters spent more time feeding than mothers of sons (Mann-Whitney U-test, Ucal = 16, p < 0.05).

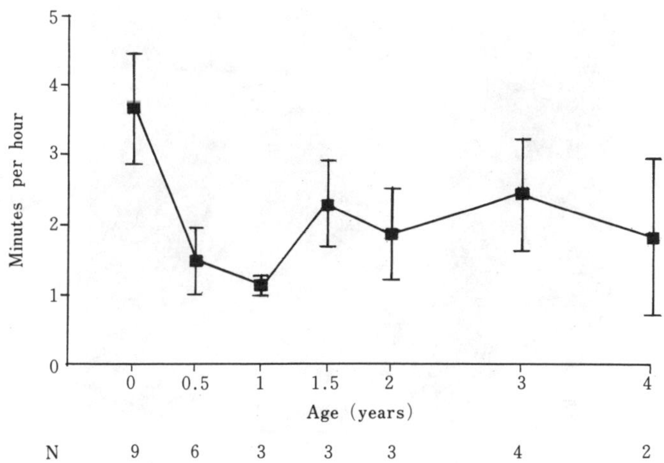

Fig. 13.3 Age changes in mean minutes of nipple contact per hour.

3. Transport of Infants

A chimpanzee infant was carried by its mother from birth until just before its younger sibling was born. When an infant was less than one year old, it was carried by its mother ventrally more often than dorsally (Wilcoxon's $T = 0$, $p < 0.05$). After one year, dorsal carrying became more common, although there was no significant difference in the percentage of the two types of transport during an infant's second and fourth year of life (Table 13.2, Fig. 13.4).

Independent travelling by an infant during long treks was occasionally seen after 6 months, but was not common until about 2 years of age. When travelling long distances, an infant under 2 years of age was carried by its mother more often than it walked independently. At 2 and 3 years

Table 13.2 Comparison of ventro-ventral transport, dorso-ventral transport and independent travelling for infants of different ages.

Age blocks	N	VV or DV	Wilcoxon's T	TR or IT	Wilcoxon's T
0	(7, 7)	VV > DV	$T = 0, p < 0.05$	TR > IT	$T = 0, p < 0.05$
0.5	(5, 5)	VV > DV	$T = 0, p < 0.05$	TR > IT	$T = 0, p < 0.05$
1	(9, 9)	VV = DV	$T = 11$, ns	TR > IT	$T = 0, p < 0.05$
2	(8, 8)	VV < DV	$T = 6, p < 0.05$	TR = IT	$T = 5$, ns
3	(7, 7)	VV = DV	$T = 5$, ns	TR = IT	$T = 6$, ns
4	(6, 6)	VV < DV	$T = 0, p < 0.05$	TR < IT	$T = 2, p < 0.05$

VV: Ventro-Ventral Transport.
DV: Dorso-Ventral Transport.
TR: All Forms of Transport.
IT: Independent Travel.

Fig. 13.4 A 17-months-old female (AM) playing on her mother's back. (Photo by T. Hasegawa)

Table 13.3 Percentage of mother-infant transport during treks for male and female infants of different ages (Mann-Whitney U-test).

Sex	2-year olds		3-year olds		4-year olds	
	No. of individuals	Mean	No. of individuals	Mean	No. of individuals	Mean
F	4	70.9	3	40.5	3	26.4
M	4	63.7	4	50.1	3	19.3
Ucal		4, ns		5, ns		3, ns

of age, percentage of transport and independent travelling were similar, while after 4 years of age, infants travelled independently more often than they were carried (Table 13.2, Fig. 13.5).

There was no significant difference between male and female infants of any age in the amount of transport versus independent travel during long distance treks (Table 13.3).

4. Grooming

An infant was seldom groomed by individuals other than its mother during its first year. The amount of time infants were groomed by their mothers did not vary with infant age (Kruskal-Wallistest, $\chi^2 = 3.1165$, $p > 0.05$). Single infants less than 2 years of age received more grooming from their mothers than infants who had older siblings (Mann-Whitney U-test, Ucal = 1, $p < 0.05$). However, no significant difference was present when the infant was older than 2 years (Mann-Whitney U-test, Ucal = 12.5, $p > 0.05$).

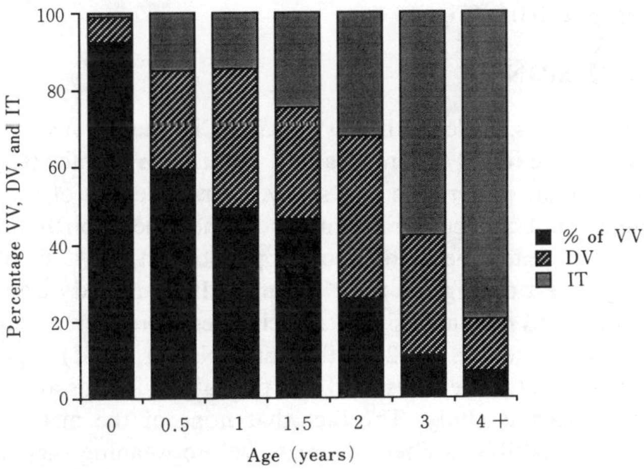

Fig. 13.5 Age changes in percentage of ventral transport, dorsal transport, and independent walking during long distance travel.

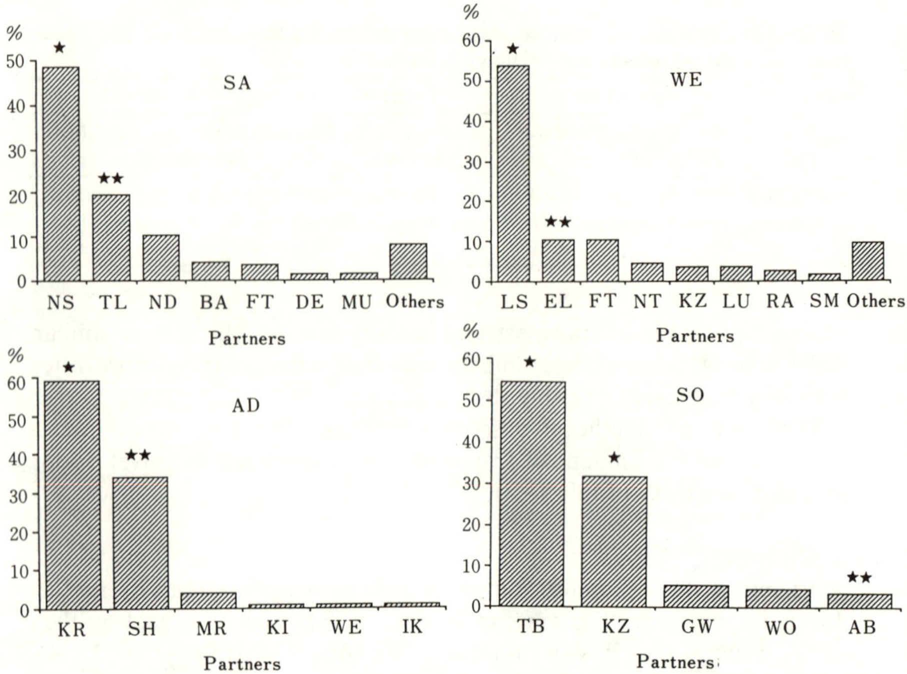

Fig. 13.6 Percent time spent grooming different partners by four mothers with weaned and unweaned offspring. ★ Weaned offspring, ★★ Unweaned offspring.

Mothers always groomed weaned offspring significantly more than they did unweaned offspring (Wilcoxon signed rank test, N = 6 pairs, Tcal = 0, $p < 0.05$) (Fig. 13.6).

IV. DISCUSSION

In chimpanzees, little change is apparent in the amount of maternal care given to the infant before weaning. A study in Gombe (Clark, 1977) showed a similar pattern. In this study, average length of a nipple contact bout did not change significantly from the time of birth until the end of the second year. The amount of nipple contact per hour of observation was greatest during the first 6 months. It decreased during the next 6 months but did not show significant changes thereafter. Captive studies (Horvat and Kraemer, 1982; Miller and Nadler, 1981) report similar results, although the methods used and the subject infants are not directly comparable to this study. The fact that none of the measurements of maternal care exhibits a linear decrease before weaning suggests that, in chimpanzees, maternal investment is fairly evenly distributed throughout the 5 year period before weaning. This may be a behavioral consequence of certain aspects of lactation in non-human primates. Primate milk production is characterized by low protein levels, very low energy yield, low peak yield, and very low energy yield per litter metabolic mass (Buss and and Voss, 1971; Oftedal, 1984).

The theory of parental investment predicts that, under certain conditions, offspring of one sex will receive greater maternal investment than offspring of the other. This occurs when increased investment before weaning has a greater effect on overall fitness of individuals of one sex (Clutton-Brock and Albon, 1982). For example, in many sexually dimorphic, polygynous mammals, variation in reproductive success among males is greater than that among females. When male reproductive success depends on body size and fighting ability, mothers invest more on individual sons than on individual daughters (red deer: Clutton-Brock, Guinness, and Albon, 1982; elephant seal: Reiter, Stinson and Le Bouef, 1978; elephant: Lee and Moss, 1986; Galapagos fur seal: Trillmich, 1986; American bison: Wolff, 1988).

Chimpanzees are sexually dimorphic and a male mates with multiple females. In Mahale, mean body weight for adult males is 42.0 kg (N=6), while that for adult females is 35.2 kg (N=8) (Uehara and Nishida, 1987). However, in this study, sex differences in maternal investment before weaning were by no means conspicuous. In addition, there was no significant difference detected in the interbirth intervals after a mother successfully reared a son versus a daughter (see Chap. 3). Although reports on sex differences in maternal care and infant development in chimpanzees are sparse and provide little quantitative data (Clark, 1977; Miller and Nadler, 1981; Nadler and Braggio, 1974), none of the studies which do exist demonstrate heavier investment toward sons than toward daughters.

Studies on developmental patterns of dimorphic mammals demonstrate heavier birth weight and faster growth rate before weaning in males than in females (Clutton-Brock, Guinness, and Albon, 1982; Lee and Moss, 1986; Reiter, Stinson, and LeBouef, 1978; Trillmich, 1986). In contrast, there is no significant sex difference reported in birth weight in chimpanzees (Gavan, 1953), and growth rate before adolescence is slightly higher in females than in males which need more time than females to reach full adult size (Gavan, 1971). Given the higher female growth rate, maternal investment might be expected to be slightly heavier in females than in males before weaning. Also, chimpanzee females usually disperse from their natal group but males typically remain. Because of the strong local mate/resource competition among males (Hamilton, 1967; Johnson, 1988), the effect of maternal investment on reproductive success might reach the point of diminishing returns more rapidly for sons than for daughters. Thus, mothers might not invest as heavily as would be expected in sons.

In this study, no data were collected which would allow comparison of post-weaning investment in sons versus daughters. However, it is unlikely that post-weaning investment in sons exceeds that in daughters; first, because the competitive ability of a male does not necessarily depend on the male's body size, and second, because the mother's support of her son in competitive situations does not greatly influence his success.

V. SUMMARY

1. The effects of infant age and sex on maternal care before weaning were investigated in chimpanzees of Mahale.

2. Patterns of suckling did not show a significant change with infant age.

3. Infants were carried by the mother until immediately before the birth of the next offspring. They started to travel independently in their third year and did so more often than they were carried from the fifth year.

4. The time spent by mothers on grooming their infants did not vary significantly according to infant age or sex. Single infants received more grooming from the mother than infants which had elder siblings.

5. Female infants suckled longer than male infants when they were younger than 6 months old, and mothers of female infants spent more time feeding when the infants were 2 to 4 years old. There was no sex difference in the amount of time the infants were carried during travel.

6. This study did not provide evidence for heavier maternal investment toward sons before weaning, a phenomenon which has been demonstrated in many dimorphic, polygynous mammals.

Acknowledgements

I am deeply grateful to T. Nishida, who was my supervisor and introduced me to the chimpanzees of Mahale, for his encouragement and criticism throughout the study; T. H. Clutton-Brock, K. J. Stewart, A. H. Harcourt and K. McComb for their discussion and criticism; Gareth S. Egarr for his assistance with English revision. The research was financed by Japan International Cooperation Agency and grants under the Monbusho International Scientific Research Program (#404120 and 56041019).

14

Role of Food Sharing between Mother and Infant in the Ontogeny of Feeding Behavior

Mariko HIRAIWA-HASEGAWA

Department of Anthropology, Faculty of Science, University of Tokyo

I. INTRODUCTION

Food sharing among chimpanzees has attracted attention since its original discovery (Goodall, 1968). It is known to occur between unrelated adults (Busse, 1978; Teleki, 1973; Wrangham, 1977), as well as among close relatives (McGrew, 1975). Food sharing between mother and infant differs from that among adults in its higher frequency of occurrence and the wider variety of food shared (McGrew, 1975). It has been argued that mother-infant food sharing has a different adaptive value from that which occurs among adults (McGrew, 1975; Silk, 1978, 1979).

This paper describes the patterns of food sharing between mother and infant chimpanzees observed in Mahale and assesses the importance of food sharing on infant feeding behavior.

II. METHODS

For a detailed description of the subjects of this study, the period of observation, and the sampling method, see Chapter 13.

During each focal sample, I measured the length of feeding of mothers and of infants and identified the species and the part of the plant eaten (fruits, leaves, flowers, piths, barks, resin, etc.).

The following definitions and measures were used in this study:

Solicitation: any attempts by the infant to obtain a food object from its mother, such as extending a hand to mother's food or mother's mouth, attempting to take the object directly from its mother's hand, or nuzzling its mouth to the mother's mouth while she eats (Fig. 14.1). Any single gesture of solicitation was counted as one.

Sharing: any solicitation resulting in the infant obtaining the desired object from its mother.

Table 14.1 Patterns of solicitation and sharing of individual food items.

Food[1]	Part eaten[2]	Easy/ Dif[3]	Min eaten[4]	Feeding time %[5]	No. sol[6]	Sol. rate[7]	Success %[8]
Albizia glaberrima	res	D	6	0.09	16	160.0	0.0
Aframomum mala	fr	D	17	0.26	7	24.7	71.4
banana	pith	D	14	0.21	5	21.4	0.0
Myrianthus holstii	fr	D	33	0.51	10	18.2	80.0
Uvaria angolensis	fr	E	26	0.40	7	16.1	85.7
meat		D	147	2.27	38	15.5	31.6
Vernonia subuligera	pith	E	9	0.13	2	13.3	50.0
Myrianthus holstii	pt	E	25	0.38	5	12.0	60.0
Grewia platyclada	fr	E	34	0.52	4	7.1	50.0
Parkia filicoidea	fr	D	19	0.29	2	6.3	0.0
Crematogaster spp.		D	286	4.42	27	5.6	85.2
Ampelocissus cavicaulis	fr	E	13	0.20	1	4.6	0.0
Strychnos innocua	fr	D	13	0.20	1	4.6	0.0
Tinospora caffra	l	E	81	1.25	6	4.4	50.0
Camponotus spp.		D	117	1.81	8	4.1	75.0
Ampelocissus cavicaulis	pith	E	32	0.49	2	3.8	50.0
Acacia sieberiana	fr	E	83	1.28	5	3.6	80.0
Landolphia owariensis	pith	E	34	0.52	2	3.5	0.0
Dalbergia boehmii	fl	E	51	0.78	3	3.5	66.7
Chlorophora excelsa	l, g	E	108	1.67	6	3.4	16.7
Ficus exasperata	l, fr	E	135	2.08	7	3.1	0.0
Aframomum mala	pith	D	65	1.00	3	2.8	33.3
Saba florida	fr	D	598	9.25	24	2.4	83.3
Ficus vallis-choudae	fr	E	477	7.38	18	2.3	88.9
Cordia millenii	fr	E	292	4.51	11	2.3	72.7
Diplorhynchus condylocarpon	fr	D	402	6.22	15	2.2	66.7
Sterculia quinqueloba	fr	E	55	0.85	2	2.2	0.0
Ficus sp.	fr	E	39	0.60	1	1.6	0.0
Ficus capensis	fr	E	316	4.88	7	1.3	14.3
Baphia capparidifolia	l, fl	E	144	2.22	3	1.3	0.0
Sterculia tragacantha	l	E	69	1.06	1	0.8	0.0
Garcinia huillensis	fr	E	380	5.87	5	0.8	40.0
Pennisetum purpureum	pith	D	108	1.67	1	0.5	0.0
Harungana madagascariensis	fr	E	248	3.83	2	0.5	0.0
Pycnanthus angolensis	fr	E	291	4.50	2	0.4	50.0
Pterocarpus tinctorius	l, fl	E	804	10.80	1	0.1	0.0

[1] Name of food.
[2] Part of food eaten (fl: flower, fr: fruits. g: galls, l: leaf pt. petiole res: resin)
[3] Easy or difficult.
[4] Number of minutes during which food was eaten by mothers when their infants were within 5 m.
[5] Percentage of time during which food was eaten relative to the total time of feeding.
[6] Number of solicitations recorded for the food.
[7] Solicitation rate per hour of feeding on the food.
[8] Percentage of successful solicitations.

Food items are listed in order of solicitation rate per hour.

Difficult foods: any food item that infants were not able to procure or process independently because of lack of dexterous manipulation of tools, active pursuit of prey, or sufficient strength.

Easy foods: any food item that infants were able to procure or process independently.

Solicitation rate per hour: number of an infant's solicitations divided by the total number of minutes that mothers fed on a food item while the infant was within 5 m.

Percent successful solicitations: the proportion of an infant's total solicitations that resulted in sharing.

Sharing rate per hour: number of an infant's successful solicitations for a food item divided by the total number of minutes that its mother fed on that item while the infant was within 5 m.

Analysis focused on 32 food types on which the target mothers fed for a combined total of more than 30 minutes. Since food sharing was a rare phenomenon and also because of the small number of infants in each age class, data were pooled in each age class. Individual and sex differences among infants could not be considered.

III. RESULTS

Of the 82 food types taken by the mothers, 36 were solicited by the infants and 31 were shared. Table 14.1 summarizes the list of food items and the information on solicitation and sharing.

During the first year, infants solicited food from their mothers almost once every 3 minutes when their mothers were feeding. Solicitation decreased rapidly over time and after the third year did not exceed once every 30 minutes (Fig. 14.2).

As infants matured, they became more capable of obtaining food independently. Therefore, mothers might be expected to react differently toward solicitations by infants of different ages. However, the percentage

Fig. 14.1 A 2-year-old female (PN) begging her mother (FT) for sugarcane. (Photo by T. Hasegawa)

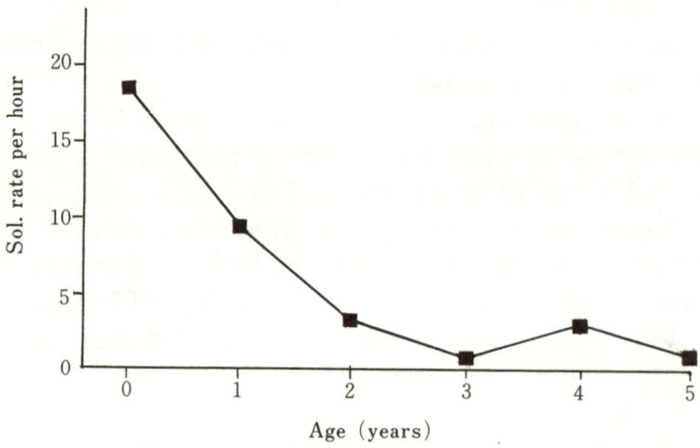

Fig. 14.2 Age changes in solicitation rate per hour of feeding (all foods combined).

of successful solicitations did not exhibit remarkable change but remained rather constant over infant age (Fig. 14.3). As solicitation rate decreased with infant age, sharing rate also decreased significantly, especially at the beginning of the infant's third year (Fig. 14.3). Food sharing occurred about once every 12 minutes of an infant's feeding during its first and second year but occurred only once every 50 minutes after its third year (Mann-Whitney U test, Ucal = 4, $p < 0.05$).

As infants grew, their food-getting skill improved. Thus, one can expect

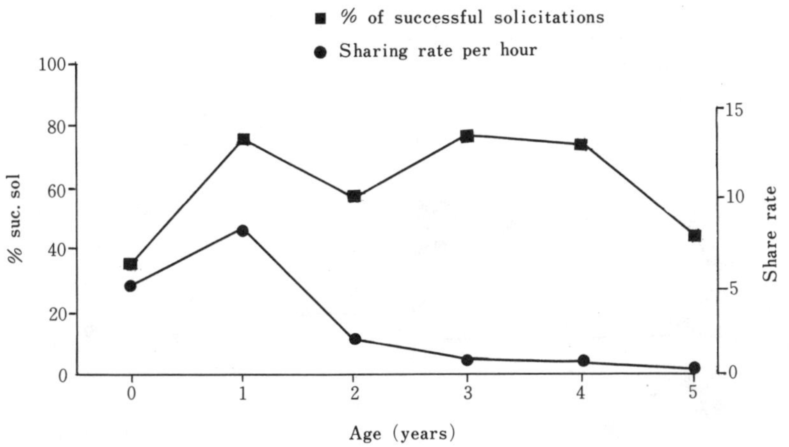

Fig. 14.3 Age changes in percent of successful solicitations relative to total number of solicitations, and in the sharing rate per hour of feeding (all food items).

an effect of age on patterns of solicitation over difficult and easy foods. The age changes in solicitation rates per hour of feeding on difficult and easy foods are indicated in Fig. 14.4. From the first year through the sixth year, the solicitation rate for difficult foods was higher than that for easy foods. An infant virtually ceased to solicit easy foods after its fourth year while it still continued to solicit difficult foods until the sixth year.

The percentage of an infant's total feeding time that was spent on shared food was calculated and related to infant age (Fig. 14.5). During

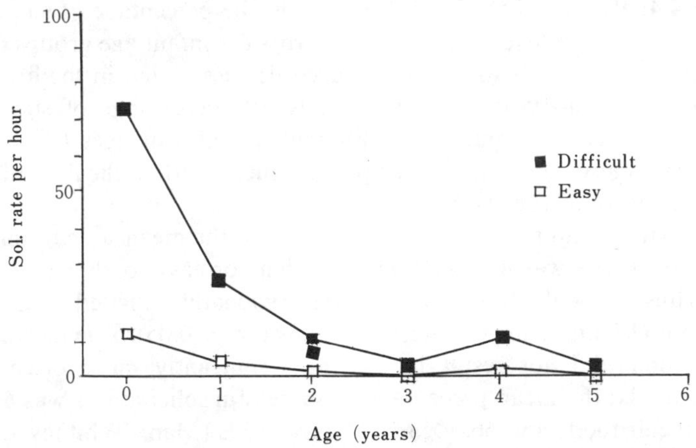

Fig. 14.4 Age changes in solicitation rate per hour of feeding on difficult and easy foods.

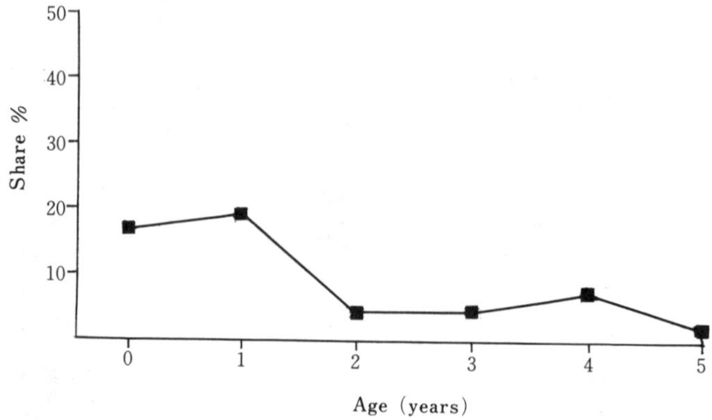

Fig. 14.5 Age changes in percent time spent feeding on shared foods relative to total feeding time.

the first and second year of life, nearly 20% of the infants' total solid food intake was provided by their mothers, whereas after the second year, shared food was only 2 to 5% of total intake.

Mothers might be expected to share difficult foods more willingly than easy foods because this might have an important effect on the ability of their offspring to acquire feeding skills. From 0 to 4 years of age, there was no significant difference between difficult and easy food in the percentage of successful solicitations (Mann-Whitney U-test, 0-year olds, Ucal = 30.5; 1-year olds, Ucal = 55; 2, 3-year-olds, Ucal = 12.5; 4-year olds, Ucal = 4.4; all $p > 0.05$). For difficult foods, the percentage of successful solicitations was not evenly distributed across the infant age groups ($\chi^2 = 31.72$, $p < 0.01$). Such solicitations succeeded less often in the first year than in the second year. For easy foods, the percentage of successful solicitations was not significantly different at different ages ($\chi^2 = 6.69$, $p > 0.05$). Age changes in the degree of success for difficult and easy foods are shown in Fig. 14.6.

When data for all target infants were pooled, the mean solicitation rate for difficult foods was 4.2 per hour and that for easy foods was 1.2 per hour. Thus, difficult foods were more frequently solicited than were easy foods (Mann-Whitney U-test, $z = 2.494$, $p < 0.05$). Solicitations for difficult foods did not result in sharing significantly more often than for easy foods; the mean percentage of successful solicitations was 64.1% for 9 difficult foods and 56.6% for 23 easy foods (Mann-Whitney U-test, Ucal = 74, $p > 0.05$). However, as infants solicited difficult foods more frequently than they did easy foods, the proportion of food acquired by sharing was significantly higher for difficult foods. Of the time spent feeding on the 9 difficult foods (741 minutes), an average of 20.6% was

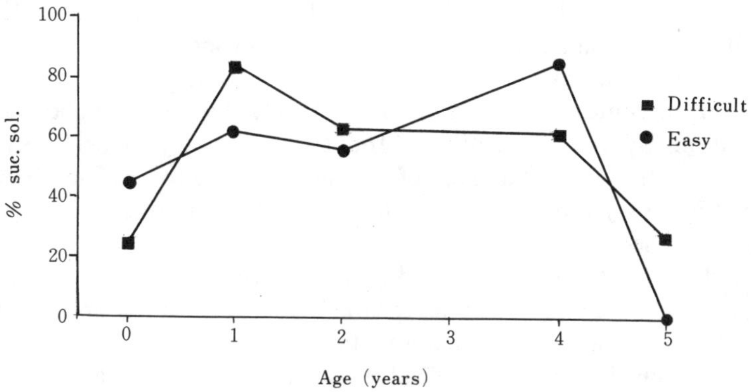

Fig. 14.6 Age changes in percentage of successful solicitations for difficult and easy foods.

spent on shared foods, while in the case of the 18 easy foods (1976 minutes), this figure was 3.9% (Mann-Whitney U-test, Ucal = 9, $p < 0.01$).

IV. DISCUSSION

It has been suggested that food sharing between mother and infant chimpanzee is a maternal strategy to facilitate transition from nursing to independent feeding (McGrew, 1975; Silk, 1978, 1979). From the data comparing food acquired by sharing to the total food intake, it can be concluded that food sharing is most important during the first and second year, providing almost 20% of the infant's total solid food intake. Therefore, food sharing may have an important role in decreasing the mother's lactation load in this period. After the second year, shared food accounted for less than 5% of total intake. However, the number of minutes spent feeding on a food item provides only a rough estimate of the nutritional and caloric intake. Difficult foods, in general, are richer in nutrition than easy foods because they include meat, insects, and hard-shelled fruits which have high nutrient content. Therefore, it is possible that the nutrition and energy provided through sharing difficult foods may be greater than is indicated solely in terms of feeding time.

Silk (1978) showed that difficult foods were shared more often and over longer periods of time than easy foods among the chimpanzees of Gombe. She argued that, as the costs to the mother of continuing nursing increase with the infant's age, mothers could minimize their own costs by sharing foods selectively, and by being increasingly discriminating about which foods they will share with their offspring.

This study also showed that difficult foods were shared over longer periods of time than easy foods, and that the proportion of shared food to the total food intake of infants was higher in the former than the latter. However, in contrast to Silk's results, this was due to the higher rate of solicitations by infants for difficult foods than for easy foods. There was no evidence that the willingness of mothers to share food varied between difficult and easy foods or that it changed with infant age, except during the first and second year.

Whether or not a particular food item is difficult for an infant to process independently is not the only factor likely to influence the probability of food sharing. Food abundance and the availability of alternative food items of equivalent nutritional value are also important factors. When a food item is abundant, the cost for a mother in sharing the food might be trivial, whether or not the food is difficult for an infant. When the food is easy, there will not be an urgent need for an infant to beg. However, if an infant does beg for an easy food, the likelihood that the mother will share is great since the cost of sharing for her is trivial. Similarly, if a difficult food item is abundant, infant solicitation will be likely to prove successful since the cost of sharing for the mother is again trivial.

When a particular food item is rare, the frequency of food sharing will depend on the relative availability of alternative foods of equivalent value for mother and infant. The interplay of numerous factors may cause great variation in the sharing rate of individual food items, and may make it difficult to discern whether mothers are willing to share difficult foods.

Food abundance and the availability of alternative foods for chimpanzees fluctuates both within a year and over several years. Patterns of food sharing during the season when food is limited may be compared with those when food is abundant, in order to clarify the importance of food sharing to the infant's survival.

V. SUMMARY

1. Patterns of food sharing between mother and infant chimpanzees in Mahale were analyzed with regard to an infant's rate of solicitations per hour of feeding, the percentage of successful solicitations, and the time spent feeding on shared food relative to total feeding time. Variation in these factors in relation to infant age and the characteristics of food items was examined.

2. Infants solicited foods once every 3 minutes of mother's feeding

time during their first and second years but only once in 30 minutes thereafter.

3. The percentage of successful solicitations did not show a remarkable change in the course of infant development.

4. Difficult foods were solicited more often and shared for a longer period of time than were easy foods; although solicitations for difficult foods did not result in sharing significantly more often than did those for easy foods.

5. The proportion of food intake that consisted of shared food was highest during an infant's second year but seemed negligible thereafter.

Acknowledgements

I am deeply grateful to T. Nishida, who was my supervisor and introduced me to the chimpanzees of Mahale, for his encouragement and criticism throughout the study; T. Nishida, T. Hasegawa, Y. Takahata, H. Hayaki, and H. Takasaki, for their cooperation in the field; T. H. Clutton-Brock, K. J. Stewart, A. H. Harcourt, and K. McComb for their discussion and criticism; Gareth S. Egarr for his assistance with English revision. The research was financed by Japan International Co-operation Agency and grants under the Monbusho International Scientific Research Program (#404120 and 56041019).

15

A Note on the Ontogeny of Feeding

Mariko HIRAIWA-HASEGAWA

Department of Anthropology, Faculty of Science, University of Tokyo

I. INTRODUCTION

Among omnivorous animals which utilize a wide range of foods it is unlikely that food selection is genetically determined. Infants must learn how to discriminate suitable food objects among the variety of potential foods present in their environment. The acquisition of adult foraging patterns by the time of weaning is essential for an individual's survival. Surprisingly, little is known about the development of independent feeding in any primate species (Nicolson and Demment, 1982; Watts, 1985).

Chimpanzee infants typically continue to suckle for as long as 4 to 5 years after birth (Clark, 1977, see Chap. 13). Infants start eating solid food well before weaning is completed. By doing so, they may not only be supplementing their diet of milk, but may also be learning the adult pattern of foraging. This paper describes the ontogeny of feeding and discusses the transmission of knowledge from mother to infant.

II. METHODS

For detailed descriptions of the subjects of this study, the period of observation and the sampling method, see Chapters 13 and 14. The vegetation of the home ranges of the two subject groups is mostly similar, consisting mainly of a mosaic of secondary deciduous forest and riverine forest (see Nishida, 1979b).

Duration of each focal sample ranged from 30 minutes to over 4 hours. During the focal watches, I measured the length of nipple contact bouts, the amount of time spent feeding of mother and of infant, and identified the species being eaten (Nishida and Uehara, 1981), and in the case of plant food, the part eaten, as well, i.e., fruits, leaves, flowers, piths, barks, resin, etc.

There are two problems with the data used in this analysis. First, the chimpanzees were partially provisioned during the study period. We fed a 5–10 cm stalk of sugarcane or one banana to each individual chimpanzee when first encountered on each day. Because this represented a small portion of the animal's daily nutritional requirement, time spent feeding on sugarcane and bananas was not included in the analysis. Second, due to the small number of target animals in each age class, data on male and female infants was not separated, and sex differences could not be considered.

The following terms were used to describe the feeding behavior of chimpanzees.

Plant food part: part of the plant eaten; fruits, flowers, leaves, piths, petioles, barks, and resin.

Number of food types: one plant food part from one plant species was counted as one food type. When two different plant food parts from one species of plant were eaten, they were counted as 2 different food types. Ants were divided into two food types; *Crematogaster* spp. and *Camponotus* spp. Termites and vertebrate meat were each counted as one food type regardless of the prey species.

Diet component: all the foods were divided into the following 6 components, fruits, leaves, piths, other plant food types, insects and meat. Leaves were subdivided into young leaves of trees, mature leaves of trees, and leaves of herbs.

Feeding bout: a feeding bout was counted from the start of an animal's feeding on one food type until it ceased to eat that food type.

Independent feeding bout: a feeding bout of an infant in which it ate a different food type from that eaten by its mother at the same feeding time.

III. RESULTS

1. Total Feeding Time

During the first 4 months of its life, an infant is totally dependent on its mother for nutrition. Age changes in nipple contact behavior were described in Chapter 13. Around the age of 4 to 5 months, an infant starts to eat solid food.

After the first 6 months, time spent in foraging increased constantly. Age changes in the mean number of minutes of suckling and of foraging per hour of observation are shown in Fig. 15.1

2. Number of Food Types

The target mothers were observed to eat a total of 82 food types (77

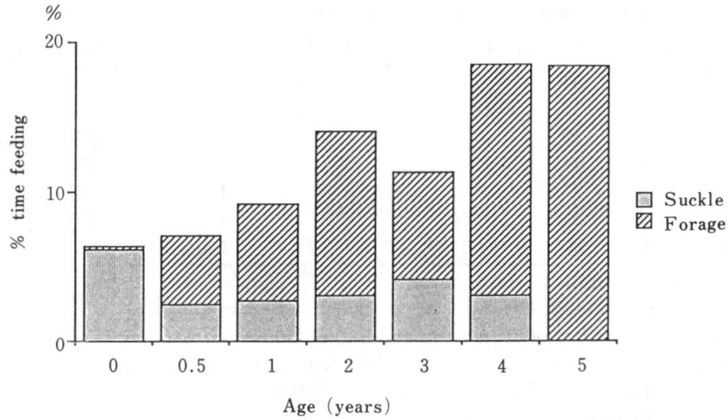

Fig. 15.1 Age changes in percentage of observation time spent feeding.

plant food parts from 65 plant species, 2 types of ants, termites, meat, and soil). Target infants were observed to eat a total of 69 food types (64 plant food parts from 56 plant species, 2 types of ants, termites, meat, and soil) (Fig. 15.2).

A comparison of the number of food types eaten by mothers and infants and the number of independent feeding bouts are shown in Table 15.1 During the first 6 months, infants ate approximately 22% of the mothers' food types. During the next 6 months this figure was 68%. After their

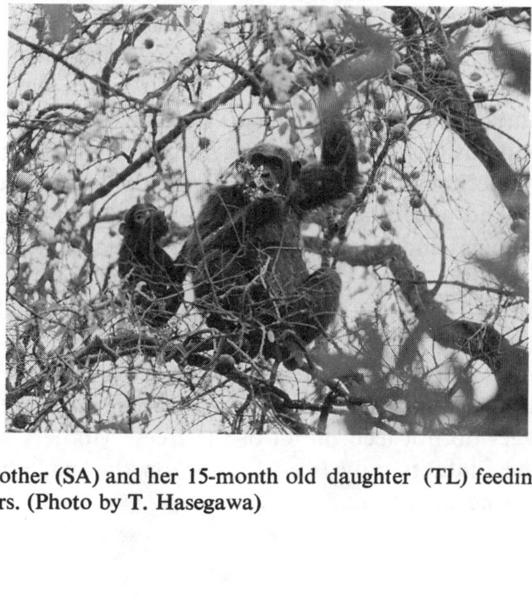

Fig. 15.2 A mother (SA) and her 15-month old daughter (TL) feeding together on buds and flowers. (Photo by T. Hasegawa)

Table 15.1 Comparison of maternal and infant feeding.

Age	N	FT(M)	FT(I)	FT(C)	FTOI	NF	FB	IFB	%IFB
0	5	31	7	7	0	0	5	0	0.0
0.5	5	35	22	22	0	0	27	6	22.2
1	7	50	40	39	1	9	167	37	22.2
2	6	41	37	28	9	1	93	19	20.4
3	5	27	23	21	2	1	69	16	23.2
4	5	37	39	32	7	1	161	55	34.2
5	4	27	32	20	12	1	109	55	50.5

N: Number of mother-infant pairs.
FT(M): Total number of food types eaten by mothers.
FT(I): Total number of food types eaten by infants.
FT(C): Number of food types commonly eaten by mothers and infants.
FTOI: Number of food types eaten only by infants.
NF: Number of non-food items eaten by infants.
FB: Total number of feeding bouts of infants.
IFB: Number of independent feeding bouts.
%IFB: ratio of IFB to FB.

second year, infants started to eat food types which were included in the food list of the M Group (Nishida and Uehara, 1983), but which were not eaten by the target mothers. The ratio of the number of food types eaten solely by infants to the total number of food types eaten by infants increased with the infant's age.

During its first and second year, an infant always ate the same food types as its mother during a feeding bout, but after the third year it often ate a different food type. The percentage of independent feeding bouts relative to the total number of feeding bouts increased with infant age and reached 50% at the age of 5 years (Table 15.1). The mean length of independent feeding bouts were usually less than 2 minutes and were shorter than bouts when infants fed on the same food types as their mothers. An infant sometimes tried a food which was not eaten by adults.

On several occasions, a mother was observed preventing her infant from eating some items. For example, when a 9-month old infant, PN, extended her hand and touched leaves of *Ficus exasperata*, her mother, FT, took PN's hand and moved it away from the leaves. As PN continued to touch and pull the leaves toward her mouth, FT took the leaves from PN's hand, plucked all the leaves within her arm's reach and dropped them to the ground. At least one other mother also exhibited the same kind of behavior towards her infant's attempt to eat the leaves of the same tree. There are many *F. exasperata* trees within the home range of the M Group, but chimpanzees do not feed on all of the trees. Mothers were observed to prohibit their infants only from feeding on the individual trees that they themselves never fed on.

Table 15.2 The percentage of total feeding time spent on each diet component.

Diet component	Mothers min. feeding	Mothers %	Offspring min. feeding	Offspring %
Fruits	3927	60.8	1984	57.9
Leaves	673	10.4	208	6.1
Piths	406	6.3	269	7.8
Other plant parts	774	12.0	553	16.1
Insects	523	8.1	382	11.1
Meat	150	2.3	22	0.6
Soil	10	0.2	10	0.3
Total	6463	100.0	3428	100.0

Table 15.3 Comparison of leaf diet between mothers and infants (percentage of time spent feeding).

	Young leaves of trees	Mature leaves of trees	Leaves of herbs
Mothers	47.4	33.4	19.2
Infants	63.9	23.3	12.8

Food sharing may facilitate an infant's learning of the diversity of adult foods. During the first year, all of the food types an infant takes are novel to it. Target infants of under 1 year old ate 27 food types; in 20 of them food was shared by their mothers. One-year-old infants were seen to eat 40 food types, of which 17 were novel and were not included in the infant's food list of the previous year. Food sharing was seen in 76% of these 40 types. Of the 37 food types taken by 2-year-old infants, 9 were novel and food sharing was seen in 3 of them. For 3- and 4-year-old infants, food sharing was present in 2 of the 7 novel food types. When the time spent feeding on food obtained through sharing was compared to that obtained independently, novel foods were shared significantly more often than familiar foods ($\chi^2 = 213.165$, df = 1, $p < 0.01$).

3. Diet Components

In Mahale, the diet of chimpanzees, in terms of time spent feeding, is composed of 57.7% fruits, 12.2% leaves, 19.8% other plant food types, 6.4% insects, 2.5% meat, and 1.3% soil. When data on all infants before weaning was pooled, the distribution of feeding time for each diet component was not significantly different from that of mothers (Table 15.2, $\chi^2 = 1.2713$, df = 4, $p > 0.05$; soil and other plant food parts combined, insects and meat combined).

Analysis of the leaf diet revealed a difference in the quality of leaves taken by mothers and by infants. Infants took relatively more young leaves of trees, and fewer other leaf types than did mothers (Table 15.3, $\chi^2 = 11.048$, df = 2, $p < 0.01$).

IV. DISCUSSION

Analysis of the time spent on feeding suggests that infants start to supplement their diet of milk with substantial amounts of solid food from their second year, although weaning is completed only during their fifth year. The adult food repertoire is acquired quickly and infants are probably familiar with most adult food types by the age of 3 years. The tendency for infants to select foods different from those eaten by their mothers during the same bout suggests that infants may acquire knowledge of their food supply partly by copying their mothers and partly from their own independent experience.

Many plants contain secondary compounds as a defense against leaf-predators, particularly in mature leaves (Stumpf and Conn, 1982), and many primates avoid such materials by feeding selectively (Glander, 1978; Milton, 1979; Oates, Waterman, and Choo, 1980). The relatively small portion of the infant diet that consisted of mature leaves of trees suggests that infants may be inefficient in detoxifying such compounds.

The prohibition by their mothers against feeding on the leaves of particular trees of *F. exasperata* is especially interesting for what it indicates about teaching by chimpanzees. Polymorphism in plant toxicity is widespread (Cooper-Driver and Swain, 1976; Maxwell, Jenkins, and Parrot, 1972). Levels of secondary compounds in individual plants show remarkable variation and are correlated with a plant's susceptibility to hervibore damage. It is possible that *F. exasperata* contains some kind of harmful secondary compound which is polymorphic in its concentration. The prohibitive behaviour of the chimpanzee mothers may be interpreted as a form of "teaching" to avoid feeding on a poisonous individual.

The preliminary results presented in this paper suggest that, along with observational learning by infants, positive transmission of knowledge from mother to infant through food sharing and prohibition from eating unsuitable items may play an important role in an infant's acquisition of adult feeding behavior.

V. SUMMARY

1. Development of independent feeding by infant chimpanzees of

Mahale was observed.

2. Infants started to supplement their diet of milk with substantial amounts of solid food during their second year.

3. The adult food repertoire was acquired rapidly, and by the age of 3 years infants seemed to be familiar with most of the food types eaten by adults.

4. Infants before weaning ate lesser amounts of mature leaves than did their mothers.

5. Active transmission of knowledge from mother to infant apparently has an important influence on the acquisition of adult feeding patterns.

Acknowledgements

I am deeply grateful to T. Nishida, who was my supervisor and introduced me to the chimpanzees of Mahale, for his encouragement and criticism throughout the study; T. Nishida, T. Hasegawa, Y. Takahata, H. Hayaki, and H. Takasaki for their cooperation in the field; T. H. Clutton-Brock, K. J. Stewart, A. H. Harcourt, and K. McComb for their discussion and criticism; Gareth S. Egarr for his assistance with English revision. The research was financed by Japan International Cooperation Agency and grants under the Monbusho International Scientific Research Program (#404120 and 56041019).

16

Deceptive Behavior in Young Chimpanzees: An Essay

Toshisada NISHIDA

Department of Zoology, Faculty of Science, Kyoto University

I. INTRODUCTION

The scientific interest in deception has increased in recent years, in part stimulated by new developments in the theoretical framework of behavioral ecology. The paradigm of the "selfish gene" (Dawkins, 1976; Dawkins and Krebs, 1978; Maynard Smith, 1982) has been particularly important in this respect.

In the early 1970s, strong evidence began to appear that nonhuman primates may manipulate and deceive their conspecifics. Perhaps most famous is the observation by Menzel (1973), who studied communicative behavior in young chimpanzees kept in an enclosure.

A juvenile female was shown where food had been hidden in the large enclosure. She was returned to the original cage, and after a while she and some other chimpanzees were released into the enclosure together. She ran straight to the hidden food and her peers followed suit, guessing from her expressions that there was something to eat there. However, when she retrieved the food, a high-ranking male snatched her prize. After several such trials with similar results, she first ran to a place where no food was hidden. Once she had confirmed that her peers were following, she suddenly doubled back, allowing them to go on ahead, while she went directly to the actual site.

Woodruff and Premack (1979) proved that a juvenile chimpanzee can intentionally deceive a "bad" trainer who consistently refused to share a banana with him. De Waal (1982, 1985) described many episodes of social manipulation and deceptive behavior in a semi-captive chimpanzee colony.

However, it is not yet clear how chimpanzees use deception in the natural environment. Therefore, although they may be anecdotal, additional data from observations in the wild will be important for understanding the origins and development of social intelligence in chimpanzees. This essay was stimulated by Byrne and Whiten (Byrne and Whiten, 1985, 1987; Whiten and Byrne, 1986, 1988), who have collected anecdotes regarding deception from field workers the world over, and who have proposed a promising technique for the evaluation of anecdotes.

II. METHODS

Observations were made on the chimpanzees of M and K Groups. Episodes described here were obtained mostly during focal sampling of mother-infant pairs.

Deceptive behavior of chimpanzees is tentatively classified into two categories: concealment and deception. Concealment is defined as the misleading of other individuals by concealing those autonomic responses, postures, gestures, facial expressions, or vocalizations which usually accompany a particular social interaction. Deception is defined as the misleading of other individuals by providing postures, gestures, facial expressions, or vocalizations that are irrelevant to the context.

III. RESULTS

1. Concealment or Withdrawal of Information

Before provisioning of the chimpanzees was ended in the early 1980s, sugarcane was cut and distributed over a wide area upon the ground. In 1975, an 8-year-old female, MW (Mwese), lost her pieces of sugarcane to older chimpanzees whenever she got one. After a time, MW ceased to go and pick up pieces of sugarcane, even when they dropped in the nearby bush. Instead, she continued to sit where she was and waited until the older chimpanzees had consumed their share and moved away. Then she entered the bush where she had seen a piece of sugarcane drop. An interesting aspect of her behavior is that she never glanced in the direction of the sugarcane when there were other chimpanzees about, thus effectively concealing information about its location from them.

2. Deception

A ritualized form of deception has been reported among the chimpanzees of Mahale. Males, both adult and adolescent, rip leaves to pieces as a courtship display as if simulating feeding, or as if making a leaf tool (Nishida, 1980a). The following are two examples of unritualized forms of deception.

In the first, a juvenile female feigned interest in one activity in order to carry out another unnoticed. In 1987, a 5-year-old female, SV (Silvy), was much interested in her 3-month-old baby sister, and tried over and over to touch her. However, her mother, SL (Silafu), prevented her from touching the baby by covering it with her arm and/or by turning around. SV apparently gave up her attempts and left SL. When she returned on the next occasion, she earnestly groomed her mother. As she was grooming, however, she slowly reached out her leg and furtively touched her sister. Her mother did not notice her daughter's surreptitious activity since SV did not even glance at her little sister while touching her.

In the second example, there are three characters; the deceiver, the deceived, and the social tool. This episode was partly reported elsewhere (Nishida, 1981).

When a chimpanzee reaches about 5 years of age, it is able to feed, walk, and perform other daily activities on its own without the help or constant reassurance of its mother. At this stage, the mother begins to resume estrus and is capable of producing her next offspring. She shakes off her youngster when it tries to ride on her back, and prevents it from suckling by folding her arms in front of her breasts or by lying prone on the ground. The weanling whimpers and displays a variety of temper tantrums in an attempt to regain maternal attention.

In 1979, a 5-year-old male, KB (Katabi), was in the process of being weaned by his mother, CH (Chausiku). Whenever his suckling attempts were discouraged, he would whimper, whimper-scream, or fall to the ground in a temper tantrum. Sometimes he would hang dangerously from a branch with one leg and shake himself, beating his head against the trunk of the tree, as if he was about to fall. Such hysterical behavior was—at least in part—a sham, since in the midst of his display KB would sometimes glance at his mother as if to check her response to his tantrum.

One day, CH and her consort, KM (Kamemanfu, who was the alpha male and the only adult male of K Group then) were grooming. KB suddenly whimpered, left CH, and lay supine on the ground. He then got up and moved still farther away. CH approached KB, but KB again moved away. KB then came towards me, and began to scream loudly, reaching his hand toward me as if pointing. Then he went round me repeatedly, still screaming and reaching out a hand. CH and KM at once glanced at me, with hair erect. I retreated a little away from KB, in order to avoid precipitating an attack by the adult chimps. Undoubtedly both KM and CH believed that KB had been attacked or teased by me, when, in fact, I had done nothing to him. Thus, it appears that KB was attempting to attract CH's attention and sympathy (and hopefully, one supposes, to be held and finally allowed to nurse) by feigning an attack by a human observer.

A similar event took place 6 minutes later. This time KB availed himself of a conspecific, rather than the human observer. While looking at CH, KB rushed away, and at the distance of 3m turned a backward-somersault along the slope, all the while continuously emitting screeches and screams. He repeated this three times, and then approached an older adolescent male MS (Masisa), who was sitting alone away from the group. KB displayed the same gestures he had to the human observer, reaching a hand to MS. MS appeared perplexed or "embarrassed." As soon as KM and CH looked at MS, he stood up and violently pant-grunted to KM. He then left KB, moving away from them all. This was the same reaction that I displayed, except, of course, I did not pant-grunt! Thus, it appeared that MS understood the dangerous situation into which he was driven by KB. KB was finally allowed to suckle 3 minutes after the second episode.

Apparently, KB had experienced various opportunities to learn that he was far more likely to be embraced and allowed to nurse by his mother as a form of reassurance when he was threatened by other chimpanzees or human observers.

IV. DISCUSSION

It appears that weanling chimpanzees possess a high degree of social awareness. The grooming behavior of weanling chimps is also intriguing (Nishida, 1988). There were four cases of weanling infants (one female, three male) whose mothers resumed sexual cycling. In all four, the rate of infantile grooming per hour of observation, and the duration of infantile relative to maternal grooming increased when their mothers were in estrus. Both the mean and the median bout-length of infantile grooming were longer when their mothers were in estrus than when they were not. The case of SV was particularly interesting. Her mother, SL, resumed estrus when SV was only 35 months old. SV did not groom her mother before SL's resumption of estrus, or during any one of the anestrous intervals between estrous periods. SV groomed her mother only when SL was in estrus and her attention began to focus on the males. This suggests that infants often do not groom their mothers to the full extent of their capability. It is possible that SV concealed her social ability in the hope that, by pretending to be less capable than she really was, she might receive more maternal attention. Thus, an infant might garner more maternal care than the mother has selected to give.

Deceptive behavior in weanling chimpanzees may originate in mother-offspring conflict. Postweaning juveniles seem to be competent social manipulaters. For example, a 5-year-old female, PN (Penelope), groomed an adult female in order to obtain access to her one-year-old infant (Nishida, 1988). An eight-year-old female, LS (Lusya), manipulated her mother by kidnapping her infant; she made off with her sister in an attempt to persuade her mother to follow her when her mother was resting or moving in a different direction (Nishida, 1988).

During the weaning period, an offspring uses deceptive behavior as a tactic to induce prolonged maternal investment. The offspring may cry as if it were very hungry when it is really not so hungry. Such a tactic may work well, since a mother who rejects the request of her hungry infant may be pursuing the more risky strategy than a mother who allows her well-fed offspring to suckle. Trivers (1974) pointed out that "an offspring will often have better knowledge of its real needs than will its parent." However, the mother who can discriminate between the real and false needs of her offspring will be more reproductively successful. The interplay between the infant's deception and the mother's attempts to discriminate may lead to an evolutionary "arms race" producing greater powers of deception in infants and discrimination in mothers.

Trivers (1974) also pointed out that species in which different, unrelated males commonly father a female's successive offspring, are expected to show stronger weaning conflict than species in which siblings are usually fathered by the same male. If the ability to engage in social manipulation originates in the weaning period, species which show more intense weaning conflict may be expected to display more social skills. This reasoning is supported by the observation that chimpanzees are far more socially adept than are gorillas or gibbons.

Deception is more likely to occur in species which are organized in large-sized social units and which exhibit a fission-fusion grouping system. This is because deception would be more likely to be detected in small and compact societies, where cheaters would soon be penalized, than in a fission-fusion society, where the possibility for deception is enhanced by the difficulty of monitering all the information on each group member. Deceptive behaviors have been suggested for baboons (Kummer, 1982; Byrne and Whiten, 1985), which live in large societies which sometimes split into subgroups. Evidence on deception in gorillas, orangs, and gibbons is scanty (Chevalier-Skolnikoff, 1985; Whiten and Byrne, 1988). Chimpanzees and bonobos, and perhaps Pleistocene humans, may have evolved deception because of their social framework.

On the other hand, the property of "self-recognition" has been documented in orangs and chimpanzees, but not in cercopithecids or gorillas (Gallup, Wallnau, and Suarez, 1980). This is curious because "self-recognition" appears to be a prerequisite to purposeful deception. The origins of primate intelligence will continue to be confusing until more observational data accumulate (see also McGrew, 1989).

Acknowledgements

The episodes described in this article were obtained during fieldwork conducted in 1975, 1979, and 1987. The study was financed by the Japan International Cooperation Agency and grants under the Monbusho International Scientific Research Program (#404120 and 62041921). I thank M.A. Huffman for revision of the English.

Appendix

Summary of Meteorological Data at Mahale Research Camps, 1973–1988

Hiroyuki TAKASAKI, Toshisada NISHIDA, Shigeo UEHARA, Kohshi NORI-KOSHI, Kenji KAWANAKA, Yukio TAKAHATA, Mariko HIRAIWA-HASEGA-WA, Toshikazu HASEGAWA, Hitoshige HAYAKI, Kenichi MASUI, and Michael A. HUFFMAN

A large portion of the meteorological data from Mahale has been available only to a limited group in the form of *Mahale Mountains Chimpanzee Research Project, Meteorological Reports* (Nishida and Norikoshi, 1976; Uehara and Norikoshi, 1977; Uehara, 1978; Kawanaka, 1979; Hasegawa, 1980; Hasegawa and Hasegawa, 1981, 1982; Takahata, 1983; Takahata, Nyundo, and Seifu, 1984; Takahata and Takasaki, 1986; Takasaki, 1988, 1989), although a one-year profile of ambient temperature (Nishida, 1968) and a brief summary of rainfall data (Uehara, 1982) have been published elsewhere. In recording the data, besides the authors and the reporters above, M. Bunengwa, H. Huseini, H. Kabukula, H. Kajugi, A. Kasulamemba, J. Katensi, S. Katensi, H. Kawanaka, S.N. Kimaro, R. Kitopeni, S. Musa, M. Masui, R. Masui, G.S. Mkombwe, M. Msenga, V.R. Mushi, H. Nishida, H.I. Sariko, H. Takahata, K. Takasaki, T. Tsukahara, and M. Uehara also cooperated.

Tables A.1 to 6 summarize the accumulated data for Kansyana, Myako, and Bilenge Camps. Although the temperature and relative humidity data were also read at 8:00 and 14:00 (East African Standard Time, GMT +6 hours) when possible, only the rainfall data and the daily maximum and minimum temperature data are summarized here. The reasons are: (1) the 14:00 data were relatively scanty, especially in recent years, thanks to the good observation of the chimpanzees, (2) the 14:00 temperature differs little from the daily maximum temperature, (3) the 8:00 temperature differs little from the daily minimum temperature, and (4) the relative humidity is almost always 85–100% at 8:00.

McGrew, Baldwin, and Tutin (1981) put Kasoje in the third position among six well studied chimpanzee habitats on a wet-dry spectrum (1.

Mt. Okorobiko, Rio Muni; 2. Budongo Forest, Uganda; 3. Kasoje; 4. Gombe, Tanzania; 5. Kasakati Basin, Tanzania; 6. Mt. Assirik, Senegal). The habitat of the Kasoje chimpanzees thus is not so dry, in particular in the Kasoje Forest, where Kansyana Camp is located. Among the chimpanzee habitats within Tanzania, "Kasoje is the most forested of all, Gombe less so, while Bilenge tends to resemble the openland habitats around Kabogo, Kasakati, and Filabanga" (Collins and McGrew, 1988).

As can be read in the rainfall data (Tables A.1–3), the year can be divided into the rainy season (October–May, with a short dry season from mid-January to February) and the dry season (June–September). As Takahata (1983) noted, interestingly, in the long-term rainfall data there is a slight biennial periodicity (Tables A.1–3).

Table A.1a Rainfall (mm) at Kansyana Camp, 1973 –1988.

	Oct.	Nov.	Dec.	Jan.	Feb.	Mar.	Apr.	May	June	July	Aug.	Sept.	Total
1973–74	88.2+	489.6	348.4	155.7	137.0								
1974–75								26.1	4.9	2.9	0.0	83.6	
1975–76	118.3	129.6	203.0	147.3	195.6	212.9+	238.3+	66.4	0.0	0.0	6.8	11.8	1330.0+
1976–77	96.3	539.5	278.1	330.6	140.3	169.1	461.8	104.1	2.1	0.0	0.0	2.8	2124.7
1977–78	101.6	310.3	384.8	209.0	140.3+	227.6+	187.2+	14.8		0.0	37.4	61.2	1674.2+
1978–79	288.7	336.8	400.5	212.1	176.3	305.8	447.1	170.6	0.0	0.0	0.0	0.0	2337.9
1979–80	78.3	247.4	259.0	320.9	335.2	131.9	210.3	78.7	0.0	0.0	0.0	87.9	1749.6
1980–81	82.3	268.7	379.0	184.9	153.0	380.4	255.8	78.7	0.0	0.0	13.5	0.0	1796.3
1981–82	39.5	110.3	318.9	147.1	199.4	259.9	267.0	146.7	0.0	0.0	0.0	30.4	1519.2
1982–83	239.8	395.0	377.4	169.4	100.0	546.5	381.4	20.0	0.0	0.0	54.4	2.6	2286.5
1983–84	222.0	279.2	276.2	218.8	168.8	287.4	212.9	22.5	0.0	0.0	0.0	1.7	1689.5
1984–85	86.5	406.8	251.5	306.0	322.1	345.5	255.8	56.1	0.0	0.0	5.0	8.0	2043.3
1985–86	3.6	256.8	354.3	203.3	322.7	359.6	245.8	135.5+	2.5	0.0		68.6	1892.1+
1986–87	130.4	318.1	322.4	389.0	155.4	395.1+	405.0	48.8	0.0	0.0	0.0	87.8	2252.0+
1987–88	173.5	193.7	255.0	211.4	146.1	202.3	264.1	0.0	2.1	0.0	5.1	5.1	1458.4
1988	110.5	240.9	307.4										
Mean	124.0	301.5	307.7	229.0	196.3	298.8	309.7	64.1	0.9	0.2	9.4	32.3	1835.5*

* Column mean.

Table A.1b Number of rainy days at Kansyana, 1973 –1988.

	Oct.	Nov.	Dec.	Jan.	Feb.	Mar.	Apr.	May	June	July	Aug.	Sept.	Total
1973–74		20	17	0+	16								
1974–75									2	0	0	7	
1975–76	7	14	17	13	16	15	13	6	0	0	3	6	110
1976–77	13	16	18	19	15	22	28	13	1	0	0	1	146
1977–78	6	18	18	15	13	15	16+	3	1	0	1	8	114
1978–79	9	26	24	25	12	17	24	10	0	0	0	0	147
1979–80	5	21	25	11	20	19	19	14	0	0	0	5	139
1980–81	9	22	22	20	12	21	18	7	0	0	3	0	134
1981–82	9	11	27	11	11	21	19	14	0	0	0	2	125
1982–83	24	29	21	20	13	25	22	6	0	0	3	2	165
1983–84	19	17	28	24	20	28	25	8	0	1	2	1	173
1984–85	13	27	29	26	24	20+	20	7	0	0	1	6	173+
1985–86		20	23	17	13	29	26	18+	1	0		4	
1986–87	9	23	24	22	11+	22	28	12	1	0	1	12	165+
1987–88	16	20	22	25	26	20	29	3	2	1	8	5	177
1988	16	27	30										
Mean	11.9	20.7	23.0	19.1	16.2	21.2	22.6	8.6	0.6	0.1	1.7	4.2	143.0*

* Column mean.

Table A.2a Rainfall (mm) at Myako Camp, 1976–1984.

	Oct.	Nov.	Dec.	Jan.	Feb.	Mar.	Apr.	May	June	July	Aug.	Sept.	Total
1975–76				112.3	295.1	204.2	293.6	40.2	5.0	0.0	19.1	0.6	
1976–77	52.4	457.4	213.9	288.3	109.0	142.3	350.7	152.1	0.0	0.0	0.0	0.0	1766.1
1977–78	108.2	250.5	289.8	167.9	168.0	338.9	185.4	6.1	0.5	0.0	74.8	35.8	1625.9
1978–79	154.8	367.8	492.9	185.2	157.3	308.0	343.8	143.0	0.0	0.0	0.0	0.0	2152.8
1979–80	89.5	161.2	220.6	156.8	440.0	58.8	143.9	71.7	0.0	0.0	0.0	236.4	1578.9
1980–81	39.8		355.1	176.3	121.8	320.8	283.5	130 0	0.0	0.0	4.3	2.1	
1981–82	54.4	110.2	230.1		127.7	234.3	334.5	128.1	0.0	0.0	0.0	34.3	
1982–83	322.1	326.6	197.8	130.3	61.1	341.9	381.5	18.5	0.0	0.3	40.3	1.0	1821.4
1983–84	266.8	296.8	271.8	240.0	170.8	179.5	259.4	6.6	4.2	0.0	0.0	2.9	1698.8
1984	93.9	284.4	199.5										
Mean	131.3	281.9	274.6	182.1	183.4	286.3	246.5	77.4	1.0	0.03	15.4	34.8	1774.0*

* Column mean.

Table A.2b Number of rainy days at Myako, 1976–1984.

	Oct.	Nov.	Dec.	Jan.	Feb.	Mar.	Apr.	May	June	July	Aug.	Sept.	Total
1976				11	20	14	21	8	1	0	2	1	
1976–77	9	17	18	19	14	25	27	14	0	0	0	0	143
1977–78	6	22	20	15	16	23	21	5	1	0	1	5	135
1978–79	10	26	22	22	11	17	21	7	0	0	0	0	136
1979–80	6	17	17	12	12	16	15	10	0	0	0	4	109
1980–81	8		17	22	11	21	7	8	0	0	3	1	
1981–82	7	8	20		8	15	17	12	0	0	0	7	
1982–83	18+	24	16	15	7	20	24	8	0	1	4	1	138+
1983–84	16	15+	19	11+	9+	17	18	2	1	0	0	1	109+
1984	7	23	16										
Mean	8.6	19.6	18.3	16.6	12.4	18.7	19.0	8.2	0.3	0.1	1.1	2.2	130.8*

* Column mean.

Table A.3a Rainfall (mm) at Bilenge, 1978–1982.

	Oct.	Nov.	Dec.	Jan.	Feb.	Mar.	Apr.	May	June	July	Aug.	Sept.	Total
1977–78			101.2+	161.7	92.8	305.0	76.8	6.2	0.0	0.0	8.5	35.9	
1978–79	138.4	295.8	211.6	260.8	114.0	268.3	312.0	42.2	0.0	0.0	0.0	0.0	1643.1
1979–80	83.4	130.1	213.8	253.7	293.3	65.4	109.4	43.7	0.0	0.0	0.0	162.0	1354.8
1980–81	41.5	205.5	266.3	235.0	215.2	170.4+	50.0+	71.1	0.0	0.0	2.3	1.9	1259.2
1981–82	20.2	36.5	170.6		92.8	134.0	80.1	53.1	0.0	0.0	0.0	5.6	
1982	106.7	288.6	124.9										
Mean	78.0	191.3	197.4	227.8	161.6	193.2	144.6	43.3	0.0	0.0	2.2	41.1	1419.0*

* Column mean.

Table A.3b Number of rainy days at Bilenge, 1978–1982.

	Oct.	Nov.	Dec.	Jan.	Feb.	Mar.	Apr.	May	June	July	Aug.	Sept	Total
1977–78			9+	16	19	21	18	2	0	0	3	5	
1978–79	6	26	20	16	13	22	22	4	0	0	0	0	129
1979–80	9	15	13	13	10	9	9	5	0	0	0	3	86
1980–81	5	12	15	10+	7	7+	3+	5	0	0	2	1	67+
1981–82	3	5	11		5	9	9	6	0	0	0	3	
1982	12	14	6										
Mean	7.0	14.4	13.0	15.0	10.8	15.3	14.5	4.4	0.0	0.0	1.0	2.4	107.5*

* Column mean.

Table A.4a Monthly mean daily maximum temperature (°C) at Kansyana, 1973 –1988.

	Oct.	Nov.	Dec.	Jan.	Feb.	Mar.	Apr.	May	June	July	Aug.	Sept.	Mean
1973–74			28.4	28.8	29.1								
1974–75													
1975–76													
1976–77			27.3	27.5	28.2	27.9	26.7	27.1	27.5	28.1	29.3	31 4	
1977–78	32.7	30.2	28.7	28.2	28.0	28.6	27.7	28.5	28.7	28.4	30.0	31.4	29.3
1978–79	31.6	27.3	27.3	28.1	28.5	28.9	26.9	28.4	27.7	28.5	31.1	33.8	29.0
1979–80	31.6	29.1	28.8		26.1	27.0	26.7	27.5	28.9	28.3	28.5	28.4	
1980–81	28.6	26.8	25.9	26.6	27.5	25.5	26.3	26.7	27.0	26.8	28.3	30.3	27.2
1981–82	31.1	29.4	26.8	27.5	26.8	27.0	26.6	26.9	28.2	28.7	31.0	31.9	28.5
1982–83	29.0	27.0	27.4	28.2	29.9	28.9	28.3	28.3	29.5	30.0	29.3	32.1	29.0
1983–84	29.4	29.0	27.7	27.8	28.0	28.3	28.8	30.3	29.3	29.4	31.1	31.6	29.2
1984–85	30.2	27.0	26.9	27.1	27.3	28.1	28.9	31.0	31.7	31.0	32.1	30.6	29.3
1985–86				27.8	28.5	28.1	30.1	30.6	30.9	30.7		31.6	
1986–87	29.5	26.8	24.7	27.0	27.9	27.9	29.0	31.5	32.8	32.7	33.0	33.0	29.7
1987–88		28.5	27.9	27.3	28.8	30.4	31.0	33.5	33.6	32.6	32.3	33.5	
1988	32.5	27.7	26.1										
Mean	30.6	28.1	27.2	27.7	28.0	28.1	28.1	29.2	29.7	29.6	30.5	31.6	29.0*

* Row mean.

Table A.4b Monthly mean daily minimum temperature (°C) at Kansyana, 1973 –1988.

	Oct.	Nov.	Dec.	Jan.	Feb.	Mar.	Apr.	May	June	July	Aug.	Sept.	Mean
1973–74			18.1	18.2	18.5								
1974–75													
1975–76													
1976–77			18.6	18.4	18.7	19.2	19.1	18.4	16.5	15.6	17.0	18.5	
1977–78	19.8	20.3	19.2	19.5	19.3	19.3	19.6	18.5	16.3	15.4	16.6	18.1	18.5
1978–79	18.3	18.8	18.8	19.5	19.1	18.7	19.4	18.5	16.3	15.3	16.3	17.4	18.0
1979–80	19.0	19.5	19.3	19.5	19.3	19.4	19.7	19.0	17.0	16.3	17.6	18.4	18.7
1980–81	19.2	18.9	18.9	19.1	18.9	19.0	19.3	18.3	16.7	16.0	17.3	18.3	18.3
1981–82	19.7	19.3	18.9	19.0	19.0	18.8	19.2	19.2	17.4	15.8	16.6	18.2	18.4
1982–83	19.2	18.6	19.4	19.3	19.9	19.8	19.9	19.5	17.5	16.8	17.5	17.6	18.8
1983–84	18.7	18.5	18.4	18.4	18.7	18.7	18.8	18.0	16.3	16.3	16.3	17.2	17.9
1984–85	18.6	18.7	18.6	18.8	18.7	19.0	19.1	18.3	16.1	15.4	16.5	17.8	18.0
1985–86				18.9	18.3	18.5	19.3	18.6	15.9	14.7		17.4	
1986–87	20.0	19.5	19.1	18.7	19.0	19.4	19.4	19.4	16.5	15.9	17.3	19.0	18.6
1987–88		19.3	19.3	19.0	19.2	19.3	19.5	18.2	17.4	16.5	17.4	17.8	
1988	19.1	18.8	18.5										
Mean	19.2	19.1	18.6	18.9	19.0	19.1	19.4	18.7	16.7	15.8	16.9	18.0	18.3*

* Row mean.

Table A.5a Monthly mean daily maximum temperature (°C) at Myako, 1977–1984.

	Oct.	Nov.	Dec.	Jan.	Feb.	Mar.	Apr.	May	June	July	Aug.	Sept.	Mean
1976–77												34.4	
1977–78	33.8	30.7	30.7	31.1	31.3	31.0	30.4	33.3	32.5	31.5	31.8	31.4	31.6
1978–79	31.6	27.3	27.3	31.7	31.1	30.7	30.5	31.6	31.1			34.0	
1979–80	33.6	30.4	31.2			28.9	29.2	28.8	30.0	30.3	31.0	31.6	
1980–81	29.7		25.9	26.4	28.4	26.8	27.6	29.2	30.8	30.7	30.8	33.3	
1981–82					28.8	28.3	28.5	29.4	31.6	31.2	32.3	32.4	
1982–83	31.1	26.4	26.7	27.2	28.9	28.4	28.4	30.7	32.0	32.2	31.7	33.3	29.8
1983–84	30.9	28.3	26.8	27.0	27.3	27.7	28.5	31.4	31.1	30.8	32.1	34.0	29.7
1984	33.0	27.1	27.0										
Mean	32.0	28.4	27.9	28.7	29.3	28.8	29.0	30.6	31.3	31.1	31.6	33.1	30.2*

* Row mean.

Table A.5b Monthly mean daily minimum temperature (°C) at Myako, 1977–1984.

	Oct.	Nov.	Dec.	Jan.	Feb.	Mar.	Apr.	May	June	July	Aug.	Sept.	Mean
1976–77												22.0	
1977–78	23.0	21.1	20.8	21.2	21.3	20.5	21.2	20.7	19.6	18.6	20.1	21.1	20.8
1978–79	21.2	20.4	20.7	21.0	21.1	20.8	20.8	20.4	19.1			20.6	
1979–80	21.5	20.9	20.3	20.8	20.9	20.2	20.6	20.5	18.7	18.9	19.7	20.9	20.3
1980–81	21.2		20.5	20.6	20.6	20.9	21.2						
1981–82					21.3	21.2	21.2	20.8	20.5	19.8	20.7	21.5	
1982–83	21.9	21.0	21.8	21.3	22.0	21.8	21.4	21.6	20.7	20.1	20.7	21.4	21.3
1983–84	21.0	20.9	20.5	20.3	20.8	20.8	20.8	21.4	20.5	19.8	20.2	21.3	20.7
1984	21.7	20.7	21.0										
Mean	21.6	20.8	20.8	20.9	21.1	20.9	21.0	20.9	19.9	19.4	20.3	21.3	20.7*

* Row mean.

Table A.6a Monthly mean daily maximum temperature (°C) at Bilenge, 1977–1984.

	Oct.	Nov.	Dec.	Jan.	Feb.	Mar.	Apr.	May	June	July	Aug.	Sept.	Mean
1977–78			20.3	33.0	32.8	30.5	32.1	34.1	32.6	31.9	32.8	34.3	
1978–79	35.2	30.0	27.7	27.9	27.1	27.1	28.1	28.4	29.7				
1979–80						28.4	29.0	29.6	28.8	28.6	29.7	29.8	
1980–81	29.8	27.8	26.8	28.0		28.7	29.7				29.3		
1981–82		28.9	26.5		28.4	29.3	29.0	29.6	30.1	29.6			
1982–83			27.3										
1983–84											29.4	30.8	
1984	30.5	28.1	27.5										
Mean	31.8	28.7	26.0	29.6	29.4	28.8	29.6	30.4	30.3	30.0	30.3	31.6	29.7*

* Row mean.

Table A.6b Monthly mean daily minimum temperature (°C) Bilenge, 1977 –1984.

	Oct.	Nov.	Dec.	Jan.	Feb.	Mar.	Apr.	May	June	July	Aug.	Sept.	Mean
1977–78			20.3	19.5	20.4	20.1	20.7	19.8	19.2	18.0	18.8	20.3	
1978–79	20.2	19.7	19.7	20.5	20.4	20.3	20.4	19.9	17.9	17.6	17.9	18.8	19.4
1979–80	20.5	20.9	20.6			20.9	21.2	20.6	18.4	17.1	17.7	19.1	
1980–81	19.5	19.1	20.1	19.3	19.8	20.0	19.4	19.2	18.0	16.8	17.3	19.3	19.0
1981–82			20.3			19.9						20.2	
1982–83			20.5										
1983–84											18.3	19.0	
1984	20.1	19.5	19.8										
Mean	20.1	19.8	20.2	19.8	20.2	20.2	20.4	19.9	18.4	17.4	18.0	19.5	19.5*

* Row mean.

Bibliography

Articles marked with (*) at the end of references are unpublished semi-official reports to the government of Tanzania.

Allen, M., 1981. Individual copulatory preference and the "strange female effect" in captive group-living male chimpanzee (*Pan troglodytes*). *Primates*, **22**: 221–236.

Altmann, J., 1974. Observational study of behavior: Sampling methods. *Behaviour*, **49**: 227–267.

Altmann, J., 1980. *Baboon Mothers and Infants*. Harvard University Press, Cambridge, Mass.

Altmann, S.A. and J. Altmann, 1979. Demographic constraints on behavior and social organization. In: *Primate Ecology and Human Origins*, I.S. Bernstein and E.O. Smith (eds.), Garland, New York, pp. 47–63.

Anonymous, 1980. *Mahale: Study for the Proposed Mahale Mountains National Park, Final Report May 1980*. Japan International Cooperation Agency, Tokyo.

Avon, R.P., 1915–16. Vie sociale des Wabende au Tanganyika. *Anthropos*, **10–11**: 98–113.

Azuma, S. and A. Toyoshima, 1961–62. Progress report of the survey of chimpanzees in their natural habitat, Kabogo Point Area, Tanganyika. *Primates*, **3**: 61–70.

Bertram, B.R., 1975. Social factors influencing reproduction in wild lions. *J. Zool., Lond.*, **177**: 463–482.

Boesch, C. and H. Boesch, 1981. Sex differences in the use of natural hammers by chimpanzees: A preliminary report. *J. Human Evol.*, **10**: 585–593.

Boesch, C. and H. Boesch, 1984. Possible causes of sex differences in the use of natural hammers by wild chimpanzees. *J. Human Evol.*, **13**: 415–440.

Boesch, C. and H. Boesch, 1989. Hunting behavior of wild chimpanzees in the Tai National Park. *Am. J. Phys. Anthropol.*, **78**: 547–573.

Boesch, C. and H. Boesch, in press. Evolution of tool use and tool making in wild chimpanzees. *Folia Primatol.*

Bourne, G.H., 1971. Nutrition and diet of chimpanzees. In: *The Chimpanzee, Vol. 4*, G.H. Bourne (ed.), S.Karger, Basel, pp. 373–400.

de Bournonville, D., 1967. Contribution à l'étude du chimpanzé en République de Guinée. *Bull. Inst. Fond. Afr. Noire*, **29**: 1188–1269.

Bowden, D.M. and M.L. Jones, 1979. Aging research in nonhuman primates. In: *Aging in Nonhuman Primates*, D.M. Bowden (ed.), Van Nostrand Reinhold, New York, pp. 1–13.

Buss, D.H. and W.R. Voss, 1971. Evaluation of four methods for estimating the milk yield of baboons. *J. Nutr.*, **101**: 901–910.

Busse, C.D., 1977. Chimpanzee predation as a possible factor in the evolution of red colobus monkey social organization. *Evolution*, **31**: 907–911.

Busse, C.D., 1978. Do chimpanzees hunt cooperatively? *Am. Nat.*, **112**: 767–770.

Bygott, J.D., 1974. *Agonistic behaviour and dominance in wild chimpanzees*. Ph.D. thesis, University of Cambridge.

Bygott, J.D., 1979. Agonistic behavior, dominance, and social structure in wild chimpanzees of the Gombe National Park. In: *The Great Apes*, D.A. Hamburg and E.R. McCown (eds.), Benjamin/Cummings, Menlo Park, pp. 405–427.

Byrne, R.W. and J.M. Byrne, 1988. Leopard killers of Mahale. *Natural History*, **97**(3): 22–26.

Byrne, R.W. and A. Whiten, 1985. Tactical deception of familiar individuals in baboons (*Papio ursinus*). *Anim. Behav.*, **33**: 669–673.

Byrne, R.W. and A. Whiten, 1987. The thinking primate's guide to deception. *New Scientist*, 3 December 1987, pp. 54–57.

Caldecott, J., 1987. Medicine and the fate of tropical forests. *Brit. Med. J.*, **295**: 229–230.

Cameron, V.L., 1877. *Across Africa*. Daldy & Isbister, London.

Cane, L.B., 1947. S.S. "Liemba." *Tanganyika Notes and Records*, No. 23, pp. 31–46.

Carcason, R. H., 1966. Lepidoptera Rhopalocera collected by the Kyoto University African Anthropoid Expedition in the Kigoma area of western Tanganyika. *Kyoto Univ. Afr. Studies*, **1**: 11–72.

Cheney, D.L., R.M. Seyfarth, and B. Smuts, 1986. Social relationships and social cognition in nonhuman primates. *Science*, **234**: 1361–1366.

Chevalier-Skolnikoff, S., 1985. An exploration of the ontogeny of deception in human beings and nonhuman primates. In: *Deception*, R.W. Mitchell and N.S. Thompson (eds.), SUNY Press, Albany, pp. 205–220.

Chevalier-Skolnikoff, S. and F.E. Poirier (eds.), 1977. *Primate Bio-social Development*. Garland, New York.

Clark, C.B., 1977. A preliminary report on weaning among chimpanzees of Gombe National Park, Tanzania. In: *Primate Bio-social Development*, S. Chevalier-Skolnikoff and F.E. Poirier (eds.), Garland, New York, pp. 235–260.

Clutton-Brock, T.H. and S.D. Albon, 1982. Parental investment in male and female offspring in mammals. In: *Current Problems in Sociobiology*, King's College Sociobiology Group (ed.), Cambridge University Press, Cambridge, pp. 195–237.

Clutton-Brock, T.H., F.E. Guinness, and S.D. Albon, 1982. *Red Deer*. University of Chicago Press, Chicago.

Collins, D.A. and W.C. McGrew, 1985. Chimpanzees' (*Pan troglodytes*) choice of prey among termites (Macrotermitinae) in Western Tanzania. *Primates*, **26**: 375–389.

Collins, D.A. and W.C. McGrew, 1987. Termite fauna related to differences in tool-use between groups of chimpanzees (*Pan troglodytes*). *Primates*, **28**: 457–471.

Collins, D.A. and W.C. McGrew, 1988. Habitats of three groups of chimpanzees (*Pan troglodytes*) in western Tanzania compared. *J. Human Evol.*, **17**: 553–574.

Comfort, A., 1956. *The Biology of Senescence*. Rinehart, New York.

Cooke, J.A.L., 1964. A revisionary study of some spiders of the rare family Prodidomidae. *Proc. Zool. Soc. London*, **142**: 257–305.

Cooper-Driver, G.A. and T. Swain, 1976. Cyanogenic polymorphism in bracken in relation to herbivore predation. *Nature*, **260**: 604.

Davis, R., 1959. University of Oxford Second Tanganyika Expedition. *Nature*, **184**: 1543.

Davis, R., 1961. The fitness of names to drawings: A cross-cultural study in Tanganyika. *Brit. J. Psychol.*, **52**: 259–268.

Dawkins, R., 1976. *The Selfish Gene*. Oxford University Press, Oxford.

Dawkins, R. and J.R. Krebs, 1978. Animal signals: Information or manipulation? In: *Behavioural Ecology: An Evolutionary Approach*, J.R. Krebs and N.B. Davies (eds.), Blackwell, Oxford, pp. 282–309.

DeRousseau, C.J., L.Z. Bitop, and P.L. Kaufman, 1986. Age-dependent impairments of the rhesus monkey visual and musculoskeletal system and apparent behavioral consequences. In: *The Cayo Santiago Macaques, History, Behavior and Biology*, R.G. Rawlins and M.J. Kessler (eds.), SUNY Press, Albany, pp. 233–251.

Dittus, W.P.J., 1977. The social regulation of population density and age-sex distribution in the toque monkey. *Behaviour*, **63**: 281–322.

Dollman, G., 1935. The occurrence of the chimpanzees in Tanganyika Territory. *Proc. Linn. Soc. Lond.*, 21 November 1935, pp. 15–16.

Donisthorpe, J.H., 1958. A pilot study of the mountain gorilla (*Gorilla gorilla beringei*) in southwest Uganda, February to September 1957. *South Afr. J. Sci.*, **54**: 195–217.

Dunbar, R.I.M., 1980. Demographic and life history variables of a population of gelada baboons (*Theropithecus gelada*). *J. Anim. Ecol.*, **49**: 485–506.

Dunbar, R.I.M., 1988. *Primate Social Systems*. Croom Helm, London.

Eaton, R.K., 1978. Why some felids copulate so much: A model for the evolution of copulation frequency. *Carnivore*, **1**: 42–51.

Fairman, C., F.I. Reyes, J.S.D. Winter, and W.C. Hobson, 1981. Endocrinology of pregnancy in apes. In: *Reproductive Biology of the Great Apes*, C.E. Graham (ed.), Academic Press, New York, pp. 45–68.

Foley, R.A., 1988. *Another Unique Species*. Longman Scientific & Technical, Essex.

Freeland, W.J., 1976. Pathogens and the evolution of primate sociality. *Biotropica*, **8**: 12–24.

Fryer, G. and T.D. Iles, 1972. *The Cichlid Fishes of the Great Lakes of Africa*. Oliver & Boyd, Edinburgh.

Fujii, H., 1975. A psychological study of the social structure of a free-ranging group of Japanese monkeys in Katsuyama. In: *Contemporary Primatolgy*, S. Kondo, M. Kawai, and A. Ehara (eds.), S. Karger, Basel, pp. 428–436.

Galdikas, B.M., 1981. Orangutan reproduction in the wild. In: *Reproductive*

Biology of the Great Apes, C.E. Graham (ed.), Academic Press, New York, pp. 281–300.

Gallup, G.G. Jr., L.B. Wallnau, and S.D. Suarez, 1980. Failure to find self-recognition in mother-infant and infant-infant rhesus monkey pairs. *Folia Primatol.*, **33**: 210–219.

Gaulin, S.J.G., 1980. Sexual dimorphism in the human post reproductive lifespan: Possible causes. *J. Human Evol.*, **9**: 227–232.

Gavan, J.A., 1953. Growth and development of the chimpanzee: A longitudinal and comparative study. *Human Biol.*, **5**: 93–143.

Gavan, J.A., 1971. Longitudinal, postnatal growth in chimpanzee. In: *The Chimpanzee, Vol. 4*, G.H. Bourne (ed.), S. Karger, Basel, pp. 46–102.

Ghiglieri, M.P. 1984. *The Chimpanzees of Kibale Forest.* Columbia University Press, New York.

Glander, K., 1978. Howling monkey feeding behavior and plant secondary compounds: A study of strategies. In: *The Ecology of Arboreal Folivores,* G.C. Montgomery (ed.), Smithonian Instititution Press, Washington D.C., pp. 561–574.

Goodall, J., 1968. The behaviour of free-living chimpanzees in the Gombe Stream Reserve. *Anim. Behav. Monogr.*, **1**: 161–311.

Goodall, J., 1971. *In the Shadow of Man.* Collins, London.

Goodall, J., 1973a. Cultural elements in a chimpanzee community. In: *Precultural Primate Behavior*, E.W. Menzel, Jr. (ed.), S. Karger, Basel, pp. 144–184.

Goodall, J., 1973b. The behaviour of chimpanzees in their natural habitat. *Am. J. Psychiat.*, **130**: 1–12.

Goodall, J., 1983. Population dynamics during a 15 year period in one community of free-living chimpanzees in the Gombe National Park, Tanzania. *Z. Tierpsychol.*, **61**: 1–60.

Goodall, J., 1986. *The Chimpanzees of Gombe: Patterns of Behavior.* Belknap, Harvard University Press, Cambridge, Mass.

Graham, C.E., 1979. Reproductive function in aged female chimpanzees. *Am. J. Phys. Anthropol.*, **50**: 291–300.

Graham, C.E., 1986. Endocrinology of reproductive senescence. In: *Comparative Primite Biology, Vol. 3: Reproduction and Development*, W.R. Dukelow and J. Erwin (eds.) Alan R. Liss, New York, pp.93–99.

Grant, C.H.B., 1946. The distribution of the chimpanzee in Tanganyika Territory. *Tanganyika Notes and Records*, No. 21, pp. 110–111.

Grant. C.H.B., 1949. Notes on Uvinza, Tongwe and Ubende, Western Tanganyika. *Tanganyika Notes and Records*, No. 27, pp. 68–73.

Hamilton, W.D., 1966. The moulding of senescence by natural selection. *J. Theoret. Biol.*, **12**: 12–45.

Hamilton, W.D., 1967. Extraordinary sex ratios. *Science*, **156**: 477–488.

Harcourt, A.H., D. Fossey, and J. Sabater-Pi, 1981. Demography of *Gorilla gorilla. J. Zool., Lond.*, **195**: 215–233.

Harcourt, A.H., P.H. Harvey, S.G. Larson, and R.V. Short, 1981. Testes weight, body weight and breeding system in primates. *Nature*, **293**: 55–57.

Harcourt, A.H. and K.J. Stewart, 1983. Interactions, relationships and social structure: The great apes. In: *Primate Social Relationships: An Integrated Approach*, R.A. Hinde (ed.), Blackwell, Oxford, pp. 307–314.

Harvey, P.H. and A.H. Harcourt, 1984. Sperm competition, testes size, and breeding system in primates. In: *Sperm Competition and the Evolution of Animal Mating Systems*, R.L. Smith (ed.), Academic Press, New York, pp. 589–600.

Harvey, P.H., R.D. Martin, and T.H. Clutton-Brock, 1987. Life histories in comparative perspective. In: *Primate Societies*, B.B. Smuts, D.L. Cheney, R.M. Sayfarth, R.W. Wrangham, and T.T. Struhsaker (eds.), University of Chicago Press, Chicago, pp. 181–196.

Hasegawa, T., 1980. Precipitation, atmospheric temperature, relative humidity and some other observations in Kasoje Chimpanzee Research Station, January–December 1979. *Mahale Mountains Chimpanzee Research Project, Meteorological Report No. 5.* (*)

Hasegawa, T., 1989. Sexual behavior of immigrant and resident female chimpanzees at Mahale. In: *Understanding Chimpanzees*, P.G. Heltne and L.A. Marquardt (eds.), Harvard University Press, Cambridge, Mass., pp. 90–103.

Hasegawa, T. and M. Hiraiwa-Hasegawa, 1980. The eighth ascent of Mt. Nkungwe: Chimpanzees of the Humo-group and other larger mammals of the main ridge of Mahale Massif. *Mahale Mountains Chimpanzee Research Project, Survey Report No. 10.* (*)

Hasegawa, T. and M. Hiraiwa-Hasegawa, 1981. Precipitation, atmospheric temperature, and relative humidity in Mahale Mountains Wildlife Research Center, January–December 1980. *Mahale Mountains Chimpanzee Research Project, Meteorological Report No. 6.* (*)

Hasegawa, T. and M. Hiraiwa-Hasegawa, 1982. Precipitation, atmospheric temperature, and relative humidity of January–December 1981. *Mahale Mountains Chimpanzee Research Project, Meteorological Report No. 7.* (*)

Hasegawa, T. and M. Hiraiwa-Hasegawa, 1983. Opportunistic and restrictive matings among wild chimpanzees in the Mahale Mountains, Tanzania. *J. Ethol.*, **1**: 75–85.

Hasegawa, T. and M. Hiraiwa-Hasegawa, 1984. Seasonal changes in ranging and grouping patterns of wild chimpanzees in Mahale, Tanzania. *Int. J. Primatol.*, **5**: 345.

Hasegawa, T., M. Hiraiwa, T. Nishida, and H. Takasaki, 1983. New evidence on scavenging behavior in wild chimpanzees. *Curr. Anthropol.*, **24**: 231–232.

Hasegawa, T. and T. Nishida, 1984. Progress report on Mahale National Park. *IUCN/SSC Primate Specialist Group Newsletter, 4 March 1984*, pp. 37–38.

Hata, N., 1968. Ndagaa fishing and settlement formation along Lake Tanganyika. *Kyoto Univ. Afr. Studies*, **2**: 31–50.

Hatchell, G.W., 1941. Some account of the people living under the protection of Mount Mkungwe. *Tanganyika Notes and Records*, No. 11, pp. 41–46.

Hauser, M.D. and G. Tyrrell, 1984. Old age and its behavioral manifestations: A study on two species of macaque. *Folia Primatol.*, **43**: 24–35.

Hayaki, H., 1985a. Copulation of adolescent male chimpanzees, with special reference to the influence of adult males, in the Mahale National Park, Tanzania. *Folia Primatol.*, **44**: 148–160.

Hayaki, H., 1985b. Social play of juvenile and adolescent chimpanzees in the Mahale Mountains National Park, Tanzania. *Primates*, **26**: 343–360.

Hayaki, H., 1988. Association partners of young chimpanzees in the Mahale Mountains National Park, Tanzania. *Primates*, **29**: 147–161.

Hayaki, H., M.A. Huffman, and T. Nishida, 1989. Dominance among male chimpanzees in the Mahale Mountains National Park, Tanzania: A preliminary study. *Primates*, **30**: 187–197.

Hinde, R.A., 1974. *Biological Bases of Human Social Behaviour*. McGraw-Hill, New York.

Hinde, R.A. (ed.), 1984. *Primate Social Relationships: An Integrated Approach*. Blackwell, Oxford.

Hiraiwa-Hasegawa, M., 1986. *Mother-infant relationships in the free-ranging chimpanzees of the Mahale National Park, Tanzania*. Dissertation, University of Tokyo. (in Japanese)

Hiraiwa-Hasegawa, M., 1989. Sex differences in the behavioral development of chimpanzees at Mahale. In: *Understanding Chimpanzees*, P.G. Heltne and L.A. Marquardt (eds.), Harvard University Press, Cambridge, Mass., pp. 104–115.

Hiraiwa-Hasegawa, M., R.W. Byrne, H. Takasaki, and J.E. Byrne, 1986. Aggression toward large carnivores by wild chimpanzees of Mahale Mountains National Park, Tanzania. *Folia Primatol.*, **47**: 8–13.

Hiraiwa-Hasegawa, M. and T. Hasegawa, 1988. A case of offspring desertion by a female chimpanzee and the behavioral changes of the abandoned offspring. *Primates*, **29**: 319–330.

Hiraiwa-Hasegawa, M., T. Hasegawa, and T. Nishida, 1984. Demographic study of a large-sized unit-group of chimpanzees in the Mahale Mountains, Tanzania: A preliminary report. *Primates*, **25**: 401–413.

Honjo, S., 1985. A scientific and ethical view of the use of nonhuman primates for bio-medical experiment. *Primate Research*, **1**: 31–38. (in Japanese)

van Hooff, J.A.R.A.M., 1973. A structural analysis of the social behaviour of a semi-captive group of chimpanzees. In: *Social Communication and Movement*, M. von Cranach and I. Vine (eds.), Academic Press, London, pp. 75–162.

Hori, M., 1985a. Hunting success of three piscivorous species of the genus *Lamprologus* (Cichlidae). *Ecol. Limnol. Tanganyika*, **3**: 50–51.

Hori, M., 1985b. Hunting success of two scale-eaters, *Perissodus microlepis* and *P. straeleni* (Pisces: Cichlidae). *Ecol. Limnol. Tanganyika*, **3**: 52–53.

Hori, M., 1985c. Faunal surveys of terrestrial and aquatic insects. *Ecol. Limnol. Tanganyika*, **3**: 58.

Hori, M., 1987. Mutualism and commensalism in the fish community of Lake Tanganyika. In: *Evolution and Coadaptation in Biotic Communities*, S. Kawano, J.H. Conell and T. Hidaka (eds.), University of Tokyo Press, Tokyo, pp. 219–239.

Horvat, J.R. and H.C. Kraemer, 1982. Behavioural changes during weaning in captive chimpanzees. *Primates*, **23**: 488–499.

Howell, N., 1979. *Demography of the Dobe ! Kung*. Academic Press, New York.

Hrdy, S., 1981. "Nepotists" and "altruists": The behavior of old females among macaques and langur monkeys. In: *Other Ways of Growing Old*, P. Amos and S. Harrell (eds.), Stanford University Press, Stanford, pp. 59–76.

Hutchins, M. and D.P. Barash, 1976. Grooming in primates: Implications for its utilitarian function. *Primates*, **17**: 145–150.

Huffman, M.A. and M. Seifu, 1989. Observations on the illness and consumption of a possibly medicinal plant *Vernonia amygdalina* (Del.), by a wild chimpanzee in the Mahale Mountains National Park, Tanzania. *Primates*, **30**: 51–63.

Imanishi, K., 1951. *Prehuman Societies*. Iwanami-shoten, Tokyo. (in Japanese)

Imanishi, K., 1957. Identification: A process of enculturation in the subhuman society of *Macaca fuscata*. *Primates*, **1**: 1–29. (in Japanese)

Imanishi, K., 1958. Gorillas: A preliminary survey in 1958. *Primates*, **1**: 73–78.

Imanishi, K., 1960. *The Gorilla*. Bungeishunjyu-shinsha, Tokyo. (in Japanese)

Imanishi, K., 1961. The origin of human family: A primatological approach. *Jpn. J. Ethnol.*, **25**: 110–130. (in Japanese)

Itani, J., 1961. *The Forest of Gorillas and the Pygmies*. Iwanami-shoten, Tokyo. (in Japanese)

Itani, J., 1964. *Animals of Africa*. Kawade-shobo, Tokyo. (in Japanese)

Itani, J., 1966. The social organization of chimpanzees. *Shizen*, **21** (8): 17–30. (in Japanese)

Itani, J., 1967. An artificial feeding of wild chimpanzees and their social organization. *Kagaku-Asahi*, **27** (2): 79–85. (in Japanese)

Itani, J., 1970. *In the Pursuit of Chimpanzees*. Chikuma-shobo, Tokyo. (in Japanese)

Itani, J., 1974. African great apes: A comparative study of ecology and social structure and their future problems. *Africa-Kenkyu* (*Journal of African Studies*), **14**: 1–13. (in Japanese)

Itani, J., 1977a. Ethnozoology of the Batongwe. In: *The Natural History of Man*, J. Itani and R. Harako (eds.), Yuzan-kaku, Tokyo, pp. 441–537. (in Japanese)

Itani, J., 1977b. *The Savanna of Chimpanzees*. Heibon-sha, Tokyo. (in Japanese)

Itani, J., 1977c. Introduction. In: *The Chimpanzees*, J. Itani (ed.), Kodan-sha, Tokyo, pp. 3–56. (in Japanese)

Itani, J., 1977d. A travel to Ugalla. In: *The Chimpanzees*, J. Itani (ed.), Kodansha, Tokyo, pp. 393–435. (in Japanese)

Itani, J., 1979. Distribution of chimpanzees in an arid area. In: *The Great Apes*, D.A. Hamburg and E.R. McCown (eds.), Benjamin/Cummings, Menlo Park, pp. 55–71.

Itani, J., 1980a. Social structure of African great apes. *J. Reprod. Fert., Suppl.*, **28**: 33–41.

Itani, J., 1980b. Individual and group aging in primates. *J. Japanese Geriatric Society*, **17** (2): 137–141. (in Japanese)

Itani, J., 1984a. The natural villages of the Batongwe. In: *Studies on African Culture*, J. Itani and T. Yoneyama (eds.), Akademia-shuppankai, Kyoto, pp. 699–728. (in Japanese)

Itani, J., 1984b. Inequality versus equality for coexistence in primate societies. In: *Absolute Values and the New Cultural Revolution, New York, 12th Conference on the Unity of the Sciences*, I.C.U.S. Books, New York, pp. 161–189.

Itani, J., 1985. The evolution of primate social structures. *Man* (*n.s.*), **20**: 593–611.

Itani, J., 1988. The origin of human equality. In: *Social Fabrics of the Mind*, M.R.A. Chance (ed.), Lawrence Erlbaum Associates, London, pp. 137–156.

Itani, J., T. Nishida, and M. Kakeya, 1973. *The Eastern Hinterland of Lake Tanganyika*. Chikuma-shobo, Tokyo. (in Japanese)

Itani, J. and A. Suzuki, 1967. The social unit of chimpanzees. *Primates*, **8**: 355–381.

Itoigawa, N., 1982. Aging and behavior in Japanese monkeys. *Jpn. J. Ethnol.*, **46**: 376–389. (in Japanese)

Izawa, K., 1970. Unit groups of chimpanzees and their nomadism in the savanna woodland. *Primates*, **11**: 1–46.

Izawa, K., 1975. *Good bye Brucy*. NHK Publishing Association, Tokyo. (in Japanese)

Izawa, K., 1977. The chimpanzees of Kasakati Basin: (II) Unit-groups and their nomadic life. In: *The Chimpanzees*, J. Itani (ed.), Kodan-sha, Tokyo, pp. 187–248. (in Japanese)

Izawa, K. and J. Itani, 1966. Chimpanzees in Kasakati Basin, Tanganyika: (1) Ecological study in the rainy season, 1963–1964. *Kyoto Univ. Afr. Studies*, **1**: 73–156.

Johnson, C.N., 1988. Dispersal and sex ratio at birth in primates. *Nature*, **332**: 726–728.

Juniper, B.E., 1959. University of Oxford Tanganyika Expedition. *Nature*, **183**: 726 –727.

Kakeya, M., 1976. Subsistence ecology of the Tongwe, Tanzania. *Kyoto Univ. Afr. Studies*, **10**: 143–212.

Kakeya, M., 1978. The materials and logics of Sikomelo: Magical medicines used by the Batongwe tribe. *Africa-Kenkyu* (*Journal of African Studies*), **17**: 1–33. (in Japanese)

Kano, T., 1971a. The chimpanzee of Filabanga. *Primates*, **12**: 229–246.

Kano, T., 1971b. Distribution of the primates on the eastern shore of Lake Tanganyika. *Primates*, **12**: 281–304.

Kano, T., 1972. Distribution and adaptation of the chimpanzee in the open country on the eastern shore of Lake Tanganyika. *Kyoto Univ. Afr. Studies*, **7**: 37–129.

Kano, T., 1977. The distribution of chimpanzees in western Tanzania. In: *The Chimpanzees*, J. Itani (ed.), Kodan-sha, Tokyo, pp. 337–391. (in Japanese)

Kano, T., 1987. A population study of a unit group of pygmy chimpanzees of Wamba—with a special reference to the possible lack of intraspecific killing.

In: *Animal Societies: Theories and Facts*, Y. Ito, J.L. Brown, and J. Kikkawa (eds.), Japan Scientific Societies Press, Tokyo, pp. 159–172.

Kawabata, M., 1975. Fish community near Mukuyu, the east coast of Lake Tanganyika. *Kyoto Univ. Afr. Studies*, **9**: 21–31.

Kawabata, M. and T. Doi, 1972. Non-Cichlidae fish community in Mukuyu and Karago areas, the east coast of Lake Tanganyika. *Kyoto Univ. Afr. Studies*, **7**: 171–179.

Kawabe, M., 1966. One observed case of hunting behavior among wild chimpanzees living in the savanna woodland of western Tanzania. *Primates*, **7**: 393–396.

Kawabe, M., 1977. The chimpanzees of the Kasakati Basin: (I) The grouping pattern. In: *The Chimpanzees*, J. Itani (ed.), Kodan-sha, Tokyo, pp. 153–186. (in Japanese)

Kawai, M., 1958. On the rank system in a natural group of Japanese monkey (I)—The basic and the dependent rank. *Primates*, **1**: 111–130. (in Japanese)

Kawai, M., 1961. *The Gorilla Expedition*. Kobun-sha, Tokyo. (in Japanese)

Kawai, M. and H. Mizuhara, 1959. An ecological study on the wild mountain gorilla (*Gorilla gorilla beringei*)—Report of the Japan Monkey Center Second Gorilla Expedition 1959. *Primates*, **2**: 1–42.

Kawanabe, H., 1983. Further observation of territoriality of *Tropheus moorei* and *Petrochromis polyodon* (Osteichthyes: Cichlidae). *Ecol. Limnol. Tanganyika*, **2**: 10–11.

Kawanaka, K., 1979. Precipitation, atmospheric temperature, relative humidity and some other observations in Kasoje Chimpanzee Research Station, January–December 1978. *Mahale Mountains Chimpanzee Research Project, Meteorological Report No.* 4. (*)

Kawanaka, K., 1981. Infanticide and cannibalism in chimpanzees, with special reference to the newly observed case in the Mahale Mountains. *Afr. Study Monogr.*, **1**: 69–99.

Kawanaka, K., 1982a. Further studies on predation by chimpanzees of the Mahale Mountains. *Primates*, **23**: 364–384.

Kawanaka, K., 1982b. A case of inter-unit-group encounter in chimpanzees of the Mahale Mountains. *Primates*, **23**: 558–562.

Kawanaka, K., 1984. Association, ranging, and the social unit in chimpanzees of the Mahale Mountains, Tanzania. *Int. J. Primatol.*, **5**: 411–434.

Kawanaka, K., 1988. Social structure of African apes: A review. *Africa-Kenkyu* (*Journal of African Studies*), **32**: 69–88. (in Japanese)

Kawanaka, K., 1989. Age differences in social interactions of young males in a chimpanzee unit-group of the Mahale Mountains National Park, Tanzania. *Primates*, **30**: 285–305.

Kawanaka, K. and T. Nishida, 1975. Recent advances in the study of inter-unit-group relationships and social structure of wild chimpanzees of the Mahali Mountains. *Proc. Symp. 5th Cong. Int'l. Primat. Soc.*, S. Kondo, M. Kawai, A. Ehara, and S. Kawamura (eds.), Japan Science Press, Tokyo, pp. 173–186.

Kenagy, G.L. and S.C. Trombulak, 1986. Size and function of mammalian testes in relation to body size. *J. Mammal.*, **67**: 1–22.

Kielland, J., 1968. Rhopalocera from the eastern side of Lake Tanganyika. *Atalanta*, **2**: 53–73.

Kielland, J., 1978. A provisional checklist of the Rhopalocera of the eastern side of Lake Tanganyika. *Tijdschrift voor Entomologie*, **121**: 147–237.

Kimura, D. and H. Ihobe, 1985. Feral cattle (*Bos taurus*) on Kuchinoshima, southwestern Japan: Their stable ranging and unstable grouping. *J. Ethol.*, **3**: 39–47.

Kirschstein, E.F., 1937. Some tales of Tanganyika natives. *Tanganyika Notes and Records*, No. 4, pp. 82–83.

Kummer, H., 1982. Social knowledge in free-ranging primates. In: *Animal Mind —Human Mind*, D.R. Griffin (ed.), Springer-Verlag, Berlin, pp. 113–130.

Kummer, H., W. Gotz, and W. Angst, 1974. Triadic differentiation: An inhibitory process protecting pair bonds in baboons. *Behaviour*, **49**: 62–87.

Kurland, J.A., 1977. *Kin Selection in the Japanese Monkey*. S. Karger, Basel.

Kuwamura, T., 1985a. Breeding ecology of cichlid fishes in Lake Tanganyika. *Ecol. Limnol. Tanganyika*, **3**: 54–55.

Kuwamura, T., 1985b. Reproductive biology and general ecology of the world's largest cichlid *Boulengerochromis microlepis*. *Ecol. Limnol. Tanganyika*, **3**: 56–57.

Kuwamura, T., 1986a. Substratum spawning and biparental guarding of the Tanganyikan cichlid *Boulengerochromis microlepis* with notes on its life history. *Physiol. Ecol. Jpn.*, **23**: 31–43.

Kuwamura, T., 1986b. Parental care and mating systems of cichlid fishes in Lake Tanganyika: A preliminary field survey. *J. Ethol.*, **4**: 129–146.

Kuwamura, T., 1987a. Distribution of fishes in relation to the depth and substrate at Myako, east-middle coast of Lake Tanganyika. *Afr. Study Monogr.*, **7**: 1–14.

Kuwamura, T., 1987b. Male mating territory and sneaking in a maternal mouthbrooder, *Pseudosimochromis curvifrons* (Pisces: Cichlidae). *J. Ethol.*, **5**: 203–206.

Lee, P.C. and C.J. Moss, 1986. Early maternal investment in male and female African elephant calves. *Behav. Ecol. Sociobiol.*, **18**: 353–361.

Leslie, P.W. and P.H. Fry, 1989. Extreme seasonality of births among nomadic Turkana pastoralists. *Am. J. Phys. Anthropol.*, **79**: 103–115.

McConell, R.B., 1945. A further note on chimpanzee in the Kigoma district. *Tanganyika Notes and Records*, No. 19, p. 18.

McGinnis, P.A., 1979. Sexual behavior in free-living chimpanzees: Consort relationships. In: *The Great Apes*, D.A. Hamburg and E.R. McCown (eds.), Benjamin/Cummings, Menlo Park, pp. 429–439.

McGrew, W.C., 1974. Tool-use by wild chimpanzees in feeding upon driver ants. *J. Human Evol.*, **3**: 501–508.

McGrew, W.C., 1975. Patterns of plant food sharing by wild chimpanzees. In: *Contemporary Primatology*, S. Kondo, M. Kawai, and A. Ehara (eds.), S. Karger, Basel, pp. 304–309.

McGrew, W.C., 1977. Socialization and object manipulation of wild chimpanzees. In: *Primate Bio-social Development:* S. Chevalier-Skolnikoff and F.E. Poirier (eds.), Garland, New York, pp. 261–288.

McGrew, W.C., 1979. Evolutionary implications of sex difference in chimpanzee predation and tool use. In: *The Great Apes*, D.A. Hamburg and E.R. McCown (eds.), Benjamin/Cummings, Menlo Park, pp. 440–463.

McGrew, W.C., 1983. Animal foods in the diet of wild chimpanzees (*Pan troglodytes*): Why cross-cultural variation? *J. Ethol.*, **1**: 46–61.

McGrew, W.C., 1989. Why is ape tool use so confusing? In: *Comparative Socioecology: The Behavioural Ecology of Humans and Other Mammals*, V. Standen and R.A. Foley (eds.), Blackwell, Oxford, pp. 457–472.

McGrew, W.C., P.J. Baldwin, and C.E.G. Tutin, 1981. Chimpanzees in a hot, dry and open habitat: Mt. Assirik, Senegal, West Africa. *J. Human Evol.*, **10**: 227–244.

McGrew, W.C., P.J. Baldwin, and C.E.G. Tutin, 1988. Diet of wild chimpanzees (*Pan troglodytes verus*) at Mt. Assirik, Senegal: I Composition. *Am. J. Primatol.*, **16**: 213–226.

McGrew, W.C. and D.A. Collins, 1985. Tool use by wild chimpanzees (*Pan troglodytes*) to obtain termites (*Macrotermes herus*) in the Mahale Mountains, Tanzania. *Am. J. Primatol.*, **9**: 47–62.

McGrew, W.C. and C.E.G. Tutin, 1978. Evidence for a social custom in wild chimpanzees? *Man (n.s.)*, **13**: 234–251.

McGrew, W.C. and J. Webster, 1989. Birth seasonality in cotton-top tamarins (*Saguinus oedipus*) despite constant food supply and body weight. Unpublished manuscript.

von Majerus, P., 1911. Brautwerbung und Hochzeit bei den Wabende (Deutsch-Ostafrika). *Anthropos*, **6**: 893–900.

von Majerus, P., 1915–6. Das Eherecht der Wabende (Deutsch-Ostafrika). *Anthropos*, **10–11**: 781–788.

Martin, D.E., 1981. Breeding great apes in captivity. In: *Reproductive Biology of the Great Apes*, C.E. Graham (ed.), Academic Press, New York, pp. 343–373.

Martin, D.E., C.E. Graham, and K.G. Gould, 1978. Successful artificial insemination in the chimpanzee. *Symp. Zool. Soc. Lond.*, **43**: 249–260.

Martin, P. and P. Bateson, 1986. *Measuring Behaviour*. Cambridge University Press, Cambridge.

Masui, K., 1986. An infant chimpanzee was captured and lethally injured by the first-ranking male: The seventh case of infant-killing at Mahale Mts. National Park, west Tanzania. *Mahale Mountains Chimpanzee Research Project, Ecological Report No. 46.* (*)

Maxwell, F.G., J.N. Jenkins, and W.L. Parrot, 1972. Resistance of plants to insects. *Advances in Agronomy*, **24**: 187–265.

Mayer, P.J., 1982. Evolutionary advantage of the menopause. *Human Ecol.*, **10**: 477–494.

Maynard Smith, J., 1982. Do animals convey information about their intentions? *J. Theoret. Biol.*, **97**: 1–5.

Menzel, E.W., Jr., 1973. Leadership and communication in young chimpanzees.

In: *Precultural Primate Behavior*, E.W. Menzel, Jr. (ed.), S. Karger, Basel, pp. 192–225.

Miller, L.C. and R.D. Nadler, 1981. Mother-infant relations and infant development in captive chimpanzees and organgutans. *Int. J. Primatol.*, **2**: 247–261.

Milton, K., 1979. Factors influencing leaf choice by howler monkeys: A test of some hypothesis of food selection by generalist herbivores. *Am. Nat.*, **114**: 362–378.

Missakian, E.A., 1973. Genealogical mating activity in free-ranging groups of rhesus monkeys (*Macaca mulatta*) on Cayo Santiago. *Behaviour*, **45**: 224–240.

Møller, A.P., 1988. Ejaculate quality, testes size and sperm competition in primates. *J. Human Evol.*, **17**: 479–488.

Moore, J., 1986. Arid chimp survey, Ugalla (Tanzania), Nov. 1–5, 1985. *Report to the L.S.B. Leakey Foundation.*

Moreau, R.E., 1942. The distribution of the chimpanzee in Tanganyika Territory. *Tanganyika Notes and Records*, No. 14, pp. 52–55.

Moreau, R.E., 1943. A contribution to the ornithology of the east side of Lake Tanganyika. *Ibis*, **85**: 377–412.

Mori, A., 1975. Signals found in the grooming interactions of wild Japanese monkeys of the Koshima troop. *Primates*, **16**: 107–140.

Mori, A., 1982. An ethological study on chimpanzees at the artificial feeding place in the Mahale Mountains, Tanzania—with special reference to the booming situation. *Primates*, **23**: 45–65.

Mori, A., 1983. Comparison of the communicative vocalizations and behaviors of group ranging in eastern gorillas, chimpanzees and pygmy chimpanzees. *Primates*, **24**: 486–500.

Myers, N., 1979. *The Sinking Ark*. Pergamon Press, Oxford.

Nadler, R.D. and J.T. Braggio, 1974. Sex and species differences in captive-reared juvenile chimpanzees and orang-utans. *J. Human Evol.*, **3**: 541–550.

Nakagawa, N., 1989. Feeding strategies of Japanese monkeys against deterioration of habitat quality. *Primates*, **30**: 1–16.

Nakamichi, M., 1984. Behavioral characteristics of old female Japanese monkeys in a free-ranging group. *Primates*, **25**: 192–203.

Nakamichi, M., 1988. Aging and behavioral changes of female Japanese macaques. In: *Research Reports of the Arashiyama West and East Groups of Japanese Monkeys Published by the Laboratory of Ethological Studies, Osaka University*, pp. 87–97.

Narita, T., 1983. Species composition, vertical distribution and density of zooplanktons and some limnological features of the coast of Mahale Mountains in Lake Tanganyika. *Ecol. Limnol. Tanganyika*, **2**: 12–14.

Narita, T., N. Mulimbwa, and T. Mizuno, 1985. Vertical distribution and seasonal abundance of zooplanktons in Lake Tanganyika. *Afr. Study Monogr.*, **6**: 1–16.

Newton, P. and T. Nishida, in press. Possible buccal administration of herbal drugs by wild chimpanzees (*Pan troglodytes*). *Anim. Behav.*

Nicolson, N., 1987. Infants, mothers and other females. In: *Primate Societies*, B.B. Smuts, D.L. Cheney, R.M. Seyfarth, R.W. Wrangham, and T.T. Struhsaker (eds.), University of Chicago Press, Chicago, pp. 330–342.

Nicolson, N. and M.W. Demment, 1982. The transition from suckling to independent feeding in wild baboon infants. *Int. J. Primatol.*, **3**: 318.

Nishida, T., 1967. The society of wild chimpanzees. *Shizen*, **22** (8): 31–41. (in Japanese)

Nishida, T., 1968. The social group of wild chimpanzees in the Mahali Mountains. *Primates*, **9**: 167–224.

Nishida, T., 1970. Social behavior and relationship among wild chimpanzees of the Mahali Mountains. *Primates*, **11**: 47–87.

Nishida, T., 1972. A note on the ecology of the red-colobus monkeys (*Colobus badius tephrosceles*) living in the Mahali Mountains. *Primates*, **13**: 57–64.

Nishida, T., 1973a. *The Children of the Mountain Spirits.* Chikuma-shobo, Tokyo. (in Japanese)

Nishida, T., 1973b. The ant-gathering behaviour by the use of tools among wild chimpanzees of the Mahali Mountains. *J. Human Evol.*, **2**: 357–370.

Nishida, T., 1974. Ecology of wild chimpanzees. In: *Human Ecology*, R. Ohtsuka, J. Tanaka, and T. Nishida (eds.), Kyoritsu-shuppan, Tokyo, pp. 15–60. (in Japanese)

Nishida, T., 1975. A general survey of the southern part of the Mahale Mountains. *Mahale Mountains Chimpanzee Research Project, Survey Report No. 1.* (*)

Nishida, T., 1976a. The bark-eating habits in primates, with special reference to their status in the diet of wild chimpanzees. *Folia Primatol.*, **25**: 277–287.

Nishida, T., 1976b. The sixth ascent of Mt. Nkungwe, the paramount peak of the Mahale Massif. *Mahale Mountains Chimpanzee Research Project, Survey Report No. 4.* (*)

Nishida, T., 1977a. Chimpanzees of the Mahale Mountains: (I) Ecology and the structure of a unit-group. In: *The Chimpanzees*, J. Itani (ed.), Kodansha, Tokyo, pp. 543–638. (in Japanese)

Nishida, T., 1977b. The social structure of chimpanzees. In: *Anthropology Series 2: Ecology*, J. Itani (ed.), Yuzan-kaku, Tokyo, pp. 277–314. (in Japanese)

Nishida, T., 1978. Revolution in the Lake Tanganyika Region. *Tenbo*, No. 232, pp. 103–124. (in Japanese)

Nishida, T., 1979a. Infant-rearing in chimpanzees. In: *Children*, UTP (ed.), University of Tokyo Press, Tokyo, pp. 157–185. (in Japanese)

Nishida, T., 1979b. The social structure of chimpanzees of the Mahale Mountains. In: *The Great Apes*, D.A. Hamburg and E.R. McCown (eds.), Benjamin/Cummings, Menlo Park, pp. 73–121.

Nishida, T., 1979c. A general survey to the Muhela Plateau. *Mahale Mountains Chimpanzee Research Project, Survey Report No. 7.* (*)

Nishida, T., 1980a. The leaf-clipping display: A newly-discovered expressive gesture in wild chimpanzees. *J. Human Evol.*, **9**: 117–128.

Nishida, T., 1980b. Local differences in responses to water among wild chimpanzees. *Folia Primatol.*, **33**: 189–209.

Nishida, T., 1980c. On chimpanzee culture. *Kagaku*, **50**: 146–154. (in Japanese)

Nishida, T., 1981. *The World of Wild Chimpanzees*. Chuokoron-sha, Tokyo. (in Japanese)

Nishida, T., 1982. Japanese aid to the establishment of Mahale National Park. *Gekkan-Apikku*, No. 51, pp. 28–33. (in Japanese)

Nishida, T., 1983a. Alpha status and agonistic alliance in wild chimpanzees (*Pan troglodytes schweinfurthii*). *Primates*, **24**: 318–336.

Nishida, T., 1983b. Alloparental behavior in wild chimpanzees of the Mahale Mountains, Tanzania. *Folia Primatol.*, **41**: 1–33.

Nishida, T., 1984. The social structure of chimpanzees and bonobos. In: *Anthropology: Its Diverse Development*, Anthropological Society of Nippon (ed.), Nikkei-saiensu-sha, Tokyo, pp. 240–246. (in Japanese)

Nishida, T., 1985. The Mahale Mountains declared a National Park. *IPPL Newsletter*, **12** (3): 9.

Nishida, T., 1986a. Development of reciprocity in grooming between chimpanzee mother and offspring. *Primate Report*, No. 14, p. 186.

Nishida, T., 1986b. Reciprocity in grooming among wild chimpanzees: A preliminary report. *Proceedings of the 23rd Congress of Japan Association for African Studies, May 31–June 1, 1986*, p. 4.

Nishida, T., 1987a. The present condition of the behavioral ecology of chimpanzees in the Mahale Mountains. *Gakujyutu-geppou*, **40**: 417–422. (in Japanese)

Nishida, T., 1987b. Local traditions and cultural transmission. In: *Primate Societies*, B.B. Smuts, D.L. Cheney, R.M. Seyfarth, R.W. Wrangham, and T.T. Struhsaker (eds.), University of Chicago Press, Chicago, pp. 462–474.

Nishida, T., 1988. Development of social grooming between mother and offspring in wild chimpanzees. *Folia Primatol.*, **50**: 109–123.

Nishida, T., 1989a. A note on the chimpanzee ecology of the Ugalla area, Tanzania. *Primates*, **30**: 129–138.

Nishida, T., 1989b. Social interactions between resident and immigrant female chimpanzees. In: *Understanding Chimpanzees*, P.G. Heltne and L.A. Marquardt (eds.), Harvard University Press, Cambridge, Mass., pp. 68–89.

Nishida, T., in press. Social structure and dynamics of the chimpanzee: A review. In: *Perspectives in Primate Biology*, P. Seth and S. Seth (eds.), Today's and Tomorrow's Publisher, Delhi.

Nishida, T. and M. Hiraiwa, 1982. Natural history of a tool-using behavior by wild chimpanzees in feeding upon wood-boring ants. *J. Human Evol.*, **11**: 73–99.

Nishida, T. and M. Hiraiwa-Hasegawa, 1984. Behavior of an adult male in the Mahale Mountains, Tanzania. *Int. J. Primatol.*, **5**: 367.

Nishida, T. and M. Hiraiwa-Hasegawa, 1985. Responses to a stranger mother-son pair in the wild chimpanzee: A case report. *Primates*, **26**: 1–13.

Nishida, T. and M. Hiraiwa-Hasegawa, 1987. Chimpanzees and bonobos: Cooperative relationships among males. In: *Primate Societies*, B.B. Smuts,

D.L. Cheney, R.M. Seyfarth, R.W. Wrangham, and T.T. Struhsaker (eds.), University of Chicago Press, Chicago, pp. 165–177.

Nishida, T., M. Hiraiwa-Hasegawa, T. Hasegawa, and Y. Takahata, 1985. Group extinction and female transfer in wild chimpanzees in the Mahale National Park, Tanzania. *Z. Tierpsychol.*, **67**: 284–301.

Nishida, T., J. Itani, M. Hiraiwa, and T. Hasegawa, 1981. A newly-discovered population of *Colobus angolensis* in East Africa. *Primates*, **22**: 557–563.

Nishida, T. and K. Kawanaka, 1972. Inter-unit-group relationships among wild chimpanzees of the Mahali Mountains. *Kyoto Univ. Afr. Studies*, **7**: 131–169.

Nishida, T. and K. Kawanaka, 1985. Within-group cannibalism by adult male chimpanzees. *Primates*, **26**: 274–284.

Nishida, T., K. Kawanaka, S. Uehara, H. Hayaki, and H. Takasaki, 1984. *The University of Tokyo Ape Expedition to Africa (1983–84): Interim Report.* (*)

Nishida, T. and K. Norikoshi, 1975. Distribution of the unit-groups of chimpanzees in the Kasoje area in the *Garcinia huillenis* season of 1975. *Mahale Mountains Chimpanzee Research Project, Survey Report No. 3.* (*)

Nishida, T. and K. Norikoshi, 1976. Precipitation, atmospheric temperature, relative humidity and some other observations in Kasoje Chimpanzee Research Station, June 1975–May 1976. *Mahale Mountains Chimpanzee Research Project, Meteorological Report No. 1.* (*)

Nishida, T. and S. Uehara, 1980. Chimpanzees, tools, and termites: Another example from Tanzania. *Curr. Anthropol.*, **21**: 671–672.

Nishida, T. and S. Uehara, 1981. Kitongwe names of plants: A preliminary listing. *Afr. Study Monogr.*, **1**: 109–131.

Nishida, T. and S. Uehara, 1983. Natural diet of chimpanzees (*Pan troglodytes schweinfurthii*): Long-term record from the Mahale Mountains, Tanzania. *Afr. Study Monogr.*, **3**: 109–130.

Nishida, T., S. Uehara, and R. Nyundo, 1979. Predatory behavior among wild chimpanzees of the Mahale Mountains. *Primates*, **20**: 1–20.

Nishida, T., R.W. Wrangham, J. Goodall, and S. Uehara, 1983. Local differences in plant-feeding habits of chimpanzees between the Mahale Mountains and Gombe National Park, Tanzania. *J. Human Evol.*, **12**: 467–480.

Nishimura, A. and S. Azuma, 1977. The chimpanzees of Kabogo. In: *The Chimpanzees*, J. Itani (ed.), Kodan-sha, Tokyo, pp. 59–151. (in Japanese)

Noë, R., F.B.M. de Waal, and J.A.R.A.M. van Hooff, 1980. Types of dominance in a chimpanzee colony. *Folia Primatol.*, **34**: 90–110.

Norikoshi, K., 1976. Investigation of the Kasiha Valley. *Mahale Mountains Chimpanzee Research Project, Survey Report No. 5.* (*)

Norikoshi, K., 1982. One observed case of cannibalism among wild chimpanzees of the Mahale Mountains. *Primates*, **23**: 66–74.

Norikoshi, K., 1983. Prevalent phenomenon of predation observed among wild chimpanzees of the Mahale Mountains. *J. Anthrop. Soc. Nippon*, **91**: 475–479.

Oates, J.F., P.G. Waterman, and M. Choo, 1980. Food selection by the south Indian leaf monkey, *Presbytis johnii*, in relation to leaf chemistry. *Oecologia*, **45**: 45–56.

Oftedal, O.T., 1984. Milk composition, milk yield, and energy output at peak lactation: A comparative review. *Symp. Zool. Soc. Lond.*, **51**: 33–85.

O'Hagan, C.C., 1940. Mkungwe Mountain. *Tanganyika Notes and Records*, No. 9, pp. 53–60.

Ohigashi, H., T. Takagaki, K. Koshimizu, T. Nishida, M.A. Huffman, H. Takasaki, and M. Kaji, 1989. On the chemical constituents and its bioactivity of *Vernonia amygdalina*, a primate medicinal plant. *Proceedings of the 26th Congress of Japan Association for African Studies, June 3–4, 1989*, p. 15. (in Japanese)

Okada, T., N. Asada, and K. Kawanaka, 1988. A result of Drosophilid survey in Tanzania. *Afr. Study Monogr.*, **8**: 159–163.

Pain, T. and B. Verdcourt, 1962. On the occurrence of *Achatina tavaresina* Morelet in Tanganyika Territory. *J. Conchology*, **25**: 82.

Parker, G.A., 1984. Sperm competition and the evolution of animal mating strategies. In: *Sperm Competition and the Evolution of Animal Mating Systems*, R.L. Smith (ed.), Academic Press, New York, pp. 1–60.

Peacock, L.J. and C.M. Rogers, 1959. Gestation period and twinning in chimpanzees. *Science*, **129**: 959.

Pinhey, E., 1959. Odonata collected by Oxford University Tanganyika Expedition: And a west African species. *Occ. Pap. Nat. Mus. S. Rhod.*, **24B**: 509–515.

Procter, J., 1958. The second ascent of Kungwe Mountain. *Tanganyika Notes and Records*, No. 51, pp. 239–245.

Pusey, A.E., 1978. *The physical and social development of wild adolescent chimpanzees* (*Pan troglodytes schweinfurthii*). Ph.D. thesis, Stanford University.

Pusey, A.E., 1980. Inbreeding avoidance in chimpanzees. *Anim. Behav.*, **28**: 543–552.

Reiter, J., N.L. Stinson, and B.J. LeBouef, 1978. Northern elephant seal development: The transition from weaning to nutritional independence. *Behav. Ecol. Sociobiol.*, **3**: 337–367.

Riss, D. and J. Goodall, 1977. The recent rise to the alpha-rank in a population of free-living chimpanzees. *Folia Primatol.*, **27**: 134–151.

Rodriguez, E., M. Aregullin, T. Nishida, S. Uehara, R.W. Wrangham, Z. Abramowski, A. Finlayson, and G.H.N. Towers, 1985. Thiarubrine-A, a bioactive constituent of *Aspilia* (Asteracea) consumed by wild chimpanzees. *Experientia*, **41**: 419–420.

Rowell, T.E., 1966. Hierarchy in the organization of a captive baboon group. *Anim. Behav.*, **14**: 430–443.

Sabater Pi, J., 1979. Feeding behaviour and diet of chimpanzees (*Pan troglodytes troglodytes*) in the Okorobiko Mountains of Rio Muni (West Africa). *Z. Tierpsychol.*, **50**: 265–281.

Sade, D., 1968. Inhibition of mother-son mating among free-ranging rhesus monkeys. *Sci. Psychoanal.*, **12**: 18–38.

Schaller, G.B., 1963. *The Mountain Gorillas*. University of Chicago Press, Chicago.

Short, R.V., 1976. The evolution of human reproduction. *Proc. R. Soc. Lond., B.*, **195**: 3–24.

Short, R.V., 1979. Sexual selection and its component parts, somatic and genital selection, as illustrated by man and the great apes. In: *Advances in the Study of Behavior, Vol. 9*, J.S. Rosenblatt, R.A. Hinde, C. Beer, and M. Bosnel (eds.), Academic Press, London, pp. 131–158.

Silk, J.B., 1978. Patterns of food sharing among mother and infant chimpanzees at Gombe National Park, Tanzania. *Folia Primatol.*, **29**: 129–141.

Silk, J.B., 1979. Feeding, foraging, and food sharing behavior of immature chimpanzees. *Folia Primatol.*, **31**: 123–142.

Silk, J.B., 1986. Eating for two: behavioural and environmental correlates of gestation length among free-ranging baboons (*Papio cynocephalus*). *Int. J. Primatol.*, **7**: 583–602.

Simkin, P., 1958. To the Mountain. *Corona, Sept.*, pp. 349–350.

Simkin, P., 1959a. Letter from the lakeside. *Corona, Febr.*, pp. 64–66.

Simkin, P., 1959b. Oxford University Tanganyika Expedition 1958, Anthropological Report: The Waholoholo Tribe (Tanganyika). 13pp. Mimeo.

Simkin, P. and B. Juniper, 1961. Report of the University of Oxford Expedition to Tanganyika, 1958. *Oxford Univ. Exploration Club Bull.*, **10**: 7–12.

Simpson, M.J.A., 1973. The social grooming of male chimpanzees. In: *Comparative Ecology and Behavior of Primates*. R.P. Michael and J. H. Crook (eds.), Academic Press, London, pp. 411–505.

Small, M., 1984. Aging and reproductive success in female *M. mulatta*. In: *Female Primates: Studies by Women Primatologists*, M. Small (ed.), Alan R. Liss, New York, pp. 249–259.

Stanley, H.M., 1878. *Through the Dark Continent*. Sampson Low, Marston, Searle & Rivington, London.

Stevens, T.E., 1962. Oxford University Tanganyika Expeditions 1958–9. *Tanganyika Notes and Records*, No. 58–59, pp. 111–115.

Stumpf, P.K. and E.E. Conn (eds.), 1982. *Biochemistry of Plants: A Comprehensive Treatise, Vol. 7. Secondary Plant Products*. Academic Press, London.

Sugiyama, Y., 1968. Social organization of chimpanzees in the Budongo Forest, Uganda. *Primates*, **9**: 225–258.

Sugiyama, Y., 1989. Population dynamics of chimpanzees at Bossou, Guinea. In: *Understanding Chimpanzees*, P.G. Heltne and L.A. Marquardt (eds.), Harvard University Press, Cambridge, Mass., pp. 134–145.

Sugiyama, Y. and J. Koman, 1987. A preliminary list of chimpanzees' alimentation at Bossou, Guinea. *Primates*, **28**: 133–147.

Sugiyama, Y., J. Koman, and M.B. Sow, 1988. Ant-catching wands of wild chimpanzees at Bossou, Guinea. *Folia Primatol.*, **51**: 56–60.

Suzuki, A., 1966. On the insect-eating habits among wild chimpanzees living in the savanna woodland of western Tanzania. *Primates*, **7**: 481–489.

Suzuki, A., 1969. An ecological study of chimpanzees in a savanna woodland. *Primates*, **10**: 103–148.

Suzuki, A. 1977. The society of chimpanzees and their adaptation. In: *The*

Chimpanzees, J. Itani (ed.), Kodan-sha, Tokyo, pp. 251–336. (in Japanese)

Suzuki, T., 1972. The houses and the life of the Tongwe tribe in Tanzania, East Africa. *Kyoto Univ. Afr. Studies*, **7**: 269–310.

Tachibana, T., 1987. Dialogue with Kinji Imanishi: Scholars studying non-human primates, No. 1. *Anima*, **14** (12): 50–56. (in Japanese)

Takada, H. and S. Uehara, 1987. Drosophilid flies (Diptera) in the Mahale Mountains National Park, Tanzania: A preliminary report. *Afr. Study Monogr.*, **7**: 15–19.

Takahata, H. and Y. Takahata, 1989. Inter-unit-group transfer of an immature male common chimpanzee and his social interactions in the non-natal group. *Afr. Study Monogr.*, **9**: 209–220.

Takahata, Y. 1980. A preliminary report on distribution, habitat preference, and food habits of sympatric primates in the Mahale Mountains. *Mahale Mountains Chimpanzee Research Project, Ecological Report No. 16.* (*)

Takahata, Y., 1983. Precipitation, atmospheric temperature, and relative humidity of January–December 1982. *Mahale Mountains Chimpanzee Research Project, Meteorological Report No. 8.* (*)

Takahata, Y., 1985. Adult male chimpanzees kill and eat a male newborn infant: Newly observed intragroup infanticide and cannibalism in Mahale National Park, Tanzania. *Folia Primatol.*, **44**: 161–170.

Takahata, Y., 1986. The chimpanzees of the Mahale Mountains National Park, Tanzania: Their ecology and sociology. In: *Anthropology of Natural Societies in Africa*, J. Itani and J. Tanaka (eds.), Akademia-shuppankai, Kyoto, pp. 7–41. (in Japanese)

Takahata, Y., T. Hasegawa, H. Hayaki, and H. Takasaki, 1982. A brief survey on the home range of the N group chimpanzees. *Mahale Mountains Chimpanzee Research Project, Survey Report No. 11.* (*)

Takahata, Y., T. Hasegawa, and T. Nishida, 1984. Chimpanzee predation in the Mahale Mountains from August 1979 to May 1982. *Int. J. Primatol.*, **5**: 213–233.

Takahata, Y., M. Hiraiwa-Hasegawa, H. Takasaki, and R. Nyundo, 1986. Newly acquired feeding habits among the chimpanzees of the Mahale Mountains National Park, Tanzania. *Human Evol.*, **1**: 277–284.

Takahata, Y., R. Nyundo, and M. Seifu, 1984. Precipitation, atmospheric temperature, and relative humidity between January and December 1983. *Mahale Mountains Chimpanzee Research Project, Meteorological Report No. 9.* (*)

Takahata, Y. and H. Takasaki, 1986. Precipitation, atmospheric temperature, and relative humidity in 1984. *Mahale Mountains Chimpanzee Research Project, Meteorological Report No. 10.* (*)

Takamura, K., 1983a. Fish fauna on Mahale Mountains coast and a discussion on its characteristics. *Ecol. Limnol. Tanganyika*, **2**: 7–8.

Takamura, K., 1983b. Further investigation on the interspecific relationships among Aufwuchs-eating fishes of Lake Tanganyika. *Ecol. Limnol. Tanganyika*, **2**: 9.

Takamura, K., 1984. Interspecific relationships of Aufwuchs-eating fishes in Lake Tanganyika. *Env. Biol. Fishes*, **10**: 225–241.

Takamura, K., 1987. An observation of the breeding of *Lamprologus callipterus* in Lake Tanganyika. *Newsl. Intern. Assoc. Fish Ethologists*, **10**: 7–8.

Takasaki, H., 1983a. Seed dispersal by chimpanzees: A preliminary note. *Afr. Study Monogr.*, **3**: 105–108.

Takasaki, H., 1983b. Mahale chimpanzees taste mangoes—Toward acquisition of a new food item? *Primates*, **24**: 273–275.

Takasaki, H., 1985. Female life history and mating patterns among the M group chimpanzees of the Mahale National Park, Tanzania. *Primates*, **26**: 121–129.

Takasaki, H., 1988. Precipitation and atmospheric temperature at Kansyana Camp in 1987. *Mahale Mountains Chimpanzee Research Project, Meteorological Report No. 13.* (*)

Takasaki, H., 1989. Precipitation and atmospheric temperature at Kansyana Camp in 1988. *Mahale Mountains Chimpanzee Research Project, Meteorological Report No. 14.* (*)

Takasaki, H., M. Hiraiwa-Hasegawa, Y. Takahata, R.W. Byrne, and T. Kano, 1986. A case of unusually early postpartum resumption of estrous cycling in a young female chimpanzee in the wild. *Primates*, **27**: 517–519.

Takasaki, H. and K. Hunt, 1987. Further medicinal plant consumption in wild chimpanzees? *Afr. Study Monogr.*, **8**: 125–128.

Takasaki, H. and S. Uehara, 1984. Seed dispersal by chimpanzees: Supplementary note 1. *Afr. Study Monogr.*, **5**: 91–92.

Takeda, J., 1976. An ecological study of the honey-collecting activities of the Tongwe, western Tanzania, East Africa. *Kyoto Univ. Afr. Studies*, **10**: 213–247.

Tanaka, J., 1969. The ecology and social structure of Central Kalahari Bushmen: A preliminary report. *Kyoto Univ. Afr. Studies*, **3**: 1–26.

Tanaka, J., 1980. *The San, Hunter-Gatherers of the Kalahari: A Study in Ecological Anthropology*. University of Tokyo Press, Tokyo.

Teleki, G., 1973. *The Predatory Behavior of Wild Chimpanzees.* Bucknell University Press, Lewisburg, Pennsylvania.

Teleki, G., 1975. Primate subsistence patterns: Collector-predators and gatherer-hunters. *J. Human Evol.*, **4**: 125–184.

Teleki, G., 1981. The omnivorous diet and eclectic feeding habits of chimpanzees in Gombe National Park, Tanzania. In: *Omnivorous Primates: Gathering and Hunting in Human Evolution*, R.O. Harding and G. Teleki (eds.), Columbia University Press, New York, pp. 303–343.

Teleki, G., 1987. Current threats to the survival of chimpanzees in the equatorial Africa. *A report by the Committee for Conservation and Care of Chimpanzees.*

Teleki, G., E.E. Hunt, Jr., and J.H. Pfifferling, 1976. Demographic observations (1963–1973) on the chimpanzees of Gombe National Park, Tanzania. *J. Human Evol.*, **5**: 559–598.

Tilford, B., 1981. Nondesertion of a post-reproductive rhesus female by adult male kin. *J. Mammal.*, **62**: 638–639.

Tomita, K., 1966. The sources of food of the Hadzapi tribe: The life of a hunting tribe in east Africa. *Kyoto Univ. Afr. Studies*, **1**: 157–171.

Trillmich, F., 1986. Maternal investment and sex-allocation in the Galapagos fur seal, *Arctocephalus galapagoensis. Behav. Ecol. Sociobiol.*, **19**: 157–164.

Trivers, R.L., 1972. Parental investment and sexual selection. In: *Sexual Selection and the Descent of Man 1871–1971*, B. Campbell (ed.), Aldine, Chicago, pp. 136–179.

Trivers, R.L., 1974. Parent-offspring conflict. *Amer. Zool.*, **14**: 249–264.

Turnquist, J.E., 1986. Joint mobility as a function of age in free-ranging rhesus monkeys (*Macaca mulatta*). In: *The Cayo Santiago Macaques, History, Behavior and Biology*, R.G. Rawlins and M.J. Kessler (eds.), SUNY Press, Albany, pp. 243–262.

Tutin, C.E.G., 1979. Mating patterns and reproductive strategies in a community of wild chimpanzees (*Pan troglodytes schweinfurthii*). *Behav. Ecol. Sociobiol.*, **6**: 29–38.

Tutin, C.E.G., 1980. Reproductive behaviour of wild chimpanzees in the Gombe National Park, Tanzania. *J. Reprod. Fert., Suppl.*, **28**: 43–57.

Tutin, C.E.G. and P.R. McGinnis, 1981. Chimpanzee reproduction in the wild. In: *Reproductive Biology of the Great Apes*, C.E. Graham (ed.), Academic Press, New York, pp. 239–264.

Tutin, C.E.G. and W.C. McGrew, 1973. Chimpanzee copulatory behaviour. *Folia Primatol.*, **19**: 237–256.

Uehara, S., 1978. Precipitation, atmospheric temperature, relative humidity and some other observations in Kasoje Chimpanzee Research Station, January –December, 1977. *Mahale Mountains Chimpanzee Research Project, Meteorological Report No. 3*. (*)

Uehara, S., 1981. The social unit of wild chimpanzees: A reconsideration based on the diachronic data accumulated at Kasoje in the Mahale Mountains, Tanzania. *Africa-Kenkyu* (*Journal of African Studies*), **20**: 15–32. (in Japanese)

Uehara, S., 1982. Seasonal changes in the techniques employed by wild chimpanzees in the Mahale Mountains, Tanzania, to feed on termites (*Pseudacanthotermes spiniger*). *Folia Primatol.*, **37**: 44–76.

Uehara, S., 1984. Sex difference in feeding on *Camponotus* ants among wild chimpanzees in the Mahale Mountains, Tanzania. *Int. J. Primatol.*, **5**: 389.

Uehara, S., 1986. Sex and group differences in feeding on animals by wild chimpanzees in the Mahale Mountains National Park, Tanzania. *Primates*, **27**: 1–13.

Uehara, S. and T. Nishida, 1987. Body weights of wild chimpanzees (*Pan troglodytes schweinfurthii*) of the Mahale Mountains National Park, Tanzania. *Am. J. Phys. Anthropol.*, **72**: 315–321.

Uehara, S. and K. Norikoshi, 1977. Precipitation, atmospheric temperature, relative humidity and some other observations in Kasoje Chimpanzee Research Station, June 1976–May 1977. *Mahale Mountains Chimpanzee Research Project, Meteorological Report No. 2*. (*)

Uehara, S. and R. Nyundo, 1983. One observed case of temporary adoption of an infant by unrelated nulliparous females among wild chimpanzees in the Mahale Mountains, Tanzania. *Primates*, **24**: 456–466.

Ulfstrand, S., 1960a. The juvenile plumage of *Apalis argentea* Moreau 1941 and a note on the habitat of the species. *Bull. Brit. Ornithol. Club*, **80**: 2–3.

Ulfstrand, S., 1960b. A new subspecies of *Anthoscopus caroli* (Sharpe 1871) from western Tanganyika Territory. *Bull. Brit. Ornithol. Club*, **80**: 11–13.

Ulfstrand, S., 1960c. *Guttera edouardi seth-smithi* Neumann 1908 in western Tanganyika Territory. *Bull. Brit. Ornithol. Club*, **80**: 100–101.

Ulfstrand, S., 1960d. Migration of African kites, *Milvus migrans* (Bodd.) *aegypticus* Sharpe & Bouvier, and Wahlberg's eagles, *Aquila wahlbergi* Sund., at Lake Tanganyika. *Kungl. Fysiografiska Sallskaperts I Lund Forhandlinger*, Bd 30, Nr 4, pp. 31–39.

Ulfstrand, S. and H. Lamprey, 1960. On the birds of the Kungwe-Mahari area in western Tanganyika Territory. *J. East Afr. Nat. Hist. Soc.*, **23**: 223–232.

Verdcourt, B., 1961. Further notes on the genus *Ptychotrema* Morch in Eastern Africa (Mollusca, Stretaxidae). *Arch. Moll.*, **90**: 155–161.

de Waal, F.B.M., 1982. *Chimpanzee Politics: Power and Sex among Apes*. Harper & Row, New York.

de Waal, F.B.M. 1984. Sex differences in the formation of coalitions among chimpanzees. *Ethol. Sociobiol.*, **5**: 239–255.

de Waal, F.B.M., 1985. Deception in the natural communication of chimpanzees. In: *Deception*, R.W. Mitchell and N.S. Thompson (eds.), SUNY Press, Albany, pp. 221–244.

de Waal, F.B.M., 1986a. The integration of dominance and social bonding in primates. *Quat. Rev. Biol.*, **61**: 459–479.

de Waal, F.B.M., 1986b. The brutal elimination of a rival among captive male chimpanzees. *Ethol. Sociobiol.*, **7**: 237–251.

de Waal, F.B.M., 1987. Dynamics of social relationships. In: *Primate Societies*, B.B. Smuts, D.L. Cheney, R.M. Seyfarth, R.W. Wrangham, and T.T. Struhsaker (eds.), University of Chicago Press, Chicago, pp. 421–429.

de Waal, F.B.M. 1989. Dominance 'style' and primate social organization. In: *Comparative Socioecology: The Behavioural Ecology of Humans and Other Mammals*, V. Standen and R.A. Foley (eds.), Blackwell, Oxford, pp. 243–263.

de Waal, F.B.M., J.A.R.A.M. van Hooff, and W.J. Netto, 1976. An ethological analysis of types of agonistic interaction in a captive group of Java-monkeys (*Macaca fascicularis*). *Primates*, **17**: 257–290.

Wadsworth, W.J., 1963. The Kapalagulu layered intrusion of western Tanganyika. *Mineralogical Society of America, Special Paper*, **1**: 108–115.

Wallis, J., 1985. Synchrony of estrous swelling in captive group-living chimpanzees (*Pan troglodytes*). *Int. J. Primatol.*, **6**: 335–350.

Waser. P.M., 1978. Postreproductive survival and behavior in a free-ranging female mangabey. *Folia Primatol.*, **29**: 142–160.

Watt, J.M. and M.G. Breyer-Brandwijk, 1962. *The Medicinal and Poisonous Plants of Southern and Eastern Africa* (2nd ed.), E. & S. Livingstone, Edinburgh.

Watts, D.P., 1985. Observations on the ontogeny of feeding behavior in mountain gorillas (*Gorilla gorilla beringei*). *Am. J. Primatol.*, **8**: 1–10.

Whiten, A. and R.W. Byrne, 1986. A survey of deception in primates. *Primate Report*, No. 14, p. 60.

Whiten, A. and R.W. Byrne, 1988. Tactical deception in primates. *Behavioral and Brain Sciences*, **11**: 233–273.

Whitten, P.L., 1983. Diet and dominance among female verevt monkeys (*Cercopithecus aethiopus*). *Am. J. Primatol.*, **5**: 139–159.

Wolfe, L.D. and M.J. Sabra Noyes, 1981. Reproductive senescence among female Japanese macaques (*Macaca fuscata fuscata*). *J. Mammal.*, **62**: 698–705.

Wolff, J.O., 1988. Material investment and sex ratio adjustment in American bison calves. *Behav. Ecol. Sociobiol.*, **23**: 127–133.

Woodruff, G. and D. Premack, 1979. Intentional communication in the chimpanzee: The development of deception. *Cognition*, **7**: 333–362.

Wrangham, R.W., 1974. Artificial feeding of chimpanzees and baboons in their natural habitat. *Anim. Behav.*, **22**: 83–93.

Wrangham, R.W., 1975. *The behavioural ecology of chimpanzees in Gombe National Park, Tanzania*. Ph. D. thesis, University of Cambridge.

Wrangham, R.W., 1977. Feeding behaviour of chimpanzees in Gombe National Park, Tanzania. In: *Primate Ecology*, T.H. Clutton-Brock (ed.), Academic Press, London, pp. 503–538.

Wrangham, R.W., 1979a. On the evolution of ape social systems. *Soc. Sci. Information*, **18**: 335–368.

Wrangham, R.W., 1979b. Sex differences in chimpanzee dispersion. In: *The Great Apes*, D.A. Hamburg and E.R. McCown (eds.), Benjamin/Cummings, Menlo Park, pp. 481–489.

Wrangham, R.W., 1980. Chimpanzee protection in the Mahale Mountains—Wasted effort or golden opportunity. *Primate Eye*, No. 15, pp. 10–12.

Wrangham, R.W., 1986. Ecology and social relationships in two species of chimpanzees. In: *Ecological Aspects of Social Evolution*, D.I. Rubenstein and R.W. Wrangham (eds.), Princeton University Press, Princeton, pp. 352–378.

Wrangham, R.W. and T. Nishida, 1983. *Aspilia* spp. leaves: A puzzle in the feeding behavior of wild chimpanzees. *Primates*, **24**: 276–282.

Wrangham, R.W. and B.B. Smuts, 1980. Sex differences in the behavioural ecology of chimpanzees in Gombe National Park, Tanzania. *J. Reprod. Fert., Suppl.*, **28**: 13–31.

Yamagiwa, J., 1987. Male life history and the social structure of wild mountain gorillas (*Gorilla gorilla beringei*). In: *Evolution and Coadaptation in Biotic Communities*, S. Kawano, J.H. Conell, and T. Hidaka (eds.), University of Tokyo Press, Tokyo, pp. 31–51.

Yoshihara, M., 1987. Reproduction of great apes in Japanese zoos. In: *The Reproduction and Conservation of Apes: Proceeding of the 31th Annual Meeting of the Japanese Society for Primate Research, Japan Monkey Center, October 23–24, 1986*, pp. 21–37. (in Japanese)

Index

T~Z